Neutron Radiography

WCNR-11

11th World Conference on Neutron Radiography (WCNR-11) held the 2nd-7th September 2018 in Sydney, Australia

Editor
Ulf Garbe

Co-Editors
Filomena Salvemini and Joseph J. Bevitt

Australian Nuclear Science & Technology Organisation

Peer review statement

All papers published in this volume of "Materials Research Proceedings" have been peer reviewed. The process of peer review was initiated and overseen by the above proceedings editors. All reviews were conducted by expert referees in accordance to Materials Research Forum LLC high standards.

Published under License by **Materials Research Forum LLC**
Millersville, PA 17551, USA

Published as part of the proceedings series
Materials Research Proceedings
Volume 15 (2020)

ISSN 2474-3941 (Print)
ISSN 2474-395X (Online)

ISBN 978-1-64490-056-7 (Print)
ISBN 978-1-64490-057-4 (eBook)

This book contains information obtained from authentic and highly regarded sources. Reasonable efforts have been made to publish reliable data and information, but the author and publisher cannot assume responsibility for the validity of all materials or the consequences of their use. The authors and publishers have attempted to trace the copyright holders of all material reproduced in this publication and apologize to copyright holders if permission to publish in this form has not been obtained. If any copyright material has not been acknowledged please write and let us know so we may rectify in any future reprint.

Distributed worldwide by

Materials Research Forum LLC
105 Springdale Lane
Millersville, PA 17551
USA
http://www.mrforum.com

Manufactured in the United State of America
10 9 8 7 6 5 4 3 2 1

Table of Contents

Methods

Software

Application

Keyword Index

Editorial

The 11th World Conference on Neutron Radiography is back on the southern hemisphere. In 2014 at the WCNR-10 in Switzerland the decision was made through an election by ISNR members in favour of ANSTO hosting the next WCNR-11. At that time, we proposed to host the conference in Sydney near the imaging beam line DINGO to providing a tour to the newly built facility. Finally, the conference took place on the 2nd-7th September 2018 at the Maritime Museum in Sydney. I would like to acknowledge the Gadigal people of the Eora Nation, the traditional custodians of this land which we were on and pay my respects to the Elders both past and present. Ancient Aboriginal rock art paintings give us the opportunity to obtain a glimpse thousands of years into the past and to gain an understanding of the important objects in the lives of the traditional custodians of this land. Neutron imaging in some ways is analogous to the ancient rock paintings, giving us the pictures deep inside scientific and industrial samples to gain an understanding of the structure within.

A glance through the list of presentations offered by the community reveals the amazing diversity of scientific and industrial applications, the advancements of the state-of-art neutron facilities and the latest method developments. With more than 160 abstracts submitted we introduced parallel sessions to accommodate 99 oral presentations and 54 poster presentations. In addition we had an invited plenary talk every morning to underline the importance of some outstanding research.

I would like to thank the sponsors who supported this conference. An event like this does not happen overnight and I would like to thank the organising committees for their tireless work in ensuring this conference is a great success. A special thank you goes to Kelly Cubbin for her fantastic support. In particular her effort on getting the conference App running was a big step to go paperless and reduce the environmental impact. In addition I would like to thank my family and friends for their help at all stages of planning and running the conference.

Conferences such as this provide a valuable opportunity for research scientists, industry specialists and the next generation of scientists to share knowledge and experiences. It is this sharing that enables the community to grow with a large number of new ISNR members signing in.

Jack Brenizer

When I started my first attempts in neutron radiography (with film methods) around 1994 there was a meeting of a European working group on Neutron Radiography, where American and Canadian partners were also be invited. It took place at BAM (Bundesanstalt für Materialprüfung) in Berlin for three days with about 10 participants only, among them Jack Brenizer. Other prominent persons of that meeting I remember were J. Domanus (Riso, Denmark), persons from Petten (Netherlands), Saclay (France) and C. Fischer (HMI Berlin). They all had already much more experience than me – and I watched very carefully and with certain respect their presented details of studies. On a trip to USA, where I represented Switzerland in the program for the reduction in fuel enrichment for research reactors, about one year later, I met Jack a second time at his beam line of the Virginia University research reactor. I was very impressed by the radioscopy setup with video camera systems and light amplifiers in front. Unfortunately, this reactor was closed short time after – including the beam line for the real-time inspection of materials there. This was the reason why Jack returned to his initial education site – the Penn State University, where a professorship position was offered to him. At the same time, he got the access to a neutron imaging facility again because the reactor at Penn State continued running – until today. Maybe, Jacks activities at that reactor, in particular for neutron imaging, have been a strong argument to maintain the operation on good level into the present time – and in future. Later, we established a deeper contact for the preparation of conferences, exchange of knowledge and also performed private visits in both directions. In particular, he organized the "4th International Topical Meeting on Neutron Tomography" in 2001 with great success at Penn State and published the proceedings in good time. Jack was very active in the approach to standardize and to advertise neutron imaging in the U.S. and on te international level by his own studies and memberships in related committees. Before he retired in 2016, he was in Oak Ridge active as advisor for the upcoming neutron imaging facilities at ORNL. I was always impressed by his enthusiasm and optimism, supporting students and co-workers with pleasure.

Honorary Member Certificate

International Society for Neutron Radiography

Established

1996

Jack Brenizer

for his

Pioneering and lifelong contribution to the International community in the advancement of the Science and Technology of Neutron Radiography

Recommended and accepted by the Board of the ISNR in 2010. Honoured and presented at the 11th World Conference on Neutron Radiography.

Sydney, Australia, 2 – 7 September 2018

Ulf Garbe
ISNR President (2014-2018)

Yoshiaki Kiyanagi

Yoshiaki Kiyanagi is a pioneer in the development of pulsed neutron imaging and its application for engineering and material science.

He was born on January 1st, 1949 in Hokkaido (Japan) and studied at Hokkaido University, where he graduated from the Department of Nuclear Engineering in 1961 and got his PhD in 1993. He became an assistant professor, lecturer, associate professor and finally full professor of Hokkaido University. He had been the division head of the Nuclear radiation source engineering and educated many students and researchers. One of his contributions to scientific research is concerning the accelerator neutron source. He performed a lot of experiments at the Hokkaido University Neutron Source (HUNS) to obtain valuable experimental data which cannot be obtained by numerical simulations. With sufficient experimental data he has designed and developed a low-energy neutron moderation system. The ability of this system has been highly evaluated and employed by many institutes, like J-PARC in Japan, SNS of Oak ledge National Laboratory in the USA, and ISIS of Rutherford-Appleton Laboratory in UK.

He has developed the Accurate Neutron-Nucleus Reaction measurement Instrument (ANNRI) in the J-PARC. For this development, he won a technology development award from the Atomic Energy Society of Japan. He has also developed the pulsed neutron transmission spectroscopy, which gives spatial distribution of crystal structure in materials. For this development, he won a paper award from Japanese Society of Metals and Materials.. Then he became a project leader to develop the pulsed neutron imaging facility (RADEN) in the J-PARC. The construction of RADEN was finished in 2014 and at present we can perform various kinds experiments at the J-PARC.

When he was a high school student, he began Japanese martial art, "Kendo". Kendo is a traditional Japanese fencing. Normally we are using a Bamboo sword for practice instead of a real Japanese sword. So, he got great interest in the manufacturing process of the Japanese swords and started his investigation on the crystal structure of Japanese sword using pulsed neutron imaging. He is now Prof. Emeritus of Hokkaido University and X'ian University in China, and President of Japanese Society for Neutron Science. He is still actively developing an accelerator neutron source for radiation therapy equipment at Nagoya University.

For his great contributions in our society, the honorable membership award was given to Prof. Yoshiaki Kiyanagi.

Honorary Member Certificate

International Society for Neutron Radiography

Established

1996

Yoshiaki Kiyanagi

for his

**Pioneering and lifelong contribution to the
International community in the advancement
of the
Science and Technology of Neutron Radiography**

*Recommended and accepted by the Board of the ISNR in 2010.
Honoured and presented at the 11th World Conference on
Neutron Radiography.*

Sydney, Australia, 2 – 7 September 2018

Ulf Garbe
ISNR President (2014-2018)

Eberhard Lehmann

Eberhard Lehmann was born in Leipzig in Eastern Germany on 16th July 1952 where he also studied physics graduating on the topic of "Molecular dynamic calculations of proteins" in 1974. He received his PhD at the East German Academy of Science in East Berlin in 1983 with his thesis entitled: "Cross-section data of construction materials for the fast breeder reactor by reactivity measurements". In the years between 1976 and 1990 Dr Lehmann was active in research in reactor physics for fast breeders based on calculations of reactor parameters with different reactor codes and experimental work at different reactor stations in several countries of the Eastern hemisphere.

With the fall of the Berlin Wall and the Iron Curtain the Western world opened for the young physicist. He immigrated to Switzerland where he could apply his expertise and experience from 1991 to 1995 at the research reactor SAPHIR of the Paul Scherrer Institute.

1991 to 1995 as reactor physicist at the research reactor SAPHIR of the Paul Scherrer Institute. Given responsibility for core design and in particular neutron applications, he took his opportunity in the latter field and established the Swiss activities in neutron imaging starting in the mid 1990ties still at the SAPHIR reactor, which, however, was shut down in 1994.

As a reactor physicist without a reactor Dr Lehmann took his chances in the new spallation source project SINQ at PSI to establish neutron imaging at the new source. In an environment strongly dominated by neutron scattering, driving the source with applications of fundamental research in magnetism Dr Lehmann established a neutron imaging beamline, which became the reference for neutron imaging user service at large scale neutron sources in Europe and maybe even the world. Together with Burkhard Schillinger from TUM he introduced digital neutron imaging and tomography in Europe.

The user program at his neutron imaging instrument proved successful and Dr Lehmann was never tired to promote the technique nationally and internationally, to find fields of applications also beyond the main stream of non-destructive testing and established simultaneously a vivid scientific user program as well as a profitable service for industry.

He formed a group of experts contributing to nearly all fields of technical developments and applications as well as industrial service, which became a model for many state-of-the-art user instruments and imaging groups at large scale neutron sources as established today.

This included to add a second instrument dedicated to neutron imaging with cold neutrons just 10 years after the first one at SINQ. The new instrument ICON became a front runner in many modern developments utilizing monochromatic neutrons or other energy resolved techniques. Most notable grating interferometric imaging was pioneered by the ever growing group, which by the end of his career as an employee at PSI and group leader of the Neutron Imaging and Activation Group NIAG operated the two dedicated imaging beamlines NEUTRA and ICON, but also officially utilized up to 50% of two further instruments, BOA, a polarized testbeamline and POLDI, a time-of-flight diffractometer. In addition, his group is one of the driving forces and partners of the European Spallation Source (ESS) in establishing neutron imaging with a day-one instrument (ODIN) and to provide the software for imaging data analyses. Dr Eberhard also led his group from being a part of the Neutron Source Division to being a valuable and respected part of the Laboratory for Neutron Scattering and (now also) Imaging, which also underlines the successes in his time to prove the potential of neutron imaging not only for non-destructive testing but far beyond in material science and other fields of science, equivalent to (other) scattering techniques.

Dr. Eberhard Lehmann was not only in his active career a tiredless ambassador and forefighter of neutron imaging in particular at large scale facilities, an advisor to nearly all major imaging instrument projects at large scale sources but is still a very active and engaged member of the ISNR. After being a yearlong member of the board with the organization of the last world conference in Switzerland (WCNR-10, 2014) he became president and served between 2010 and 2014.

After his retirement in July 2017 he remains being an active member of the community, still serving on the ISNR board and as advisor in instrumentation projects, continuing his own research and being an ambassador of neutron imaging at large scale facilities.

Honorary Member Certificate

International Society for Neutron Radiography

Established 1996

Eberhard Lehmann

for his

Pioneering and lifelong contribution to the
International community in the advancement
of the
Science and Technology of Neutron Radiography

*Recommended and accepted by the Board of the ISNR in 2010.
Honoured and presented at the 11th World Conference on
Neutron Radiography.*

Sydney, Australia, 2 – 7 September 2018

Ulf Garbe
ISNR President (2014-2018)

Committees

Local Organizing Committee

U Garbe
K Cubbin
F Salvemini
J Bevitt
J Schulz
A Paradowska

Scientific Committee

E. Gilbert
C. Hall
D. Jun
S. Khaweerat
A. Kingston
H. Li
G. Prusty
B. Schillinger
P. Trtik
V. Peterson
M. Safavi-Naeini

International Committee

A. Kaestner
M. Arif
Les G.I. Bennett
T. Bücherl
D.F. Chen
F. De Beer
C. Grünzweig
D. Hussey
N. Kardyjilov
W. Kockelmann
E. H. Lehmann
M. Strobl
Y. Saito

Sponsors

Platinum

Gold

Conference Support

Instrumentation

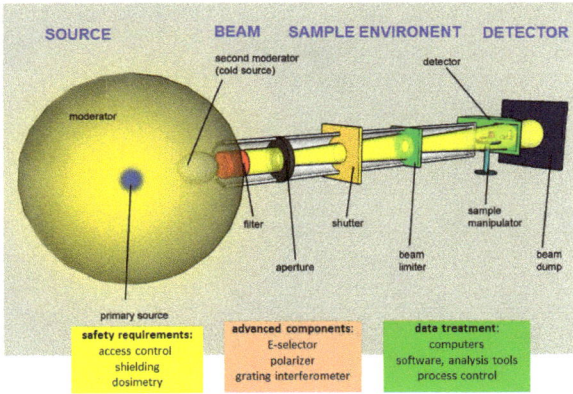

Neutron Radiography - WCNR-11   Materials Research Forum LLC
Materials Research Proceedings **15** (2020) 3-10   https://doi.org/10.21741/9781644900574-1

What Future in Neutron Imaging?

Eberhard H. Lehmann*[1], Danas Ridikas[2], Nuno Pessoa Barradas[2]

[1]Paul Scherrer Institut, Laboratory for Neutron Scattering & Imaging, CH-5232 Villigen PSI, Switzerland

[2]International Atomic Energy Agency (IAEA), Division of Physical and Chemical Sciences

eberhard.lehmann@psi.ch

* corresponding author

Keywords: Neutron Imaging, Neutron Sources, Beam Ports, Neutron Detection, Image Processing

Abstract. We describe the current situation of the neutron imaging technology, based on known "user facilities" and projects at prominent neutron sources world-wide. Although this method has become highly accepted, there is a great potential for further methodical and technical progress. Continued access to most suitable beam ports and future neutron sources are keystones for the future of neutron imaging. Promising new methods and prominent new applications are stimulating this process.

Introduction

Neutron Imaging is today a well-established technique for scientific and technical applications. It is very complementary to similar X-ray methods and can often be used symbiotically [1]. The layout of a dedicated imaging beam line should be based on state-of-the-art technologies and the valuable experience of facility operators. The future of this technique will depend on the continuous access to best suitable beam ports at present and future useful neutron sources.

This paper starts with a "generic neutron imaging facility", will reflect the situation on neutron sources and their developments, have a look onto new facility projects and upgrades of existing ones, describe options for best facility utilization and highlight the methodical progress and other new options in neutron imaging.

The generic neutron imaging facility

As shown in Fig. 1, a generic neutron imaging (NI) facility consists of four major components: the neutron source, including moderation media and filters, the beam forming equipment (collimation), the sample environment and the neutron imaging detector.

Modern NI stations [2] have become quite complex and sophisticated systems if all modern trends in this technology should be involved. Depending on complexity level and desired performance, the investments required are in the range from some ten thousands to some ten millions of Euros [3].

The technical level of a NI installation depends on the lab strategy, framed by the funding, the major applications and the demands of the user community [4]. The available facilities can be categorized roughly into four classes:

1. Operational user labs, open for an international access by scientific and industrial partners
2. Operational in-house and test facilities, mainly used by own researchers for domestic projects

3. On-going new NI installation projects or facility upgrade activities
4. Projects under considerations for potential new or upgrade facility installations

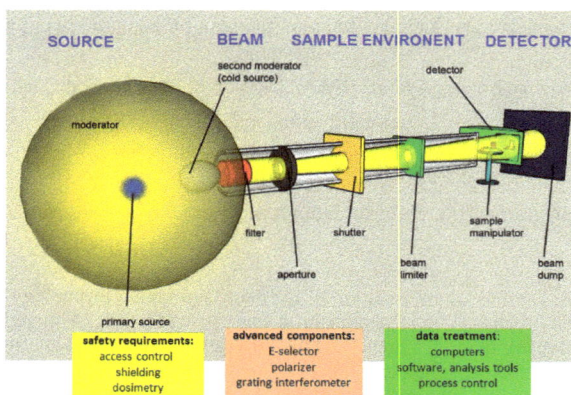

Fig. 1:   *Layout of a generic neutron imaging facility with the major components and*
supplementary features

The determining conditions underpinning the most advanced NI systems are:

- Well collimated (high L/D-ratio) neutron beam
- Beam size adequate to the sample dimensions
- High neutron beam intensity
- Narrow energy band (thermal or cold), well-known spectral conditions, necessary for quantification
- Low background from gamma rays or fast neutrons in the primary beam
- Flat beam profile
- No interference from back-scattered neutrons (and process gammas)

In general, a high intensity neutron beam is required to achieve the highest temporal, spatial or spectral resolution in a reasonable acquisition time. In addition, many more sophisticated techniques are possible in reasonable acquisition time when the intensity is suitable.

Neutron source development

From all options for the generation of neutrons, until now the research reactors remain the most common, flexible, powerful and even cost-efficient sources of neutrons. They are also by far the majority of the sources where NI stations are presently located and utilized. There exist a few intensive spallation neutron sources as well as other accelerator based neutron sources with lower output of well collimated beams of moderated neutrons. One notes separately that isotopic neutron sources have by far no chance to compete in NI performance when compared to the above mentioned accelerator or reactor based facilities.

The development of the NI technology has to be seen in a global context. In order to develop, apply and utilize modern techniques, the access to well-suited beam ports is necessary. As a general trend, the number of appropriate neutron sources is decreasing in developed countries, but more are being installed in developing countries. On the other hand, the most advanced NI facilities are still situated in a few labs in developed countries. Therefore, the knowledge transfer towards the newly implemented facilities is essential for the progress in the field and broader access for usage of the technique.

Fig. 2 describes well the world-wide situation of operational research reactors as summarized in the IAEA Research Reactor data base [5]. From the currently running research reactors with a suitable power above 100 kW, 75 facilities declare to perform "neutron radiography". Since no detailed specification is given in the data base for many facilities, it is difficult to estimate on which technological level these installations are.

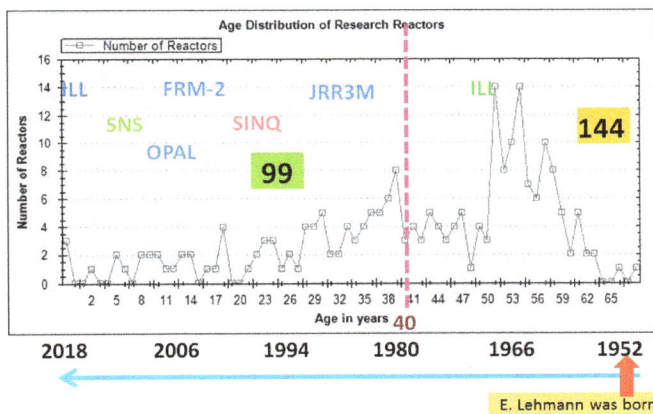

Fig. 2: *Number of newly commissioned research reactors per year in an inverse time scale: more than 140 have an age of more than 40 years. Some famous sources are highlighted (ILL is indicated twice since NI started just now – 50 years after its startup); SINQ and SNS are not reactors, but spallation neutron sources – shown for comparison. The data are taken from the IAEA RR data base, see:* https://nucleus.iaea.org/rrdb.

The show examples are by far not complete, but indicate some milestones for the imaging community.

A more pragmatic way for data achievement about NI facilities has been made by the "International Society for Neutron Radiography (ISNR)" survey as published on their homepage [6]. A list of "user facilities" can be found in [7], but an update will be given in the appendix.

Only few research reactor installations are expected to come to operation in the next years, while aged reactors will continue to shut down for different reasons. On the other hand, all spallation sources (ISIS, JPARC, SNS, ESS) involve NI as a key technology in dedicated projects, while SINQ already operates a few different stations [8] since many years with great success.

The situation within Europe is illustrated by Fig. 3 and the number of NI facilities (running and projected) is added. A similar analysis is not available for the rest of the world in the same quality.

Given the fact that new reactor based sources are not presently being built or planned in the Western world, there are initiatives to evaluate and design accelerator based neutron sources with specific performance, e.g. "high brilliance" [9] customized for specific applications. Also in such cases, NI installations are or could be foreseen from the beginning.

New installations and upgrades of NI facilities
Most of the prominent and powerful neutron sources have been "taken" by the neutron scattering community, by irradiation experiments for nuclear technologies, including isotope production and silicon doping. Therefore, only a few most suitable beam ports remained available and used for NI facilities in the past.

The situation has changed slightly after the development, at the end of last century, of digital imaging detector systems with superior performance compared to film based methods. It was possible to make very competitive installations ready in Japan, Europe and America, and more recently in Australia. Based on that progress, the family of "user facilities" has been established which have a similar operational approach now like neutron scattering instruments.

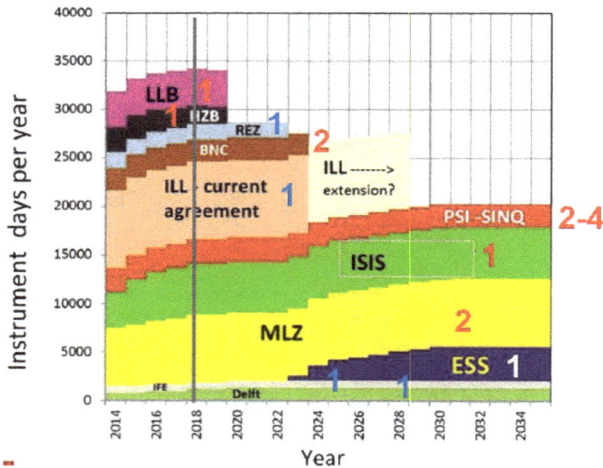

Fig. 3: *Neutron sources in Europe (data base: [6]); the numbers indicate existing (red) and planned NI facilities*

New sources tried to follow this trend already at the planning stage, including CARR in China, and planned NI facilities in reactors in Argentina, Jordan and Brazil.

Other countries, operating a research reactor for a long time, intend now to rebuild beam ports into NI stations or intend to make a major upgrade, e.g. South Africa and Indonesia. The level of performance these new setups will have depends on the financial situation, the involved (qualified) manpower and the particular source conditions. Tables 1 to 3 summarize the particular activities were the categories 2-4 as mentioned above are taken.

Table 1:   *In-house and test facilities, mainly for own projects*

country	source	situation
Thailand	reactor TTR-1, 2 MW	NI exists, limited user program
India	reactors at BARC	high potential, but limited communication with NI community
Indonesia	reactor RSG-GAS, 30 MW	NI exists, limited user program
Canada	reactor McMaster, 5MW	commercial activities, limited research
France	reactor Orphee, 14 MW	facility IMAGINE, limited user operation, shutdown after 2019
Europe	reactor @ ILL, 58 MW	a facility at cold beam port D50 under preparation, some users
Egypt	reactor ETRR-2, 22 MW	NI exists, limited user program
Brazil	reactor IEA-R1, 2 MW	NI exists, limited user program
Bangladesh	BAEC TRIGA reactor, 3 MW	NI exists, limited user program
Russia	several research reactors	IR-8, IRT-T, some NI exists
Poland	reactor MARIA	NI exists, limited user program

Table 2:   *Running NI installation projects and upgrade activities*

country	source	situation
Argentina	reactor RA-10, 30 MW	NI facility as day-one installation, completion in 2020
Brazil	new reactor like RA-10	planned and funding provided
Czech Republic	reactor LVR-15, 10 MW	NI facility with limited performance, upgrade intensions
China	reactor CARR, 60 MW	two imaging facilities under preparation
Norway	reactor JEEP-II, 2 MW	NI facility upgrade
Netherlands	reactor HOR, 5 MW	NI facility planned, preliminary setup exists
South Africa	reactor SAFARI, 20 MW	upgrade of the SANRAD facility
South Korea	reactor HANARO, 30 MW	operational again, utilization programme not yet started
USA	Idaho RR, 250 kW	upgrade program for digital NI

Table 3:   *Potential options and intentions for installations*

country	source	situation
Marocco	TRIGA reactor, 2 MW	NI facility planned
Malaysia	TRIGA reactor, 1 MW	NI facility exists, starting with digital system, limited communication
Algeria	reactor NUR, 1 MW	NI facility exists
Chile	reactor RECH 2, 10 MW	reactor out of operation, lack of manpower for NI
Peru	reactor RP-10, 10 MW	NI possible
Slovenia	TRIGA reactor, 250 kW	NI program stopped
Uzbekistan	reactor WWR-SM, 10 MW	NI not existing, but under consideration.
Russia	reactor PIK, 100 MW	operation unclear
Jordan	reactor JRTR, 5 MW	NI facility planned, reactor operational
Mexico	reactor TRIGA, 1 MW	NI considered

Best utilization

With the improvement of the detector technology also the utilization of the beam lines has been increased enormously. This gives the opportunity to perform much more investigations in shorter time while with an increased data volumes (towards the Tera-Byte region already).

In this manner, partners from different research areas can be invited for dedicated studies, with customized infrastructure and equipment. The same is valid for industrial partners on a commercial basis. Prominent cases are studies for fuel cells, batteries, particulate filters or electronic devices [10].

Since the number of such "user facilities" (see appendix) is still low and there are planned or unplanned shutdowns of these sources, a good communication and coordination between the facility operators will help to serve the increasing user community best.

Methodical developments & new applications

Starting with simple radiography studies many more advanced techniques have been developed and were introduced into the common user program in the meantime. It started with neutron tomography on a competitive level with X-rays and involved the fully quantitative analysis such as for precise water determination.

A very recent approach is "grating interferometry" which enables to study phase contrast and dark image features in samples with structures in the micro-meter range, linking to small-angle scattering studies.

In the dynamic imaging, either high frame rates are possible now (depending on the beam intensity, up to 100 Hz and more) or triggered stroboscopic investigations of repetitive processes. Due to the magnetic moment of neutrons, the separation of one of the two spin states (polarization) enables studies of magnetic properties on the macroscopic scale [11].

The separation of tiny energy bands of the initial beam has many advantages, in particular for crystalline materials with pronounced Bragg scattering behavior. It has been realized to study textures, crystal orientations and even internal stress with energy-resolved options. The use of time-of-flight techniques at pulsed sources will provide even better conditions in this respect.

Conclusions

Neutron imaging has made an enormous progress in the past years, but in limited number of labs only. Now, there is a challenging need to transfer this know-how to other suitable neutron sources and to increase the network within the neutron imaging community. The access to prominent and new installations is mandatory for the progress and strengthening of neutron imaging in the future.

References

[1] A. Kaestner et al., Combined neutron and X-ray imaging on different length scales, Proc. 6th Conference on Industrial Computed Tomography, Wels, Austria (iCT 2016)

[2] E. Lehmann, D. Ridikas, Status of Neutron Imaging – Activities in a Worldwide Context, Physics Procedia, Volume 69, 2015, Pages 10-17. https://doi.org/10.1016/j.phpro.2015.07.001

[3] IAEA TECDOC Series, Commercial Products and Services of Research Reactors, IAEA-TECDOC-1715 (2013)

[4] IAEA Nuclear Energy Series NG-T-3.16, Strategic Planning for Research Reactors (2017)

[5] https://nucleus.iaea.org/rrdb.

[6]https://ec.europa.eu/research/infrastructures/pdf/esfri/publications/esfri_neutron_landscape_group-report.pdf, page 69

[7] http://www.isnr.de/index.php/facilities/user-facilities

[8] E. Lehmann, New neutron imaging techniques to close the gap to scattering applications, Journal of Physics: Conference Series, Volume 746, Number 1. https://doi.org/10.1088/1742-6596/746/1/012070

[9] T. Cronert et al., High brilliant thermal and cold moderator for the HBS neutron source project Jülich, Journal of Physics: Conference Series, Volume 746, Number 1. https://doi.org/10.1088/1742-6596/746/1/012036

[10]   L. Donzel et al., Space-resolved study of binder burnout process in dry pressed ZnO ceramics by neutron imaging, Journal of the European Ceramic Society, Volume 38, Issue 16, Pages 5448-5453. https://doi.org/10.1016/j.jeurceramsoc.2018.08.017

[11]   N. Kardjilov et al., Three-dimensional imaging of magnetic fields with polarized neutrons, Nature Physics 4, 399 - 403 (2008). https://doi.org/10.1038/nphys912

Appendix: NI user facilities world-wide

country	site	institution	facility	neutron source	spectrum	power [MW]	status
Australia	Sydney	ANSTO	DINGO	OPAL reactor	thermal	20	operational
Germany	Munich-Garching	TU Munich	ANTARES	FRM-2 reactor	cold	25	operational
Germany	Munich-Garching	TU Munich	NECTAR	FRM-2 reactor	fast	25	operational
Germany	Berlin	HZB	CONRAD	BER-2 reactor	cold	10	operational
Hungary	Budapest	KFKI	NORMA	WWS-M reactor	cold	10	operational
Hungary	Budapest	KFKI	NRAD	WWS-M reactor	thermal	10	operational
Japan	Kyoto	Kyoto University	imaging beamline	MTR reactor	thermal	5	standby
Japan	Tokai	JAEA	imaging beamline	JRR-3M reactor	thermal	20	standby
Japan	Tokai	JAEA	RADEN	JPARC spallation	cold	0.5	operational
Korea	Daejon	KAERI	imaging beamline	HANARO reactor	thermal	30	standby
Russia	Dubna	JINR	imaging beamline	IBR-2M pulsed reactor	thermal	2	operational
Switzerland	Villigen	PSI	NEUTRA	SINQ spallation	thermal	1	operational
Switzerland	Villigen	PSI	ICON	SINQ spallation	cold	1	operational
UK	Oxfordshire	Rutherford Lab	IMAT	ISIS spallation	cold	0.3	operational
USA	Gaithersburg	NIST	BT-2	NBSR reactor	thermal	20	operational
USA	Gaithersburg	NIST	NG-6	NBSR reactor	cold	20	operational
USA	Oak Ridge	ORNL	CG-1D	HFIR reactor	cold	85	operational
South Africa	Pelindaba	NECSA	SANRAD	SAFARI reactor	thermal	20	standby

Neutron Radiography - WCNR-11
Materials Research Proceedings **15** (2020) 11-16

Materials Research Forum LLC
https://doi.org/10.21741/9781644900574-2

Overview of the Conceptual Design of the Upgraded Neutron Radiography Facility (INDLOVU) at the SAFARI-1 Research Reactor in South Africa

Frikkie de Beer[1a*], Tankiso Modise[1], Robert Nshimirimana[1], Deon Marais[1], Christo Raaths[1], Rudolph van Heerden[1], Kobus Eckard[1], Evens Moraba[1], Johann van Rooyen[1], Gerhard Schalkwyk[3], Jacoline Hanekom[2] and Gawie Nothnagel[1]

[1]Radiation Science Department, South African Nuclear Energy Corporation SOC Ltd. (Necsa), Pretoria, South Africa

[2]Engineering Department, South African Nuclear Energy Corporation SOC Ltd. (Necsa), Pretoria, South Afric

[3]SDG Nuclear Solutions, Hartbeespoort, 0216, South Africa

[a]frikkie.debeer@necsa.co.za

Keywords: Neutron Radiography, SAFARI-1, INDLOVU, Necsa

Abstract. The research and economic value of neutron beam line facilities are increasingly appreciated as is evident from the number of new facilities being planned and commissioned worldwide. In order to provide local researchers with world-class capabilities, the South African Nuclear Energy Corporation SOC Ltd. has embarked on the upgrade of the neutron beam line instruments at the SAFARI-1 nuclear research reactor which entails, inter alia, a novel, multi-functional neutron radiography (NRAD) facility, named INDLOVU (acronym for "Imaging Neutron Device to Locate the Obscure and Visualise the Unknown" and also the Zulu name for elephant). The basic design of the facility has been developed and construction will commence after the required licensing authorizations have been obtained. The INDLOVU NRAD facility comprises of a number of subsystems. However, this article describes and highlights some of the more important subsystems and components in terms of their importance, functionality and operational interconnection with other subsystems.

Introduction

The design, construction and commissioning of neutron radiography (NRAD) facilities are generally non-trivial and expensive and therefore should aim right from the start to be competitive, complementary, state of the art, optimal for a given facility, multifunctional but also flexible.

The IAEA Research Reactors database contains updated information on current neutron radiography facilities located at nuclear research reactors (RR) [1] while NRAD facilities at spallation sources (SS) are summarized in the International Society for Neutron Radiography (ISNR) database [2]. A total of 72 NRAD RR facilities in 38 countries and 5 NRAD SS facilities in 2 countries are listed of which 15 NRAD facilities are classified as operational user facilities. Also listed is development and/or upgrading of 10 RR NRAD facilities and development of 3 new SS NRAD facilities (IMAT in UK; VENUS in USA [3]; ODIN in Sweden).

Of the 18 countries that participated at the WCNR-11 (this conference) only 2 countries made presentations on existing activities for building and/or upgrading of NRAD facilities to be operational in the near future. However, there is a potential in 9 other countries for the development of new NRAD facilities. The closure of BERII in Berlin in 2020 with the subsequent loss of a highly productive NRAD facility at the Helmholtz Institute, sketch a scenario of limited available NRAD user facilities over the next 5 years. This makes the upgrade of the thermal spectrum limited South African Neutron Radiography (SANRAD) facility towards the planned multi-functional INDLOVU NRAD facility, an important capacity generating development and milestone for South Africa and the international NRAD user community.

Necsa has been planning the upgrade since 2005 starting with an IAEA supported expert mission to identify the shortcomings of the old SANRAD facility. It was followed by several IAEA supported expert missions to FRMII, Germany as well as a MCNP-X optimization study which was based on the findings of the IAEA expert mission [4]. Additionally, guidelines for the improvement, optimization and implementation of NRAD facilities as compiled and suggested by Lehmann [5] were incorporated in the design of the INDLOVU NRAD facility. This includes optimal radiation protection, large space provisioning at experimental infrastructure, options for beam limitation and energy selection, maximized radiation beam intensity, an L/D > 250 and a homogeneous radiation beam spatial distribution.

Since 2013, several ISNR related conference articles were published on the planned enhancements of the old SANRAD facility [6] towards an improved new INDLOVU NRAD facility. These included articles on the concrete characteristics of the shielding design, which was based on that of the ANTARES facility at FRMII in Germany, and experimental evidence was provided as to the shielding integrity w.r.t. radiological safety [7]. The initial envisaged scientific design philosophy, neutron optics (collimation ratios) and imaging capabilities of the facility were also presented at ITMNR-8 and described elsewhere [8].

After several engineering design changes were approved to meet local criteria and manufacturing requirements, the final design of the INDLOVU NRAD facility has been optimized and nearly concluded. The physical dimensions of INDLOVU are as follows: Average height of 3.20 m with a maximum of 5 m at the secondary shutter and collimators. The width and length are 7.411 m and 13.573 m respectively. As indicated in the schematic overview in Fig. 1, the facility has been subdivided into several subsystems which are now discussed in separation and in relation to each other.

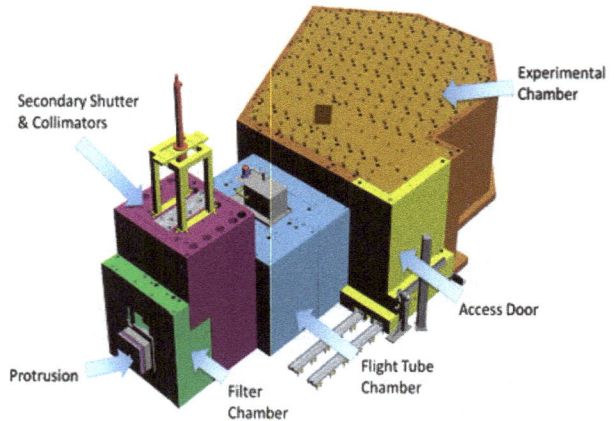

Fig. 1. Schematic model of the INDLOVU NRAD facility with its subsystems and weight (Tonnes)

Filter Exchange chamber: ~18T (Green)
Secondary Shutter: ~82T (Purple)
Flight Tube Chamber: ~78T (Blue)
Experimental Chamber: ~353T (Brown & Yellow)

Neutron Radiography - WCNR-11   Materials Research Forum LLC
Materials Research Proceedings **15** (2020) 11-16   https://doi.org/10.21741/9781644900574-2

Beam tube, Primary Shutter and Protrusion (Fixed Collimator)

The No-2 beam tube is located radial to the SAFARI-1 nuclear research reactor core thus allowing the full neutron and gamma ray spectrum to be guided through the 3.4 m long, 30 cm diameter tube towards the outside of the biological shield. Normally, during long term instrument shutdown, the tube is fitted with a concrete plug and cooled with circulating water. However, during maintenance of the filter chamber, the beam tube will only be filled with water, and, together with the primary shutter, provide adequate shielding from the reactor core. During normal operations the beam tube will be filled with He to allow uninhibited neutron transmission.

The primary shutter consists of a 0.8 m thick layered Fe and B-PE structure, which pneumatically slides horizontally and latches into discrete opened or closed positions. The primary shutter fails in a closed position through a spring mechanism. Between the shutter and the filter exchange chamber, a layered Fe and B-PE structure with a fixed cone-shaped hole is inserted to provide initial collimation of the usable radiation beam. This hole is extended to the outside of the biological shield via another Fe and B-PE laminated extension structure, which serves as the connection interface between the reactor's biological shield and NRAD shielding. Using the opening of the hole as a fixed collimator, an $L/D = 125$ is achieved.

Filter Exchange Chamber

A set of three uncooled filter materials, Bi, Sapphire (Al_2O_3) and B-PE, is located inside the filter exchange chamber. Each filter can individually be moved into or out of the radiation beam using either the Data Acquisition & Control System (DACS) or manually via chains from outside the chamber. This design concept adds to the multi-functionality of INDLOVU and allows for the radiation beam to be manipulated and thus exhibiting an applicable radiation composition for a specific experiment through a selected combination of any of the filters.

Secondary Shutter and Collimators

The collimator-shutter assembly consists of a 123 cm thick vertical sliding block with a concrete section and three fixed B-Fe collimators stacked on top of each other (Fig. 2). The collimators have fixed apertures of 12 mm, 24 mm and 42 mm and provide L/D ratios of 800, 400 and 250 respectively. The collimator-shutter assembly also acts as a secondary radiation shutter comprising of 125 cm thick high density concrete.

Flight Tube Chamber

Lead and B-10 disks are used as additional filter materials to tailor the radiation beam to the experimental requirements, and are positioned in the flight tube chamber just after the collimators. The B-10 disk is small enough to be fast moving to capture the thermal radiation beam in three modes of operation: (a) manually, to absorb the thermal neutron beam during radiography setup to minimize extreme sample activation, (b) automatically, to shut the thermal neutron beam during the tomography process to minimize extreme sample activation and (c) to be used as a thermal neutron filter in combination with the other radiation filters, to obtain either a more prominent gamma-ray, or fast neutron radiation spectrum.

Fig. 2. Three B-Fe collimators: with different aperture openings (D) of 12 mm, 24 mm and 42 mm respectively.

The radiation flight tubes are cylindrical aluminum channels, comprising of five individual sections aligned along the optical axis, extending from the fast shutter mechanism up to the translation table and sample stage (Fig. 3). The tube sections have each a different diameter,

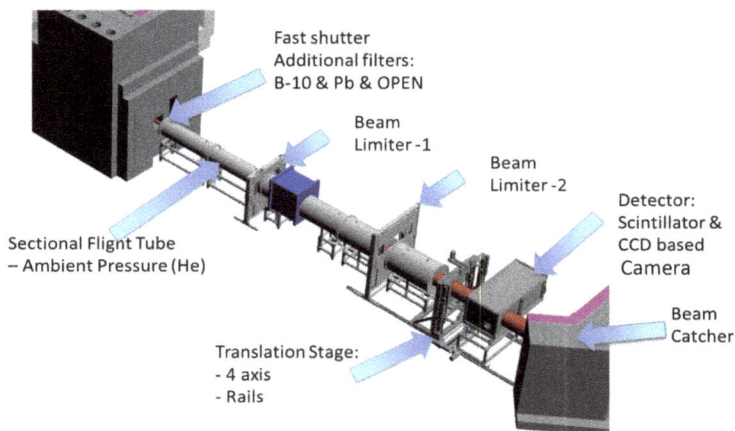

Fig. 3. Schematic diagram of the instrumentation in the flight tube chamber and experimental chamber.

increasing towards the detector to accommodate the beam divergence. He at ambient pressure as fill gas allows uninhibited neutron transport by excluding unwanted radiation scatter from air. The tube sections can be separated and removed to enable installation of additional neutron optic elements (e.g. a periscope) in the future. The mobility of the tube sections allows for the movement of the detector system towards the source for an increase in radiation flux for the envisaged addition of a micro-focus neutron camera.

Two beam limiters, a primary limiter inside the flight tube chamber and a secondary limiter fixed to the translation table in the experimental chamber, define the desired cross sectional area of the radiation beam to be dimensionally similar to the outer bounds of the sample.

Experimental Chamber

The flight tube extends into the experimental chamber and ends just before the secondary beam limiter. The experimental chamber mainly houses the sample translation table and neutron detection system. The sample table is mounted on rails which are fixed to the floor and positioned in front of the camera box. These rails enable macro set-up whereby the entire table can manually be moved parallel to the beam and locked in position. The table is equipped with high precision stepper motors to allow four degrees of freedom needed to position the sample in the center of the beam. These stages are parallel and tangential to the beam, vertical translation and rotation. During sample setup, the necessary controls are available to the operator within the experimental chamber enabling the safe initial positioning of the sample. This minimizes sample activation during operation as only minor positioning adjustments will be required while the sample is exposed to the radiation beam. The relative position of the sample w.r.t. the center of the beam and detector only becomes known during operation when the shutters are opened and a radiograph is acquired.

Neutron Radiography - WCNR-11   Materials Research Forum LLC
Materials Research Proceedings **15** (2020) 11-16   https://doi.org/10.21741/9781644900574-2

The first phase of implementation of the detection system entails the installation of a CCD-based digital camera housed in a light-tight a camera box. The box is positioned on rails which enables it to be manually moved parallel to the incoming radiation beam. A removable scintillator screen is attached to the front of the camera box. Two different sized scintillators can be accommodated namely 35×35 cm^2 and 10×10 cm^2. Various scintillator materials are available depending on the application and incident radiation beam spectrum needed. These are 0.05 mm or 0.10 mm thick ZnS:Cu/6-LiF screens, a 0.01 mm thick (GADOX) Gd_2O_2S:Tb/6LiF: 80/20 screen and a 1.5 mm thick PP30 (30% ZnS:Cu) converter. A maximum neutron flux of 1×10^9 neutrons.cm^{-2}.s^{-1} is envisaged, utilizing the full radiation spectrum – see Fig.4 from a McStas simulation of the homogeneous flat radiation field envisaged [9].

A mirror reflects the photons emanating from the scintillator at 90° towards the removable lens of the CCD camera. As the camera is mounted on a translation stage, the optimal field of view (FOV), which is dependent on the size of the sample and the required spatial resolution, can be selected through electronic means by changing the camera to scintillator distance. A focusing protocol can be initialized from NRAD DACS to enable automatic focusing of the lens. The camera box is ventilated and temperature monitored which enables automated shutdown of camera box when the temperature exceeds the upper operational limit. The second phase sees the implementation of an additional detection system positioned closer to the reactor core where the intensified neutron flux (but smaller FOV) will enable a micro neutron CT capability.

The transmitted radiation beam finally terminates in a beam catcher/dump constructed of a layered system of B-PE, Pb and Cd to minimize the radiation scatter onto the detector and thus decreases the background signal.

Safe Operation

Entry to the experimental chamber is through a manual sliding door and is constrained by an independent, separate, hard wired interlocking system that controls the operation of the shutters. A "Last man out" procedure requires the last person leaving the experimental chamber to activate a device (located at the far end within the experimental chamber) with a safety key. The door should then be locked (using the same key) in the closed position within a time limit to render the shutters operational. In addition, the door can only be unlocked when the shutters are fully closed. The

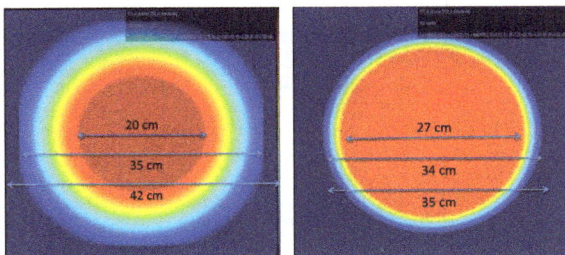

Fig. 4 McStas simulation of the Flat Field sizes and homogeneity achieved by two apertures of INDLOVU:

Aperture:	42 mm		13 mm	
L/D:	250		800	
Flat Field	20 cm		27 cm	
Total Field:		42 cm		34 cm
Screen size:		35 cm		35 cm

status of the shutters is available to the operator on the status screen of the client control computer, a mimic panel next to the access door as well through red and green indication lights placed at strategic positions on the outside of the experimental chamber. These lights are visible from the NRAD control room as well as from the SAFARI-1 control room.

Instrument Control

All motorized stages are controlled by the ANSTO implementation of the SINQ Instrument Control Software (SICS) [10]. SICS allows batch scripting of instrument parameters thereby allowing experimental sequences to be pre-programmed. This DACS functionality enables maximum facility utilization with minimum personnel.

Conclusion

The envisaged INDLOVU NRAD facility will be a unique and highly competitive analytical instrument for the neutron sciences communities of South Africa and abroad. It will offer advanced radiography/tomography functionality by exploiting the high radiation flux available at the SAFARI-1 research reactor. INDLOVU will be unique in its application as it can perform not only neutron radiography using various energy ranges, but also perform radiography utilizing a gamma-ray radiation beam. The INDLOVU beam line instrument will expand the local and international scientific and industrial user community as a multi-functionality analytical tool complementary to X-ray tomography. Future upgrades will include the installation of a periscope for monochromatic neutron radiography and tomography as well as a micro tomography capability.

Acknowledgements

The research and development for the NRAD upgrade at the SAFARI-1 nuclear research reactor was sponsored by the Department of Science and Technology's National Research Foundation (NRF 75433) and the Department of Energy (FUN-DOE-001). The authors would like to thank Dr's Grunauer, Schillinger and Schultz from FRMII, at the Technical University of Munich for the fruitful technical discussions and continuing support for the development of the future INDLOVU.

References

[1]   IAEA, "IAEA Research Reactor Data Base." [Online]. Available: http://nucleus.iaea.org/RRDB/. [Accessed: 15-Dec-2018].

[2]   "International Society for Neutron Radiography (ISNR)." [Online]. Available: https://www.isnr.de/index.php/facilities/user-facilities. [Accessed: 15-Dec-2018].

[3]   H. Bilheux, K. Herwig, S. Keener, and L. Davis, "Overview of the Conceptual Design of the Future VENUS Neutron Imaging Beam Line at the Spallation Neutron Source," *Phys. Procedia*, vol. 69, pp. 55–59, 2015. https://doi.org/10.1016/j.phpro.2015.07.007

[4]   F. Grünauer, "RS-RAD-REP-14001: Monte Carlo Simulations for the SAFARI-1 Reactor and its Instrumentation: Part C: Neutron Radiography Facility.," 2009.

[5]   E. Lehmann, Neutron Imaging Facilities in a Global Context, *J. Imaging*, vol.3, no.4, p.52, 2017. https://doi.org/10.3390/jimaging3040052

[6]   F. De Beer, "Neutron and X-ray Tomography at Necsa," *Jounal South African Inst. Min. Metall.*, vol. 108, no. OCTOBER, pp. 1–8, 2008.

[7]   F. C. De Beer, M. J. Radebe, B. Schillinger, R. Nshimirimana, and M. A. Ramushu, "Upgrading the Neutron Radiography Facility in South Africa (SANRAD): Concrete Shielding Design Characteristics," *Phys. Procedia*, vol. 69, no. October 2014, pp. 1–9, 2015. https://doi.org/10.1016/j.phpro.2015.07.017

[8]   F. C. De Beer, F. Gruenauer, M. J. Radebe, T. Modise, and B. Schillinger, "Scientific design of the new neutron radiography facility (SANRAD) at SAFARI-1 for South Africa," *Phys. Procedia*, vol. 43, pp. 34–41, 2013. https://doi.org/10.1016/j.phpro.2013.03.004

[9]   http://www.mcstas.org/ (Visited 20 Dec 2018)

[10]   H. Heer, M. Könnecke, and D. Maden, "The SINQ instrument control software system," *Physica B: Condensed Matter*, vol 241-243, pp 124-126, 1997. https://doi.org/10.1016/S0921-4526(97)00528-0

Neutron Radiography - WCNR-11   Materials Research Forum LLC
Materials Research Proceedings **15** (2020) 17-22   https://doi.org/10.21741/9781644900574-3

Reviving and Extending the Neutron Imaging Capabilities at the Penn State Breazeale Reactor

Robert Zboray [1,a*]

[1]Department of Mechanical and Nuclear Engineering, The Pennsylvania State University, 233 Reber Building, University Park, PA 16802, USA

[a]rzz65@psu.edu

Keywords: Neutron Imaging, TRIGA, Core Moderator, Bright Flash Imaging, Multi-Spectral Imaging, Fast Neutron Imaging

Abstract. The Penn State Breazeale Reactor (PSBR), a 1 MW TRIGA type reactor has been utilized successfully for neutron imaging in the past. Presently a single beam line with a thermal spectrum is utilized for imaging, however the reactor is just undergoing a major refurbishment involving the installation of a new core moderator assembly. This enables the establishment of several new neutron beam lines around the reactor including three with a cold spectrum. We report here on the ongoing developments and future plans to revive and extend neutron imaging capabilities at PSBR.

Introduction

The Breazeale reactor operated by the Radiation Science and Engineering Center (RSEC) of the Pennsylvania State University, is a 1-MW TRIGA type reactor. It has been successfully utilized in the past for neutron imaging and its imaging beam line, which has been operational until very recently (mid 2018), has been upgraded on few occasions. The layout of this old imaging beam line is shown in Fig. 1. It was an ASTM E 544 Category 1 facility with a tangential collimator. It featured a steady neutron flux of $1.7*10^7$ n/cm^2/s at full power and has an L/D collimation ratio of around 150 at the sample position. Further details on the beam line and on past imaging activities, can be found in [1] and [2].

A drawback of the old moderator and beam port design was that the beam ports, except for the imaging beam line #4 on Fig. 1, were not axially aligned with the core center therefore could not feature the highest possible flux. This was related to a change of the fuel type of PSBR from high-enriched uranium to low-enriched TRIGA fuel. Furthermore, the imaging beam line #4 had a pretty high gamma contribution. This was mainly attributed to the thermal neutron capture reaction by hydrogen in pool water which mainly takes place at the sides of the old core-moderator assembly (see Fig. 1) as pointed out in [3]. The high gamma background was not only disturbing for certain imaging applications (see below) but also for some other neutron irradiation techniques sporadically performed at the beam line.

Due to these deficiencies and in a broader context to enhance neutron science capabilities around the reactor, a conceptual design of a new core moderator assembly and new set of beam ports have been prepared using extensive Monte Carlo and thermalhydraulic simulations [3]. This has resulted in a crescent-shaped core-moderator assembly filled with D2O and a number of radial beam ports (see Fig. 2 left). The main advantages of this new design are: larger number and variety of beamlines aligned axially with the core, higher fluxes than in the existing beam line(s), improved n/γ ratios due to the reduced $^1H(n, \gamma)H^2$ reaction by the new moderator geometry and improved Cd ratios. The new thermal beam line NBP4 will be exclusively dedicated to neutron imaging. Some of the design parameters of this beam line are as follows:

the thermal neutron flux is around $7*10^7$ n/cm^2/s and the fast flux (> 2MeV) is about $6*10^6$ n/cm^2/s. These are unfiltered flux estimates. Depending on the type and thickness of the applied in-beam filters (Bi, sapphire) these values can decrease up to a factor five for thermal and a factor twenty for fast flux. Regarding flexible beam filtering and spectral shaping see the text below. The expected L/D collimation ratio is around 130 at the sample position (1 m away from the outer wall of the biological shielding). One of the new cold beam lines will also be partly used for neutron imaging. Furthermore, the exploratory free beam NBP2 could also be occasionally used for imaging related studies.

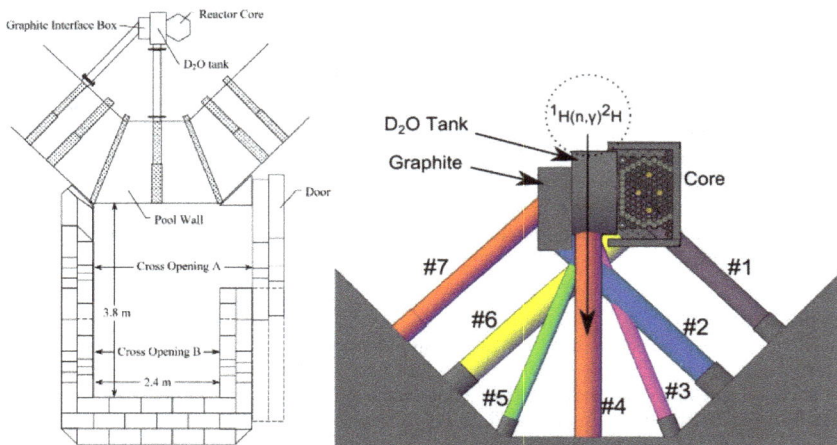

Fig. 1: Layout of the old neutron imaging beam line at the PSBR. Left: the beam cave is shown together with the beam port and the biological shield. Right: the old D2O moderator tank and the old beam ports. #4 is the imaging beam line.

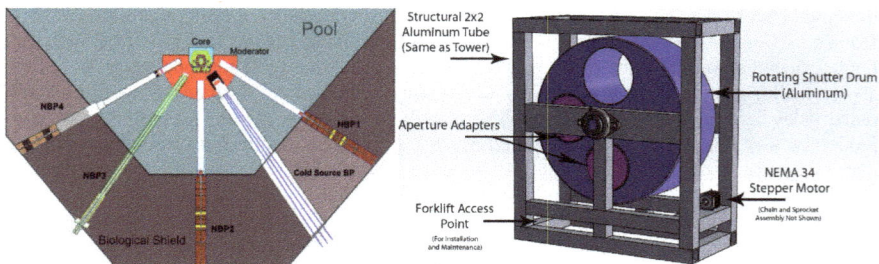

Fig. 2: Left: layout of the new core moderator assembly with the planned new radial beam port and beam lines. The image is taken from [3]. Right: Rotary beam shutter with several apertures enabling easy and fast shutting down of theNBP4 beam and also a flexible in-beam filtering.

Neutron Radiography - WCNR-11   Materials Research Forum LLC
Materials Research Proceedings **15** (2020) 17-22   https://doi.org/10.21741/9781644900574-3

Imaging technique developments at PSBR

General thermal neutron imaging. Not long before the reactor refurbishment, we have started extensive developments to enhance the imaging capabilities and activities at PSBR. First, efficient state-of-the-art digital, camera-based detectors have been introduced. One such detector for general thermal neutron imaging purposes have been obtained from NeutronOptics [4] featuring a 200×250 mm^2 field-of-view (FOV) and a CCD camera with a 1 inch Sony ICX694ALG EXview HAD CCD II chip with 2750×2200 pixels. Dark current is strongly reduced (0.002 e/pix/s @-10 °C) by thermo-electric cooling at -35C. The camera has 16 bit digital output and a high, ~75% quantum efficiency (QE). The camera is coupled with a high-resolution f/1.4 Fujinon C-mount lens. A photo of the camera box and the CCD is shown in Fig. 3.

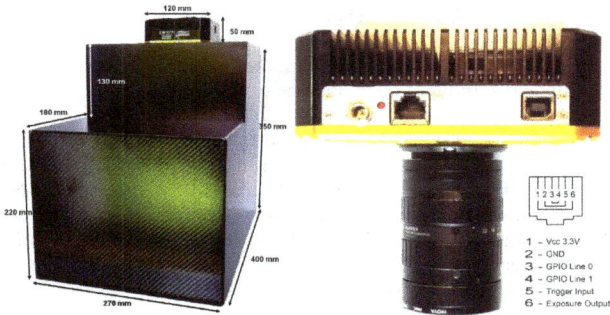

Fig. 3: Photo of the detector box showing its front (sensitive) face (left) and of the CCD camera (right).

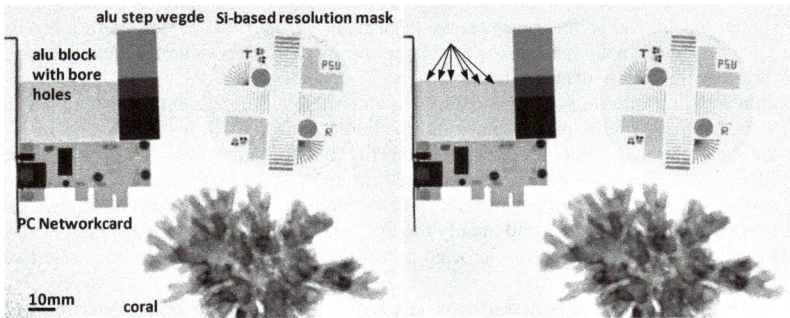

Fig. 4: Two images of several test objects obtained at the old thermal imaging beam line using the detector shown in Fig. 3 in conjunction with a high efficiency 400 um LiF/ZnS screen at only 10% power of the reactor with as short as 5s exposure (left) and a longer 60s exposure (right).

Some example neutron images of several test objects are shown in Fig. 4 taken by the above detector in combination with a 400 µm thick, high efficiency LiF/ZnS(Cu) screen from Scintacor [5] on the detector box to illustrate imaging quality at the beamline using the above detector. The high detection efficiency of the screen enabled to take reasonable quality radiographs at only 10% reactor power and with exposure as low as 5s (see Fig. 4 left). The image shows visually observable noise but the overall quality is still reasonable (SNR=22.2) only the fine bore holes in the aluminum block go undetected due to the noise as is illustrated compared to Fig. 4 on the right, which is obtained for 60s exposure and with very good quality (SNR=62.1). The arrows indicated the bore holes that are from left to right 3.5, 2, 1.5, 1 mm in diameter in a 20 mm thick aluminum block and the rightmost, 1.5 mm, is filled with a steel bar. The test objects in Fig. 4 represent different levels of structural complexity (from a simple aluminum step wedge to a PCB board of a PC network card and a highly structured coral) and cover a broad variety of material compositions (organics, metals, minerals etc.). Note that due to the thick screen the spatial resolution, is estimated based on the edge spread function of the upper part of the aluminum step wedge, is only around 500 um, however the primary purpose of the screen was to use it for bright flash imaging (see below).

Bright flash neutron imaging. TRIGA reactors, due to their special fuel composition, allow operation in pulse mode [9]. In pulse operation one of the control rods, in case of the Penn State reactor the transient rod, is rapidly ejected to a certain extent from the core with the help of a pneumatic system making the reactor prompt supercritical and engaging it on a power excursion. The power excursion is then mitigated by the strong prompt negative feedback from fuel temperature on reactivity brought about by the fuel composition (uranium mixed intimately with the zirconium hydride moderator) and the reactor power settles down to practically zero again. Peak power values can reach almost 1 GW [6]. For the definition of reactivity in $ and for details on prompt supercritical reactor physics the reader is referred to [9].

This paves the way towards high-speed neutron radiography for fast transient processes. Such processes require high acquisition rate potentially in the kilo frame per second (kfps) range to capture them with minimal motion blur. However, due the short exposure times a high neutron flux is needed to obtain images with acceptable signal-to-noise ratio and statistics.

We have recently demonstrated up to 4 kfps bright flash radiography of a two-phase bubbly flow in a simple bubbler made of aluminum using the old imaging beam line. Some illustrative results are shown in Fig. 5. For this purpose, we have replaced the CCD camera with a legacy high-speed CMOS camera with special, high brightness optics. We report on the details of our bright flash imaging developments elsewhere.

Fast neutron imaging, multi-spectral imaging. Besides thermal neutron imaging we plan to develop fast neutron or multi-spectral imaging methods. This is will be implemented by designing the new thermal imaging beam line in the most flexible way. It will feature (re)moveable in-beam filters to enable multi spectrum imaging ranging from the standard thermal to epithermal and fast energy ranges. Typically, sapphire filters are applied in thermal beam lines to reduce both the gamma and mainly the fast neutron contribution. We plan to have a removable sapphire filter seated deep in the beam port not so far away from the new moderator tank.

We have recently tried to perform fast neutron imaging using the old beam line. The test object was a massive Li-fueled power source shown in Fig.6. The detector was equipped with a 3.8 mm thick PP/ZnS:Cu fast neutron imaging screen from RC Tritec AG, Switzerland [7]. The thermal content of the beam has been removed by applying a 1-cm thick Mirrobor (a borated rubber matt [8]) in-beam filter to avoid unnecessary activation of the sample and the detector. Even exposure times as high as 12 minutes delivered suboptimal image quality (see Fig.6), though the upper oxidized area can be clearly distinguished from the lower elemental Li parts. The low image quality was likely due to the insufficient fast flux in the old tangential beam line, that was looking into the middle of the D_2O tank. Furthermore, a permanently placed sapphire filter was also present in the beam path of that beam line. For these reasons we plan the new beam line with a removable sapphire filter and the MCNP simulations also show that the new radial beam line will have a significantly higher native fast neutron content.

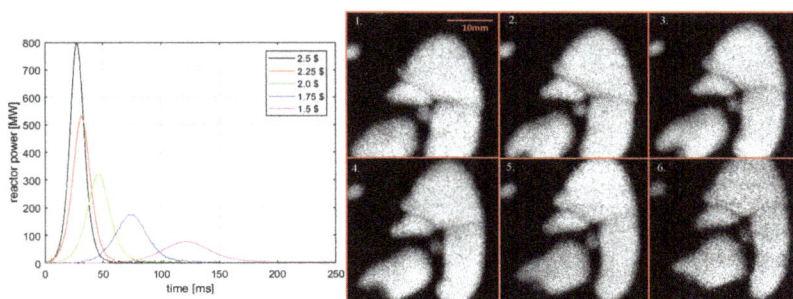

Fig. 5: Left: measured pulse shape at PSBR for a set for different reactivity insertions. Right: image sequence of bubbly, two-phase flow taken at 4 kfps for a 2.5$ pulse. The sequence 1.-6. shows the instantaneous gas volume fraction distribution (white-gas, black-liquid) for every 10th image taken, i.e. with a time gap of 2.5 ms. Image frames are colored red for better visualization.

Fig. 6: Photo (left) and fast neutron radiograph (right) of a Li-filled power source. Imaging as done at the old PSBR beam line using a 3.8 mm thick PP/ZnS:Cu screen.

As the new beam line leaves the biological shield and enters the new beam cave, we plan to place a rotary beam shutter as is shown on the right of Fig. 2. The rotary beam shutter will have several apertures enabling an easy and fast shutting down of the beam and a flexible in-beam filtering. One aperture will be open for normal beam operation and it will also give access to the removable sapphire filter seated deeper inside the beam port towards the reactor core. This aperture could also house a removable Mirrobor filter for fast neutron imaging. Another aperture will shut the beam down, while a third one will have a few mm thick Cd plate filter to enable epithermal neutron imaging. Switching easily from a thermal to an epithermal beam could have a lot of interesting applications for dual-spectrum or differential imaging of samples that contains component(s) with distinctly different attenuation coefficient e.g. strong resonance(s) in the epithermal range. Those components could be quantified by differential imaging. Epithermal imaging is also promising alternative for samples that are slightly too attenuating from thermal neutrons but not that much that only fast neutrons could be used to examine them.

Summary

The Penn State Breazeale Reactor has just recently undergone a major refurbishment. A new core-moderator assembly has been designed and built enabling a lot of new capabilities or more optimal conditions for neutron beam science. One of the major topics that is being extensively and dynamically developed is neutron imaging. New imaging beam lines are going to be established with thermal and cold spectra. A major effort is also ongoing on imaging detector and imaging methodology development. These include besides general state-of-the-art thermal neutron imaging high-speed, bright flash neutron radiography, fast neutron imaging and dual/multi-spectrum imaging such as thermal/epithermal differential imaging. First results and the corresponding developments have been reported here.

Acknowledgement

The author is indebted to Mr. Andrew Bascom for providing the results of his MCNP simulations for the flux in new imaging beam port. The author is grateful to Prof. Kenan Unlu, the director of RSEC, for his continued interest and support of his work.

References

[1] J. Brenizer, M.M. Mench, K. Ünlü, K. Heller, A. Turhan, L. Shi, J.J. Kowal, available at http://www.rsec.psu.edu (accessed July 10th, 2018)

[2] J.M. Cimbala, J.S. Brenizer, A. Po-Ya Chuang, S. Hanna, C. Thomas Conroy, A. El-Ganayni, et al. "Study of a loop heat pipe using neutron radiography" Appl Rad. Iso., 61 (2004), pp. 701-705. https://doi.org/10.1016/j.apradiso.2004.03.104

[3] D. Ucar, "Modeling and Design of a new core-moderator assembly and neutron beam ports for the Penn State Breazeale nuclear reactor" Ph.D. Thesis, The Pennsylvania State University, 2013.

[4] http://www.neutronoptics.com/cameras.html

[5] Scintacor, Neutron Screens, available at https://scintacor.com/wp-content/uploads/2015/09/Datasheet-Neutron-Screens-High-Res.pdf, (accessed on July 20th, 2018)

[6] D. L. Hetrick, Dynamics of Nuclear Reactors, University of Chicago Press, 1971

[7] RC Tritec AG, T., 2017. Scintillators. http://www.rctritec.com/en/scintillators.html.

[8] MirroBor, 2012. Radiation shielding. http://www.mirrotron.kfkipark.hu/shield.html.

[9] D. L. Hetrick, Dynamics of Nuclear Reactors, University of Chicago Press, 1971

Neutron Radiography - WCNR-11
Materials Research Proceedings **15** (2020) 23-28

Materials Research Forum LLC
https://doi.org/10.21741/9781644900574-4

PSI 'Neutron Microscope' at ILL-D50 Beamline - First Results

Pavel Trtik[1, a *], Michael Meyer[1], Timon Wehmann[1], Alessandro Tengattini[2,3],
Duncan Atkins[3], E.H. Lehmann[1], Markus Strobl[1]

[1]Paul Scherrer Institut (PSI), Laboratory for Neutron Scattering and Imaging, CH-5232 Villigen PSI, Switzerland

[2]Université Grenoble Alpes (UGA), CNRS, Grenoble INP, 3SR, Grenoble 38000, France

[3]Institut Laue-Langevin (ILL), 71 Avenue des Martyrs, Grenoble 38000, France

[a] pavel.trtik@psi.ch

*corresponding author: Pavel Trtik (PSI)

Keywords: Neutron, Microscope, High-Resolution Neutron Imaging, Cold Neutrons, Beam Intensity

Abstract. A high-resolution neutron imaging system referred to as 'Neutron Microscope' (NM) has been recently established as a piece of instrumental equipment at the Paul Scherrer Institut (PSI), Switzerland. It is providing the wide user community of the Neutron Imaging and Applied Materials Group (NIAG) with the capability of spatial image resolution below 5 μm at effective pixel sizes of 1.3 μm. The NM has been designed as a portable, self-contained system that can be moved between beamlines at PSI with only moderate effort. In this contribution, we report on the first results and experience with the Neutron Microscope externally, at a beamline of another neutron source outside the Swiss Spallation Neutron Source (SINQ). In June 2018, NM has been transported to the Institute Laue-Langevin (ILL) and was successfully installed at the D50 beamline for four days. A gadolinium based Siemens star produced at PSI has been used for the assessment of the spatial resolution. The spatial resolution achieved using the Neutron Microscope at ILL-D50 equalled 4.5 μm. Above that, several high-resolution tomographies of various samples were acquired, of which an illustrative example is presented.

Introduction

High spatial resolution neutron imaging is a fast developing area driven by the demands from the user community (for example that of electrochemistry [1]). Provided that 'high resolution neutron imaging' is loosely defined as neutron imaging with the capability to resolve about 10 μm structures or better, there are several approaches that demonstrate such capability [2],[3],[4],[5],[6],[7]. At PSI, high resolution neutron imaging has been advanced with the project 'Neutron Microscope'. In this project, a detector based on a high-numerical aperture objective [8] combined with very thin, though efficient, isotopically-enriched 157-gadolinium oxysulfide scintillator screens [9] was developed and led to the achievement of about five μm isotropic spatial resolution in 2D [10].

Despite these clear advances in the field of the high spatial resolution neutron imaging detectors, it is the available flux (even at rather powerful neutron sources) that sets limitations on both the achievable resolution and the performable experiments. Grosse & Kardjilov [11] have recently proposed a useful theoretical model that estimates the time necessary for neutron radiography/tomography to reveal unambiguously structures with a given contrast at a specific required spatial resolution. Even though the model seems to provide a conservative estimate [12], it is clear that -for certain contrast conditions- neutron tomography of very high spatial resolution requires prohibitively long exposure times even at sources as powerful as SINQ, PSI.

Neutron Radiography - WCNR-11   Materials Research Forum LLC
Materials Research Proceedings **15** (2020) 23-28   https://doi.org/10.21741/9781644900574-4

Even for favourable contrast conditions, such as in the case of the neutron tomography of a small porous gold sample [13], the exposure times of several days are required at BOA beamline [14] to achieve approximately 10 μm true spatial resolution in 3D. It is therefore clear that the high spatial resolution neutron imaging applications are bound to seek the highest neutron flux sources.

As the NM has been designed as self-contained detector that can be transported between the beamlines at PSI with only moderate effort, it can also be moved to other neutron sources. Because the neutron source of ILL provides outstanding neutron flux conditions, the NM detector system was transferred recently from PSI to the D50 beamline [15] at ILL and its performance under cutting edge flux conditions has been tested With an estimated neutron flux of 6×10^9 n/cm^{2}s^{-1}, the D50 beamline provides currently the highest available flux of cold neutrons for imaging applications. In the following we provide some results of this initial campaign.

Installation of NM at ILL-D50

The entire 'Neutron Microscope' instrumentation has been transported to and installed at ILL within one day. The NM has been installed at the most downstream position inside the current D50 beamline bunker (as shown in Figures 1a and 1b).

Figure 1 – PSI 'Neutron Microscope' at ILL-D50 beamline: (a) being installed through the open roof into the bunker, (b) positioned at the most downstream position of the ILL-D50, (c) fitted with B$_4$C sheets and other borated shielding materials. The red dashed arrow in Figure 2b indicates the neutron beam.

It is worth noting that the installation had, despite the independent control of the system, taken only about 6 hours, after which the NM was ready for the first images to be acquired. A significant part of the installation time has been spent on fitting the NM with shielding material (see Figure 1c) in order to avoid activation. The scintillator screen of the NM has been positioned at 11.13 m downstream the 30 mm-diameter pinhole, thus providing collimation ratio

Neutron Radiography - WCNR-11   Materials Research Forum LLC
Materials Research Proceedings 15 (2020) 23-28   https://doi.org/10.21741/9781644900574-4

of 371. In the next step of the installation, the NM has been focused using a standard resolution test object - a gadolinium-based Siemens star [16]. The focusing procedure was relatively quick again due to the superior neutron flux.

Results

In the first experiment, to establish the resolution capability achieved, eighty images of the Siemens star pattern, fifty open beam images and ten dark current images were acquired. All images were recorded with 30 seconds exposure time and the test object was positioned in the close vicinity of the scintillator screen (scintillator-sample distance smaller than 0.5 mm). The scintillator was an approximately 3.5 micrometres thick ^{157}Gd isotopically-enriched gadolinium oxysulfide screen. The substrate of the screen consisted of a silicon wafer coated with a 200-nm iridium layer for light output enhancement [17]. The images were acquired using a sCMOS camera (Hamamatsu ORCA Flash 4.0, pixel size 6.5 μm). Thanks to the 5-fold magnification of the NM optics, the pixel size (pixel resolution) of the acquired images was equal to 1.3 micrometers. Two separate open-beam corrected images based on 40 individual radiographs of the test object each were created. These were used for Fourier ring correlation and for the visual inspection of the achieved spatial resolution. One of these images is shown in Figure 2 (left) while the right hand side image presents an enlarged image of the Siemens star centre.

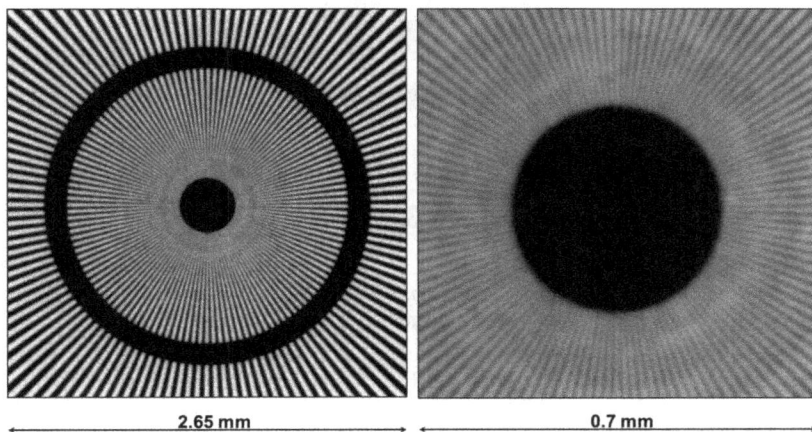

2.65 mm   0.7 mm

Figure 2 – (left) Neutron radiography of a Gd based Siemens star, (right) close-up of the centre of the Siemens star of the left hand side image clearly revealing the ends of the individual 4.5 micrometres spokes.

It is clearly visible that the thinnest ends of the individual spokes of the Siemens star can be resolved in the image, in particular in the magnified detail on the right hand side of Fig. 2. The size of the thinnest spokes is equal to 4.5 μm (line pair: 9 μm). For the quantitative analyses, Fourier ring correlation [18] was applied to the Siemens star images and resulted in a measured spatial resolution of 4.2 μm.

Subsequently, after assessing the resolution experimentally, several small static samples were tomographed. The tomographed samples included pieces of Zircaloy nuclear fuel cladding [19], bits of additively manufactured gold alloys, and a cylinder of a diameter of approximately 2.5 mm of a gold-lead alloy.

Neutron Radiography - WCNR-11
Materials Research Proceedings **15** (2020) 23-28

Materials Research Forum LLC
https://doi.org/10.21741/9781644900574-4

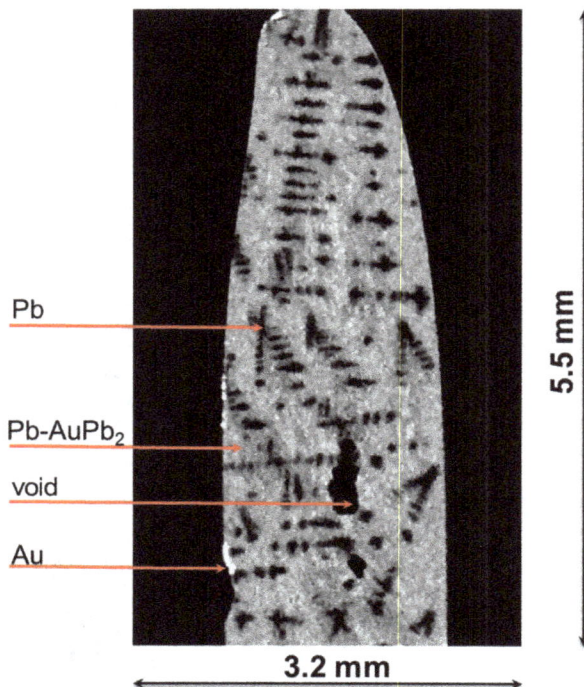

Figure 3 – Example of a neutron microtomography using the PSI 'Neutron Microscope' at the ILL-D50 beamline: A vertical slice from a neutron microtomography dataset showing dendritic microstructures of lead, voids and gold in a sample of a gold-lead alloy.

The detailed description of the results of these neutron microtomographies goes beyond the purpose and the scope of this paper. However, Figure 3 provides an example of the high quality of the resulting datasets showing a randomly chosen vertical slice from the microtomographic dataset revealing in detail the dendritic microstructure of lead in a gold-lead alloy [20] with unprecedented spatial resolution. From the point of view of the temporal resolution, the presented lead-gold alloy microtomography required approximately 12 hours of beamtime only - including the acquisition of open beam, dark current and black body [21] images.

Discussion & Outlook

From the practical and logistic point of view, the installation of NM at ILL was a rather swift procedure. One of the issues that need to be addressed when using a detector at another neutron facility is the activation of the system. Naturally, the pieces of NM that were exposed to the direct beam (i.e. scintillator screen/holder and the mirror) were expected to be activated beyond a level for immediate release. However, readily available duplicates of these pieces enable transfer and usability of the system elsewhere within about 24-48 hours.

Despite taking care with shielding the NM from any unnecessary other neutrons, few parts of the rest of the instrumentation were very slightly activated (likely due to neutrons scattered from the sample/scintillator/mirror) at the end of the campaign (~0.15 μSv/h). This activation led to the necessity to prolong the stay of NM at ILL for another 24 hours. The procurement of more efficient shielding for NM can alleviate this situation in the future.

Regarding the results themselves, it was shown that the favourable combination of NM with the superior flux at ILL-D50 can provide highest spatial resolution and quality of neutron imaging data both in 2D and in 3D with significantly reduced exposure times and hence higher efficiency. It can be concluded that highest resolution imaging capabilities at ILL are highly desirable for the neutron imaging community. This could enable not only a higher throughput of samples with scanning requirements at highest currently achievable resolutions as demonstrated in this contribution, but also achieving better than current temporal resolutions in time-resolved studies [22],[23]. In addition, there appears to be still room for (i) further pushing the spatial resolution limit towards even closer to 1 μm and (ii) combination of the detector with neutron optics (e.g. [24]).

Summary

PSI 'Neutron Microscope' was successfully installed and tested at ILL-D50 beamline in Grenoble for 4 days. The visual assessment and the Fourier ring correlation criterion of the images of the standard resolution test pattern (PSI gadolinium Siemens star) resulted in a measured spatial resolution better than 4.5 μm. Also, several high-resolution tomographies of relevant static samples were acquired within the available allocated beamtime window. They demonstrated the highest resolution neutron microtomography and underlined the capability of the fast measurements at extreme spatial resolutions at ILL-D50 with the NM technology developed at PSI.

Acknowledgement

Dariusz Gawryluk, Marc Raventós, Eric Ricardo Carreon and Marisa Medarde (all PSI) are kindly thanked for the provision of the gold-lead alloy sample.

References

[1]   P. Boillat, E. H. Lehmann, P. Trtik, and M. Cochet, "Neutron imaging of fuel cells – Recent trends and future prospects," *Curr. Opin. Electrochem.*, vol. 5, no. 1, 2017. https://doi.org/10.1016/j.coelec.2017.07.012

[2]   F. Krejci *et al.*, "Development and characterization of high-resolution neutron pixel detectors based on Timepix read-out chips," *J. Instrum.*, vol. 11, no. 12, 2016. https://doi.org/10.1088/1748-0221/11/12/C12026

[3]   A. Faenov *et al.*, "Lithium fluoride crystal as a novel high dynamic neutron imaging detector with microns scale spatial resolution," *Phys. Status Solidi Curr. Top. Solid State Phys.*, vol. 9, no. 12, 2012. https://doi.org/10.1002/pssc.201200185

[4]   A. S. Tremsin *et al.*, "High resolution neutron imaging capabilities at BOA beamline at Paul Scherrer Institut," *Nucl. Instruments Methods Phys. Res. Sect. A Accel. Spectrometers, Detect. Assoc. Equip.*, vol. 784, 2015. https://doi.org/10.1016/j.nima.2014.09.026

[5]   S. H. Williams *et al.*, "Detection system for microimaging with neutrons," *J. Instrum.*, vol. 7, no. 2, 2012. https://doi.org/10.1088/1748-0221/7/02/P02014

[6]   D. S. Hussey, J. M. LaManna, E. Baltic, and D. L. Jacobson, "Neutron imaging detector with 2 μm spatial resolution based on event reconstruction of neutron capture in gadolinium oxysulfide scintillators," *Nucl. Instruments Methods Phys. Res. Sect. A Accel. Spectrometers, Detect. Assoc. Equip.*, 2017. https://doi.org/10.1016/j.nima.2017.05.035

[7]   M. Morgano, P. Trtik, M. Meyer, E. H. Lehmann, J. Hovind, and M. Strobl, "Unlocking high spatial resolution in neutron imaging through an add-on fibre optics taper," *Opt. Express*, vol. 26, no. 2, 2018. https://doi.org/10.1364/OE.26.001809

[8]   P. Trtik *et al.*, "Improving the Spatial Resolution of Neutron Imaging at Paul Scherrer Institut

- The Neutron Microscope Project," in *Physics Procedia*, 2015, vol. 69. https://doi.org/10.1016/j.phpro.2015.07.024

[9]   P. Trtik and E. H. Lehmann, "Isotopically-enriched gadolinium-157 oxysulfide scintillator screens for the high-resolution neutron imaging," *Nucl. Instruments Methods Phys. Res. Sect. A Accel. Spectrometers, Detect. Assoc. Equip.*, vol. 788, 2015. https://doi.org/10.1016/j.nima.2015.03.076

[10]   P. Trtik and E. H. Lehmann, "Progress in High-resolution Neutron Imaging at the Paul Scherrer Institut-The Neutron Microscope Project," *J. Phys. Conf. Ser.*, vol. 746, no. 1, 2016. https://doi.org/10.1088/1742-6596/746/1/012004

[11]   M. Grosse and N. Kardjilov, "Which Resolution can be Achieved in Practice in Neutron Imaging Experiments? - A General View and Application on the Zr - ZrH2and ZrO2- ZrN Systems," in *Physics Procedia*, 2017. https://doi.org/10.1016/j.phpro.2017.06.037

[12]   M. Grosse, P. Trtik, and B. Schillinger, in preparation

[13]   P. Trtik, "Neutron microtomography of voids in gold," *MethodsX*, vol. 4, 2017. https://doi.org/10.1016/j.mex.2017.11.009

[14]   M. Morgano, S. Peetermans, E. H. Lehmann, T. Panzner, and U. Filges, "Neutron imaging options at the BOA beamline at Paul Scherrer Institut," *Nucl. Instruments Methods Phys. Res. Sect. A Accel. Spectrometers, Detect. Assoc. Equip.*, vol. 754, pp. 46–56, 2014. https://doi.org/10.1016/j.nima.2014.03.055

[15]   D. Dauti, A. Tengattini, S. Dal Pont, N. Toropovs, M. Briffaut, and B. Weber, "Analysis of moisture migration in concrete at high temperature through in-situ neutron tomography," *Cem. Concr. Res.*, no. 111, pp. 41–55, 2018. https://doi.org/10.1016/j.cemconres.2018.06.010

[16]   C. Grünzweig, G. Frei, E. Lehmann, G. Kühne, and C. David, "Highly absorbing gadolinium test device to characterize the performance of neutron imaging detector systems," *Rev. Sci. Instrum.*, vol. 78, no. 5, 2007. https://doi.org/10.1063/1.2736892

[17]   J. Crha, "Light Yield Enhancement of 157-Gadolinium Oxysulfide Scintillator Screens for the High-Resolution Neutron Imaging," *MehtodsX*, vol. 6., pp. 107-114. https://doi.org/10.1016/j.mex.2018.12.005

[18]   M. Van Heel and M. Schatz, "Fourier shell correlation threshold criteria," *J. Struct. Biol.*, vol. 151, no. 3, 2005. https://doi.org/10.1016/j.jsb.2005.05.009

[19]   W. Gong, P. Trtik, S. Valance, and J. Bertsch, "Hydrogen diffusion under stress in Zircaloy: High-resolution neutron radiography and finite element modeling," *J. Nucl. Mater.*, vol. 508, pp. 459–464, 2018. https://doi.org/10.1016/j.jnucmat.2018.05.079

[20]   M. Medarde *et al.*, "Lead–gold eutectic: An alternative liquid target material candidate for high power spallation neutron sources," *J. Nucl. Mater.*, vol. 411, no. 1, pp. 72–82, 2011. https://doi.org/10.1016/j.jnucmat.2011.01.034

[21]   P. Boillat *et al.*, "Chasing quantitative biases in neutron imaging with scintillator-camera detectors: A practical method with black body grids," *Opt. Express*, vol. 26, no. 12, 2018. https://doi.org/10.1364/OE.26.015769

[22]   P. Trtik *et al.*, "Release of internal curing water from lightweight aggregates in cement paste investigated by neutron and X-ray tomography," *Nucl. Instruments Methods Phys. Res. Sect. A Accel. Spectrometers, Detect. Assoc. Equip.*, vol. 651, no. 1, 2011. https://doi.org/10.1016/j.nima.2011.02.012

[23]   J. Terreni *et al.*, "Observing Chemical Reactions by Time-Resolved High-Resolution Neutron Imaging," *J. Phys. Chem. C*, vol. 122, no. 41, pp. 23574–23581, 2018. https://doi.org/10.1021/acs.jpcc.8b07321

[24]   M. Yamada *et al.*, "Pulsed neutron-beam focusing by modulating a permanent-magnet sextupole lens", *Progress Theoret Experim Phys*. no. 4, 2015, 043G01. https://doi.org/10.1093/ptep/ptv015

Neutron Radiography - WCNR-11   Materials Research Forum LLC
Materials Research Proceedings **15** (2020) 29-34   https://doi.org/10.21741/9781644900574-5

Wavelength-Resolved Neutron Imaging on IMAT

W. Kockelmann[1, a *], T. Minniti[1, b], R. Ramadhan[1,2, c], R. Ziesche[1,3 d],
D.E. Pooley[1, e], S.C. Capelli[1, f], D. Glaser[4, g], A.S. Tremsin[5, h]

[1]STFC-Rutherford Appleton Laboratory, ISIS Facility, Harwell, OX11 0QX, UK

[2]University of Coventry, Centre Manufacturing and Materials Engineering, Coventry, CV1 5FB, UK

[3]University College London, Torrington Place, London, WC1E 7JE, UK

[4]Council for Scientific and Industrial Research (CSIR), Pretoria, South Africa

[5]University of California at Berkeley, Space Science Laboratory, CA 94720 Berkeley, USA

[a]winfried.kockelmann@stfc.ac.uk, [b]triestino.minniti@stfc.ac.uk, [c]ramadhar@uni.coventry.ac.uk,
[d]ralf.ziesche.16@ucl.ac.uk, [e]daniel.pooley@stfc.ac.uk, [f]silvia.capelli@stfc.ac.uk,
[g]dglaser@csir.co.za, [h]ast@ssl.berkeley.edu

* corresponding author

Keywords: Neutron Imaging, Neutron Radiography, Wavelength-Resolved Imaging, Energy-Selective, Energy-Dispersive, Time of Flight, Bragg Edge

Abstract. The IMAT project is now well into its commissioning phase, and a user programme for neutron imaging has started on the new instrument at ISIS TS2. The performance parameters for white-beam tomography and energy-dispersive neutron imaging had been determined earlier. Here we report on a further evaluation of the wavelength-resolving imaging options on IMAT, including selection of neutron wavelength bands using disk choppers as well as energy-dispersive Bragg edge imaging using time-resolving detectors. We review the instrument parameters of IMAT relevant for energy-resolved imaging, and present one example of residual strain imaging.

Introduction

Energy-selective and energy-dispersive neutron imaging has been developed over the past decade as discussed at many Neutron Wavelength Dependent Imaging (NeuWave) workshops [1] and as documented by many proof-of-concept and materials science studies [2], [3] (and references therein). Energy-selective neutron imaging using crystal monochromators was explored early on at continuous neutron sources [4, 5], whilst energy-dispersive Bragg edge transmission methods were developed on time of flight (TOF) instruments [6]. Further developments of energy-selective analysis at a pulsed neutron source [7] and the development of an MCP detector with high spatial and time resolution [8] helped to advance the field. Several dedicated TOF neutron imaging beamlines were designed in the past few years, with RADEN [9] and IMAT [3, 10] having started operation already.

Here we report on results from the commissioning of the IMAT instrument at the ISIS pulsed neutron source, UK. The basic performance parameters for white-beam tomography and energy-dispersive neutron imaging have been determined earlier [10], and a more application-related characterisation of the instrument was completed recently [11]. Here we continue to discuss the wavelength-resolving imaging options on IMAT, and report on the evaluation of energy-selection related instrument parameters.

Neutron Radiography - WCNR-11   Materials Research Forum LLC
Materials Research Proceedings **15** (2020) 29-34   https://doi.org/10.21741/9781644900574-5

IMAT set-up

IMAT is installed on the coupled 18 K liquid hydrogen (LH2) moderator of beam port W5 on the 10 Hz pulsed source TS2 of ISIS. Fig. 1 shows the main components; details of the instrument setup are given in [3] and [10]. The flight path length of the beamline from the moderator centre to the centre of the sample positioner is 56 m. Two double-disk choppers, a T_0 chopper, five vanadium-foil neutron monitors and different types of TOF detectors use an external trigger indicating at the time of the neutron pulse generation.

On IMAT there are three different detector systems available that take advantage of the TOF option. A gated light-intensified CCD camera can be used with the Messina camera box [3]. A microchannel plate detector (MCP) [8] uses a 2×2 array of Timepix readout chips (512×512 pixels, each 55×55 μm^2) and has a field of view of 28×28 mm^2. An active pixel sensor (GP2) uses the PImMS-2 CMOS and a gadolinium sheet for converting neutrons to electrons for a pixel size of 70×70 μm^2 and for a field of view of 22×22 mm^2 [12]. For each pixel of a TOF neutron camera (be it the gated-CCD, the MCP or the GP2) the time of the neutron arrival relative to the external trigger is measured, with an electronic time resolution of the order of 10 ns, i.e. well below the instrument resolution $\delta\lambda$.

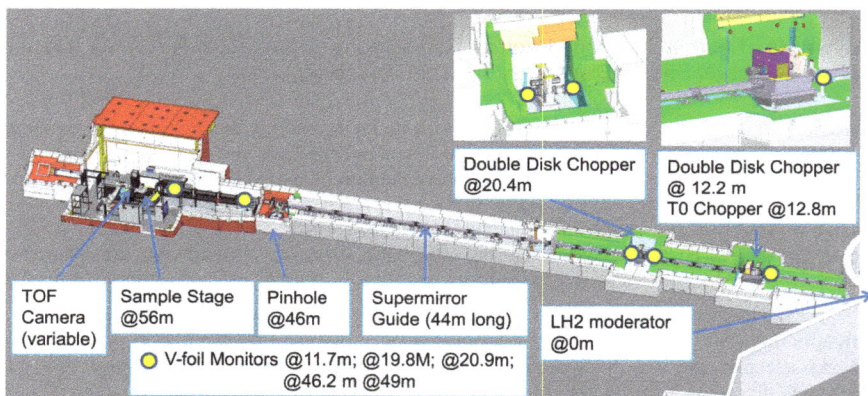

Double Disk Chopper @20.4m

Double Disk Chopper @ 12.2 m
T0 Chopper @12.8m

TOF Camera (variable)

Sample Stage @56m

Pinhole @46m

Supermirror Guide (44m long)

LH2 moderator @0m

V-foil Monitors @11.7m; @19.8M; @20.9m; @46.2 m @49m

Fig. 1: Schematic overview of the IMAT instrument.

The wavelengths of the detected neutrons are calculated from their time of flight by

$$\lambda = \frac{h(T + \Delta T_0)}{mL} = 3957 * \frac{(T + \Delta T_0)}{L} \tag{1}$$

where λ is the neutron wavelength (in Angstrom), h is Planck's constant, T is the neutron time of flight (in seconds), ΔT_0 is the time offset of the source trigger received by the data processing electronics (in seconds), m is the neutron mass, and L is the flight path from source to detector (in meters). The maximum wavelength band, defined by the need to avoid frame overlap, is about 7 Å for L=56 m and 10 Hz operation [3]. The wavelength resolution of the instrument, on the other hand, is determined by the uncertainty $\delta\lambda$ to which a given wavelength can be determined. The width and shape of the neutron pulse for a given wavelength depend on the type, geometry and physical (slowing down, storage) processes occurring within the moderator. The relative uncertainty $\delta\lambda/\lambda$, as determined recently for IMAT [10], defines the broadening of a Bragg edge or a Bragg dip in a TOF spectrum. It is this resolution function that signifies the ability of the instrument to discriminate neutron energies, not the width of a time channel (in histogram mode) or uncertainty of a TOF measurement (in event mode).

Energy selection on IMAT

Energy-selective and energy-dispersive radiography enable image contrast enhancement and the mapping of structure properties, respectively. We use the terms energy-selective/dispersive synonymously with wavelength-selection/dispersion as the kinetic energy and the wavelength of a neutron follow from the measured time-of-flight via Eq 1. We use the term 'energy-selective' if one or more (wide or narrow) wavelength bands are engaged, with monochromatic neutron imaging being a special case. Energy-selection is a prerequisite for 'energy-dispersive' neutron imaging where histogramming within a wavelength range is performed with a sufficiently fine channel width that is smaller than the instrument resolution, for example by performing a scan across a Bragg edge.

The selection of neutron wavelength bands using the two double-disk choppers on IMAT is demonstrated in Fig. 2. The four disks running at 10 Hz can be 'phased' to select broader (Fig. 2a) or narrow (Fig. 2b) wavelength bands, up to the width of the white-beam spectrum (red curves). The narrow bands are defined by the condition not to reduce the peak flux (Fig. 2b), yielding relative band widths $\Delta\lambda/\lambda$ ranging from 30% to 10%, for wavelengths from 2 to 6.5 Å, respectively. For example, the bandwidth at the flux maximum of the IMAT spectrum at 2.6 Å is $\Delta\lambda/\lambda\sim22\%$ (blue curve in Fig 2b). Given these values, and that only one energy band is engaged at a time, this method allows energy-selective rather than energy-dispersive measurements.

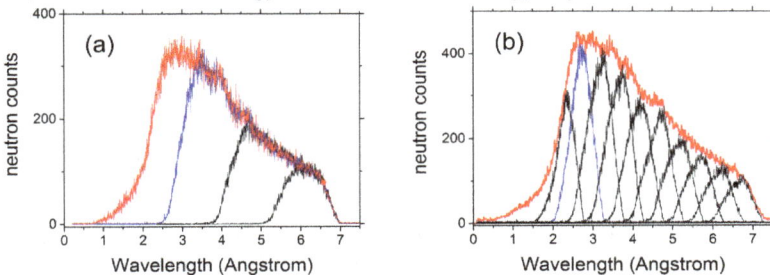

Fig. 2: Examples of wavelength band selection using choppers, for (a) broad and (b) narrow bands.

Coarse wavelength bandwidth selection is flexibly and quickly achieved via chopper dephasing from a script without changing beamline components. Experiments with a 'pink beam' (i.e. a narrow wavelength band; a term adopted from the synchrotron X-ray imaging community) allow for changing image contrasts without changing the sample set-up and for surveying energy-dependencies of attenuation coefficients for the largest field of view of IMAT (up to 20×20 cm^2) with any camera system. The neutron flux levels for the narrow wavelength bands in Fig. 2b are reduced to 3-7% of the white-beam flux (depending on wavelength) thus practically precluding tomography studies. Because of the coarse banding, detailed Bragg edge studies are not possible.

Another approach to achieve energy discrimination on a TOF instrument is to synchronise the imaging camera with the neutron source. An energy-selective radiography set-up with a gated image-intensified CCD or CMOS camera [7, 13] allows selecting a wide or narrow energy interval out of a white-beam spectrum. The energy discrimination is achieved for the maximum possible field of view, and in principle for time bins smaller than the instrument resolution. Thus, Bragg edge studies can be performed in principle; however with the gated camera only one channel is selected at a time (different from a high frame-rate camera) and only a fraction of the neutron spectrum is used at a time. Therefore, mapping of Bragg edge parameters is usually not performed on IMAT with this camera system.

With the IMAT pixel detectors that accumulate counts into multiple time-slices (3100 for the MCP and 4096 for GP2) performance is improved tremendously, in terms of acquisition times, albeit for small field of views. For a TOF instrument it is important to determine the spectral resolution, i.e. the monochromacity for a given wavelength. An energy-dispersive radiography was collected from a cylindrical CaF_2 crystal of 20 mm thickness and 35 mm diameter, at a distance of 25mm from the MCP detector and with an L/D of 250 (Fig. 3a). Fig. 3b displays Bragg dips from the CaF_2 single crystal for a selected wavelength range. Fig. 3c shows the wavelength dependence of the FWHM as an indicator of the resolution function. An alternative description was given earlier [10] using the trailing tail (asymmetry τ) of Bragg edges from a CeO_2 powder, shown in Fig. 3c for comparison. The asymmetry τ, an indicator of the storage term of the moderator process and reflecting the peak broadening, is preferred to represent the resolution function of IMAT as it is sample-independent and can be determined with a calibration measurement. The FWHM values are smaller by about a factor of two; the scatter of the values is due to the FWHM dependence on wavelength and diffraction angle.

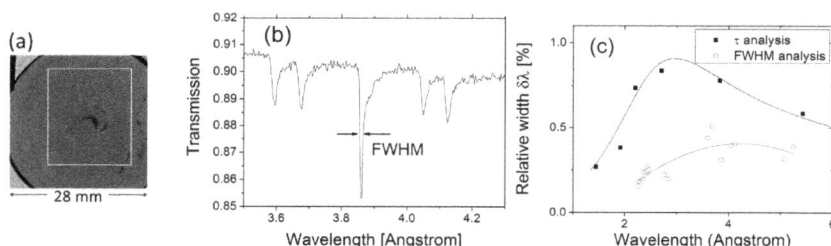

Fig. 3: (a) Radiography of a CaF_2 crystal; (b) Bragg dip spectrum for selected region of interest; (c) resolution function. "τ" data points taken from [10]. Solid curves are guides to the eye.

Table 1 summarises some of the instrument parameters of IMAT relevant for energy selective and energy-dispersive measurements for the different set-up options. The strain resolution value, demonstrating the precision with which a shift of a Bragg edge can be determined, is on the order of 0.01% for a good coherent scatterer like Fe, as determined earlier and included here for reasons of completeness [11]. The spatial resolution limit for Bragg edge analysis on IMAT is about 100 μm [11].

Strain mapping via Bragg edge analysis

A 12%Cr martensitic stainless steel sample with a laser-peened surface (see depiction in Fig. 4) relevant for steam turbine blade applications was analysed for two hours with an L/D of 125 using the MCP detector, with a time channel bin-width of 20 μs around the 110-Bragg edge. The distance of the face of the sample to the neutron-sensitive MCP was 20 mm. A transmission spectrum for a macro-pixel of 1.1 mm is shown in Fig 4. The 110-Bragg edge was fitted analytically using a Bragg edge profile (see [14] for details) based on the 'Santisteban' formula [6]. A strain map was reconstructed with a λ_0 value from the sample centre, far from the peened surface. A distinct zone of compressive residual strains ε_x is observed on the peened surface,

Neutron Radiography - WCNR-11   Materials Research Forum LLC
Materials Research Proceedings **15** (2020) 29-34   https://doi.org/10.21741/9781644900574-5

with a depth up to 1.5 mm, and a maximum compressive residual strain of about 1800 µε on the surface. The typical strain error is ± 160 µε. It should be noted that the analysed strain is averaged through the thickness of the sample along the beam direction.

Table 1: Energy-related instrument parameters of IMAT. See text for details.

White beam analysis:		Red curves in
Single-frame bandwidth	~6 Å	Fig. 2
Energy selection and 'pink beam' analysis using choppers:		Blue curve in
Energy resolution $\Delta\lambda/\lambda$	22% at 2.6 Å	Fig. 2b
Energy-selection or -dispersion using a gated CCD:		[7][13]
Energy resolution $\delta\lambda/\lambda$	<0.9%	[10] + Fig. 3c
Energy-selection or -dispersion and Bragg edge analysis using TOF detectors:		
Energy resolution: Bragg edge width, via "τ" analysis: $\delta\lambda/\lambda$	<0.9%	[10] + Fig. 3c
Energy resolution: Bragg dip width, via "FWHM" analysis: $\delta\lambda/\lambda$	<0.45%	Fig. 3c
Strain resolution: $\Delta\varepsilon = \Delta(\lambda-\lambda_0)/\lambda$	90 µε	[11]
Spatial resolution for Bragg edge mapping	>100 µm	[11]

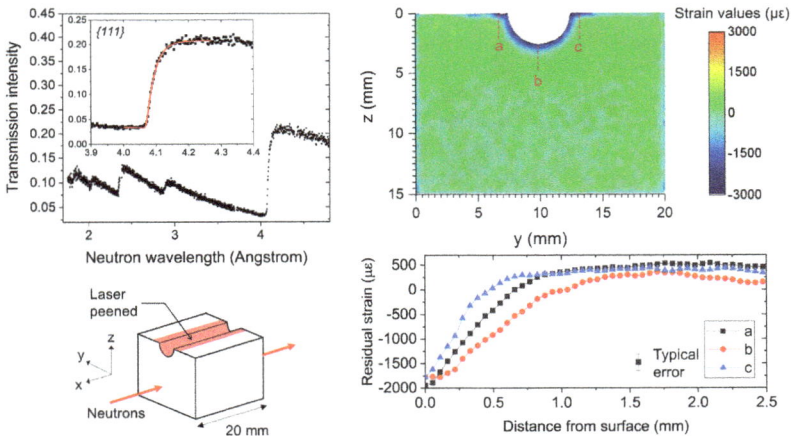

Fig. 4: Strain mapping of a laser-peened steel sample. (Top left) Bragg edge spectrum from a macro-pixel with analysed 110-Bragg edge in inset; (Top right) map of strain ε_x; (Bottom right) strain profiles for lines indicated in the strain map.

Conclusion

Coarse wavelength band selection using the IMAT disk choppers is aimed at enhancing contrasts of materials for large fields of views. For this mode, tomography scans are usually not considered because of the reduced neutron flux and low duty cycle. Much more important for the IMAT user programme are energy-dispersive measurements using the MCP or GP2 detectors given the unique possibilities for strain mapping on a TOF instrument. Other imaging methods that benefit from an energy-dispersive setup will be explored in due course, including wavelength-resolved interferometry, phase contrast imaging and grain mapping using attenuation, phase shift and/or diffractive signals.

References

[1] E.H. Lehmann, B. Schillinger, How the NEUWAVE workshop series has pushed neutron imaging developments, Neutron News 29 (2018) 25-31.
https://doi.org/10.1080/10448632.2018.1445923

[2] R. Woracek, J.R. Santisteban, A. Fedrigo, M. Strobl, Diffraction in neutron imaging - a review, Nucl Instr Meth A 878 (2018) 141–158. https://doi.org/10.1016/j.nima.2017.07.040

[3] W. Kockelmann et al., Time-of-Flight neutron imaging on IMAT@ISIS: a new user facility for materials science, J. Imaging 4 (2018) 47. https://doi.org/10.3390/jimaging4030047

[4] N. Kardjilov, B. Schillinger, E. Steichele, Energy-selective neutron radiography and tomography at FRM, Applied Radiation and Isotopes 61 (2004) 455-460.
https://doi.org/10.1016/j.apradiso.2004.03.070

[5] M. Schulz , P. Boni, E. Calzada, M. Muhlbauer, B. Schillinger, Energy-dependent neutron imaging with a double crystal monochromator at the ANTARES facility at FRM II, Nucl Instr Meth A 605 (2009) 33–35. https://doi.org/10.1016/j.nima.2009.01.123

[6] J.R. Santisteban, L. Edwards, A. Steuwer, P.J. Withers, Time-of-flight neutron transmission diffraction, J. Appl. Crystallogr. 34 (2001) 289–297. https://doi.org/10.1107/S0021889801003260

[7] W. Kockelmann, G. Frei, E. H. Lehmann, P. Vontobel, J. R. Santisteban, Energy-selective neutron transmission imaging at a pulsed source, Nucl Instr Meth A 578 (2007) 421-434.
https://doi.org/10.1016/j.nima.2007.05.207

[8] A.S. Tremsin, J.V. Vallerga, J.B. McPhate, O.H.W. Siegmund, R. Raffanti, High resolution photon counting with MCP-Timepix quad parallel readout operating at > 1 KHz frame rates. IEEE Trans. Nucl. Sci. 60 (2013) 578–585. https://doi.org/10.1109/TNS.2012.2223714

[9] Y. Matsumoto, M. Segawa, T. Kai, T. Shinohara, T. Nakatani, K. Oikawa, K. Hiroi, Y.H. Su, H. Hayashida, J. D. Parker, S.Y. Zhang, Y. Kiyanagi, Recent progress of radiography and tomography at the energy-resolved neutron imaging system RADEN, Physics Procedia 88 (2017) 162-166.
https://doi.org/10.1016/j.phpro.2017.06.022

[10] T. Minniti, K. Watanabe, G. Burca, D. Pooley, W. Kockelmann, Characterization of the new neutron imaging and materials science facility IMAT, Nucl Instr Meth A888 (2018) 184-195.
https://doi.org/10.1016/j.nima.2018.01.037

[11] R.S. Ramadhan, W. Kockelmann, T. Minniti, B. Chen, D. Parfitt, M.E. Fitzpatrick, A.S. Tremsin, Characterisation and application of Bragg edge transmission imaging for strain measurement and crystallographic analysis on the IMAT beamline, J. Appl. Cryst 52 (2019) 351-368.
https://doi.org/10.1107/S1600576719001730

[12] D.E. Pooley, J.W.L. Lee, M. Brouard, J.J. John, W. Kockelmann, N.J. Rhodes, E.M. Schooneveld, I. Sedgwick, R. Turchetta, C. Vallance, Development of the GP2 Detector: Modification of the PImMS CMOS Sensor for Energy-Resolved Neutron Radiography, IEEE TNS, 64 (2017) 2970-2981. https://doi.org/10.1109/TNS.2017.2772040

[13] T.E. McDonald Jr., T.O. Brun, T.N. Claytor, E.H. Farnum, G.L. Greene, C. Morris, Time-gated energy-selected cold neutron radiography, Nucl Instr Meth A 424 (1999) 235-241.
https://doi.org/10.1016/S0168-9002(98)01252-2

[14] A.S. Tremsin, T.Y. Yau, W. Kockelmann, Non-destructive examination of loads in regular and self-locking spiralock® threads through energy resolved neutron imaging, Strain (2016).
https://doi.org/10.1111/str.12201

Neutron Radiography - WCNR-11
Materials Research Proceedings **15** (2020) 35-41

Materials Research Forum LLC
https://doi.org/10.21741/9781644900574-6

Energy Resolved Imaging using the GP2 Detector: Progress in Instrumentation, Methods and Data Analysis

Daniel E. Pooley[1,a*], Jason W. L. Lee[2,b], Freddie A. Akeroyd[1,c], Owen Arnold[1,d],
Michael Hart[1,e], Jaya J. John[4,f], Peter M. Kadletz[3,g], Winfried Kockelmann[1,h],
Triestino Minniti[1,i], Christopher Moreton-Smith[1,j], Manuel Morgano[5,k],
Nigel J. Rhodes[1,l], Erik M. Schooneveld[1,m], Iain Sedgwick[1,n], Claire Vallance[2,o],
Robin Woracek[3,p]

[1] STFC, Rutherford Appleton Laboratory, Harwell Campus, Didcot OX11 0QX UK

[2] Department of Chemistry, Chemistry Research Laboratory, 12 Mansfield Road, University of Oxford, Oxford, OX1 3TA UK

[3] European Spallation Source ERIC, SE-221 00, Lund, Sweden

[4] Department of Physics, Denys Wilkinson Building, Keble Road, University of Oxford, Oxford OX1 3RH UK

[5] Paul Scherrer Institute, CH-5232 Villigen-PSI, Switzerland

[a]*daniel.pooley@stfc.ac.uk, [b]jason.lee@chem.ox.ac.uk, [c]freddie.akeroyd@stfc.ac.uk,
[d]owen.arnold@stfc.ac.uk, [e]michael.hart@stfc.ac.uk, [f]Jaya.John@physics.ox.ac.uk,
[g]peter.kadletz@esss.se, [h]winfried.kockelmann@stfc.ac.uk, [i]triestino.minniti@stfc.ac.uk,
[j]c.m.moreton-smith@stfc.ac.uk, [k]manuel.morgano@psi.ch, [l]nigel.rhodes@stfc.ac.uk,
[m]erik.schooneveld@stfc.ac.uk, [n]iain.sedgwick@stfc.ac.uk, [o]claire.vallance@chem.ox.ac.uk,
[p]robin.woracek@esss.se

Keywords: Neutron, Energy Resolved, Wavelength Dispersive, Imaging, Tomography, CMOS, Gadolinium, Pixel, Spallation, ToF, Time of Flight

Abstract. We report on the recent developments of the 'GP2' detector, highlighting a selection of energy resolved measurements and associated methodology. GP2 is a 100k pixel time-of-flight (ToF) neutron camera, which combines a gadolinium converter film and a CMOS (Complementary Metal Oxide Semiconductor) readout sensor. This paper describes an up-to-date specification of the detector and its variants, progress that has been made towards integration into the Imaging and Materials Science instrument (IMAT) and an independent review at the ESS test beamline, V20. Two ToF data reduction methods are detailed, namely wavelength dispersive contrast enhancement and 'wavelength frame multiplication' (WFM) reduction.

1. Introduction to GP2

The 'GP' detector series has been developed over the last six years as a collaborative project between STFC Technology Department, Oxford University and the ISIS Neutron and Muon source, all based in Oxfordshire, UK. The acronym GP derives from the use of gadolinium as a neutron converter and the implementation of an event-mode sensor developed for applications in imaging mass spectrometry, known as the 'PImMS' sensor [1]. GP1 was a scaled-down prototype detector with an active area of ≈25 mm². The second generation (GP2) detector shown in Fig 1 has a greatly increased area, 22.7×22.7 mm². Variants currently in development include the F-mode, where a full sensor readout is reduced to ~3 ms, the S-mode, where the sensor is sandwiched between two layers of gadolinium to enhance neutron detection efficiency and M, where the sensor is implemented in modular form to facilitate sparse tiling over large areas. A concise, tabulated, specification of GP2 is given in Table 1, with full detail reported in [2]. The 'register' feature is sensor-level memory allowing up to 4 events per pixel to be buffered between readout operations. This paper focuses on the recent use of the detector at two time-of-flight imaging beamlines: IMAT [3] at the ISIS source, UK, and the ESS test beamline V20 [4] located at HZB, Germany. In the case of IMAT, where the detector is available to the user programme, further detail is provided to describe integration into the existing beamline infrastructure.

Parameter	Value
Pixel size	70×70 μm
Pixel number	324×324
Active area	22.7×22.7 mm^2
Bit depth (time bins available)	12 bit (4095)
Smallest temporal bin width	12.5 ns
Registers per pixel	4
Detector mechanical size	$15 \times 15 \times 8$ cm^3
Communication standard	USB 2
Readout time using 4 registers	21.19 ms
Neutron detection efficiency (4 μm ^{NAT}Gd)	7.5 % at 2.5 Å
Gamma sensitivity	1.5×10^{-3}

Fig 1. GP2 detector with protective cover removed to expose the 22.7×22.7 mm^2 sensor.

Table 1. Specification of the GP2 detector.

2. Integration into IMAT

This section details the process of integrating GP2 into the IMAT infrastructure along with example IMAT data and ensuing contrast enhancement.

2.1 Mechanical Integration.

IMAT can be tailored to each measurement by selection of various instrument parameters, one of which is the camera system. Changes between different imaging detectors (cameras) are performed using a robotic arm, facilitating quick and accurate changes between white-beam imaging (time integrated) and time-of-flight (energy resolved) imaging. Each detector has its own plate, which provides a standardised interface (mating socket) for the robotic arm to locate and pickup. The plate for GP2 is shown below in Fig 2.

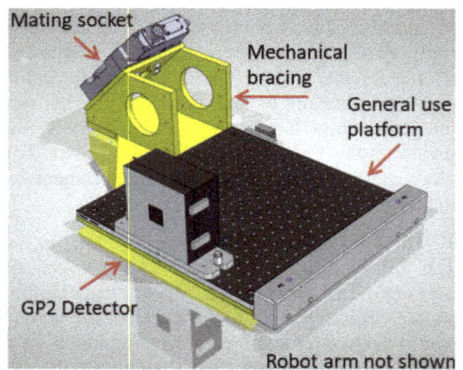

Fig 2. The robotic arm mounting plate for GP2.

2.2 Control and Data

For GP2 to be available within the IMAT user programme the detector needed to be integrated into the ISIS EPICS [5] based control system. Key objectives associated with this were:

- Synchronisation of Start/Stop/Pause commands between the local GP2 acquisition computer and the IMAT instrument control computer.
- Passing file name constructors and meta-data between the local GP2 acquisition computer and the IMAT instrument control computer.
- Stream data live to user and ISIS archive.

As the camera software is written in LabVIEW, a convenient control mechanism was network shared variables [6], for which generic EPICS support had previously been developed [7]. These network shared variables allow setting and reading of values remotely from the instrument control computer and hence scripting of the camera operation via EPICS using Python.

The data is currently saved as a binary file containing three unsigned 16 bit integers for each hit. Each frame is delimited by two 32 bit integers, providing the array size. The longer-term aim is to store the data in a HDF5 format, namely NeXus [8], which will ensure compatibility with the ISIS data archive and reconstruction software such as SAVU [9].

2.3 IMAT Data: 4 Registers.

To illustrate the 4-memory register capability of GP2 and to enumerate the typical count rate on IMAT, open beam data with a 40 mm pinhole (L/D ~250) was recorded, with GP2 synchronous to the 10 Hz source. In this configuration GP2 measured $\sim 1.1 \times 10^5$ hits per 100 ms frame, more than the total number of pixels (324×324). Fig 3 shows the hits in each register as a histogram. The number of hits in each register decreases with increasing register number since the registers are populated sequentially starting at register 1. Monitoring this distribution provides diagnostic information as to how close to saturation (missing counts) the detector is, mitigating the need for offline data correction. If the neutron flux were to increase, read-out within a frame would have to be performed at the cost of introducing time gaps in the data. In this example GP2 was operated as a 3-register camera as the fourth register was minimally occupied.

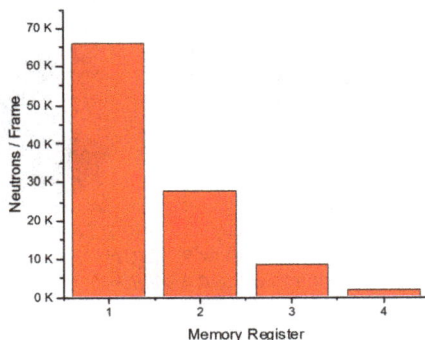

Fig 3. Histogram showing the hit-distribution across the 4 memory registers for a typical IMAT frame.

2.4 IMAT Data: Contrast Enhancement

GP2 is an event-mode detector meaning it records wavelength (energy) dispersive data. The benefits of using energy resolved neutron imaging to measure sample properties are widely reported [10] [11]. Here we emphasise that some simple diagnostic tools can be used to both ascertain the measurement quality and to provide contrast enhancement on-the-fly. A simple graphical user interface was developed which allows the user to maximise the grey-value contrast between two regions of interest (RoI) as shown in Fig 4. The program normalises the time-of-flight spectra for each RoI and displays the ratio (green). A simple thresholding of the ratio plot is used to select the time-of-flight bins (i.e. energy channels) which provide more contrast to the material in RoI-1, by applying this selection to the whole image.

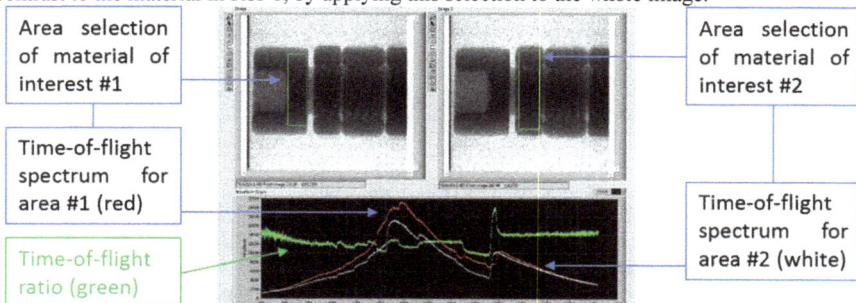

Fig 4. Screenshot of GUI used to select time channels to produce contrast enhanced image.

An example of this simple method is demonstrated using an energy resolved image of a bolt with alternating FCC and BCC steel nuts. Both materials have similar macroscopic cross sections and therefore attenuate the beam equally. In a white-beam radiograph these two materials will have the same grey value and will be indistinguishable, as seen in Fig 4 top. However, by selecting all the ToF bins that enhance one of the materials those pixels can be transported to a different grey value. This method offers a simple 'online' analysis, which can immediately distinguish features without a large overhead in data analysis.

Fig 5. LEFT: Photo of sample, alternating BCC and FCC steel. RIGHT: Example output image, where the GUI has been used to enhance the FCC bolt and nut. Time channels were selected and integrated by thresholding the ratio plot (Fig 4 green).

3. Independent Review: Measurements at the ESS Test Beamline, V20

As the GP2 detector has reached a point of technological maturity, moving from the R&D phase to deployment, impartial technology review from new users was sought. The aims of this exercise were threefold. Firstly, independent review is considered due diligence for any scientific development. Secondly new users to the technology will provide valuable feedback for future development. Thirdly, the suitability and compatibility with different source structures and flux distributions should be ascertained if GP2 is to be deployed to other facilities. To this end GP2 was tested at the dedicated ESS test beamline V20, located at the BER2 research reactor at the Helmholtz Zentrum Berlin. The beamline can mimic the time structure of the future European Spallation Source (ESS) using double-disk choppers and is equipped with optional pulse shaping choppers for 'Wavelength Frame Multiplication' (WFM). It is also dedicated to the testing and development of instrumentation, scientific methods and corresponding data reduction for the ESS [4] [12]. A number of measurements were performed, including count rate investigations, detector stability, source-structure compatibility and an in-situ tensile experiment. While the main detector parameters were found to be consistent with results previously reported and the latter experiment will be reported elsewhere in detail, we report here on two measurements particularly suited to the V20 beamline.

3.1 Detector Stability

Reactor based beamlines are inherently stable, providing an opportunity to investigate detector performance over hours and days. Fig 6 shows the variation between the integrated counts of an open-beam GP2 measurement (red) alongside the reactor power (blue) and cold source temperature (green). As the neutron flux is dependent on both the reactor power and source temperature, the observed correlation indicates that that detector stability is much better than the observed variation, i.e <2%. This type of analysis further demonstrates the advantages of both the detector and event mode data acquisition.

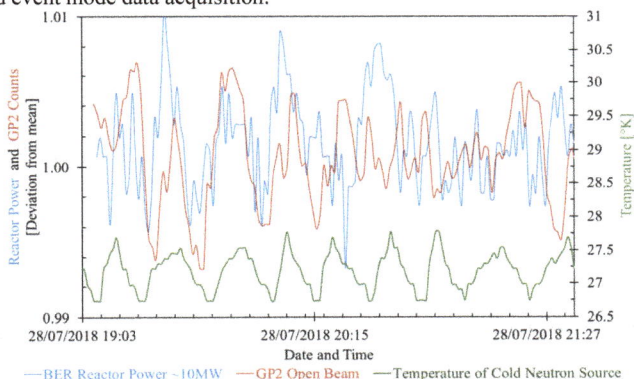

Fig 6. Stability plot showing fluctuation in reactor power (blue) and the counts measured with GP2 (red) alongside the cold source temperature (green, right-axis) variation. The neutron counts measured with GP2 is sensitive to the compound effect of both reactor power and source temperature.

3.2 Source Structure

V20 mimics the ESS pulse structure: a ~3 ms long pulse with a 14 Hz repetition rate. As at any ToF beamline, the wavelength resolution is determined by the length of the instrument, for the herein presented measurement between ~5 % (8 Å) and ~20 % (2 Å) for a flight length of ~26 m. While beamlines at the ESS will typically be much longer, several of them will have additional pulse shaping choppers to improve the resolution, similar to V20. V20 is equipped with a chopper cascade to operate in wavelength frame multiplication (WFM) mode, described fully in [4]. In this mode, six new and much narrower source pulses are created from the long pulse, all with a constant and much higher resolution. These are separated in time but overlap in wavelength and hence form a continuous wavelength band after data reduction ('stitching' using the Mantid framework [13]). The resolution is furthermore tuneable by changing the distance between the WFM choppers and the herein utilised settings provide ~2.3 %. This mode of operation offers a novel test for GP2, as the pulse structure requires a detector to handle large changes in flux with high stability. The measurement of a 10 mm thick BCC steel plate is shown in Fig 7, demonstrating both long pulse and WFM mode. By inspection of the Bragg edge features in the stitched data and observing its smooth continuous nature (Fig 7. RIGHT), the

Materials Research Forum LLC
https://doi.org/10.21741/9781644900574-6

quality of GP2 data is evident. The WFM data shows sharper, well-resolved Bragg edges (black) suitable for advanced analysis methods not possible with the long pulse data. The stitching process applied to GP2 data and its application to more complex systems will be presented in a separate and more detailed publication.

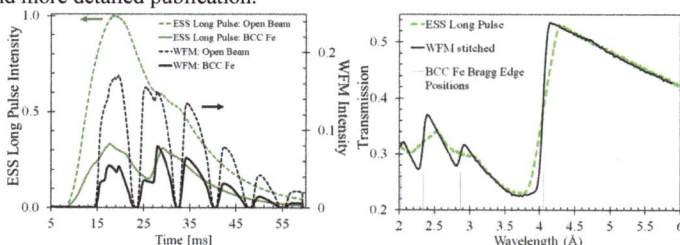

Fig 7. LEFT: Open beam and sample data, for both the 'ESS long pulse mode' and 'WFM mode'. Peak intensity of the long pulse mode data is normalised to unity (left y-axis). The same scale is applied to the WFM intensity (right y-axis). RIGHT: Transmission data, showing the difference in resolution between the two modes. The known BCC Bragg edge positions are indicated.

Summary

An overview of recent developments for GP2 has been presented, alongside two examples of ToF data reduction. With F, S and M variants currently under development and the positive feedback from the imaging community we anticipate wider deployment across the facilities.

References

[1] I. Sedgwick, A. Clark, J. Crooks, R. Turchetta, L. Hill, J. J. John, A. Nomerotski, R. Pisarczyk, M. Brouard, S. H. Gardiner, E. Halford, J. W. Lee, M. L. Lipciuc, C. Slater, C. Vallance, E. S. Wilman, B. Winter and W. H. Yuen, "PImMS: A self-triggered, 25ns resolution monolithic CMOS sensor for Time-of-Flight and Imaging Mass Spectrometry," IEEE NEWCAS (2012). https://doi.org/10.1109/NEWCAS.2012.6329065

[2] D. E. Pooley, J. W. L. Lee, M. Brouard, J. J. John, W. Kockelmann, N. J. Rhodes, E. M. Schooneveld, I. Sedgwick, R. Turchetta and C. Vallance, "Development of the "GP2" Detector: Modification of the PImMS CMOS Sensor for Energy-Resolved Neutron Radiography," IEEE TNS, 64 (2017) 2979. https://doi.org/10.1109/TNS.2017.2772040

[3] T. Minniti, K. Watanabe, G. Burca and D. E. Pooley, "Characterization of the new neutron imaging and materials science facility IMAT," NIM A, 888 (2018) 184. https://doi.org/10.1016/j.nima.2018.01.037

[4] M. Strobl, M. Bulat and K. Habicht, "The wavelength frame multiplication chopper system for the ESS test beamline at the BER II reactor-A concept study of a fundamental ESS instrument principle.," NIM A, 705 (2013) 74. https://doi.org/10.1016/j.nima.2012.11.190

[5] [Online]. Available: https://epics-controls.org

[6] [Online]. Available: http://www.ni.com/white-paper/5484

[7] [Online]. Available: http://epics.isis.stfc.ac.uk/doxygen/NetShrVar

[8] M. Könnecke, F. A. Akeroyd, H. J. Bernstein, A. S. Brewster, S. I. Campbell, B. Clausen, S. Cottrell, J. U. Hoffmann, P. R. Jemian, D. Mannicke, R. Osborn, P. F. Peterson, T. Richter, J.

Suzuki, B. Watts, E. Wintersberger and J. Wuttke, "The NeXus data format," J. Appl. Cryst,. **48** (2015) 301. https://doi.org/10.1107/S1600576714027575

[9] [Online]. Available: https://savu.readthedocs.io/en/latest/about

[10] R. Woracek, J. Santisteban, A. Fedrigo and M. Strobl, "Diffraction in neutron imaging-A review," NIM A, **878** (2018) 141. https://doi.org/10.1016/j.nima.2017.07.040

[11] N. Kardjilov, I. Manke, R. Woracek, A. Hilger and J. Banhart, "Advances in neutron imaging," Materials Today, **21** (2018) 652. https://doi.org/10.1016/j.mattod.2018.03.001

[12] R. Woracek, T. Hofmann, M. Bulat, M. Sales, K. Habicht, K. Andersen and M. Strobl, "The test beamline of the European Spallation Source – Instrumentation development and wavelength frame multiplication," NIM A, **839** (2016) 102. https://doi.org/10.1016/j.nima.2016.09.034

[13] O. Arnold, J.C. Bilheux, J.M. Borreguero, A. Buts, S.I. Campbell, L. Chapon, M. Doucet, N. Draper, R. Ferraz Leal, M.A. Gigg, V.E. Lynch, A. Markvardsen, D.J. Mikkelson, R.L. Mikkelson, R. Miller, K. Palmen, P. Parker, G. Passos, T.G. Perring, P.F. Peterson, S. Ren, M.A. Reuter, A.T. Savici, J.W. Taylor, R.J. Taylor, R. Tolchenov, W. Zhou, J. Zikovsky, "Mantid—Data analysis and visualization package for neutron scattering and μ SR experiments", NIM A, **764** (2014) 156. https://doi.org/10.1016/j.nima.2014.07.029

Neutron Radiography - WCNR-11
Materials Research Proceedings 15 (2020) 42-47

Materials Research Forum LLC
https://doi.org/10.21741/9781644900574-7

First Neutron Computed Tomography with Digital Neutron Imaging Systems in a High-Radiation Environment at the 250 kW Neutron Radiography Reactor at Idaho National Laboratory

Aaron Craft[1,a] *, Burkhard Schillinger[2,b], William Chuirazzi[1,c], Glen Papaioannou[1,d], Andrew Smolinski[1,e] and Nicholas Boutlon[1,f]

[1]Idaho National Laboratory, PO Box 1625, MS 2211, Idaho Falls, ID 83415, USA

[2]Heinz Maier-Leibnitz Center, Technische Universität München; Lichtenbergstr. 1, 85748 Garching, Germany

[a]Aaron.Craft@inl.gov, [b]Burkhard.Schillinger@frm2.tum.de, [c]William.Chuirazzi@inl.gov, [d]Glen.Papaioannou@inl.gov, [e]Andrew.Smolinski@inl.gov, [f]Nicholas.Boulton@inl.gov

Keywords: Neutron Tomography, Neutron Radiography, Computed Tomography

Abstract. The Neutron Radiography Reactor (NRAD) at Idaho National Laboratory (INL) was designed for thermal and epithermal neutron radiography for examination of highly-radioactive irradiated nuclear fuel elements. Radioactive samples are remotely lowered into the East and North Radiography Stations (ERS and NRS, respectively), and a rail transfer system remotely positions radiography cassettes into the detector position for indirect radiography. The indirect transfer method with film has been used at NRAD for around forty years, but recent efforts seek to develop digital camera-based neutron imaging systems. Two initial camera detector systems were built using an inexpensive but high-quality scientific CMOS camera with robust shielding, and tests were performed in collaboration with Heinz Maier-Leibnitz Institut of Technische Universität München. The first measurements in the ERS provided valuable experience that informed the design of an improved neutron imaging system that was tested in the NRS. The first successful digital neutron computed tomography at INL was acquired, consisting of 421 neutron radiographs acquired in 4 hours. These first tests with camera-based neutron imaging systems have demonstrated the potential to both increase the throughput of radiography by two orders of magnitude and provide higher quality spatial information with three-dimensional tomographic reconstructions compared to two-dimensional radiographic projections, which represent a significant improvement compared to current film radiography capabilities.

Introduction

Neutron radiography provides more comprehensive information about the internal geometry of irradiated nuclear fuel than any other nondestructive examination technique and has demonstrated its importance to the nuclear industry for many decades [1]. The first neutron radiography experiments at INL were first conducted in 1964, with the first dedicated neutron radiography beamline being built in 1967 at the Transient Reactor Test (TREAT) facility [2]. The construction of the NRAD reactor in 1978 included two neutron beamlines specifically designed for neutron radiography of irradiated nuclear fuels [3]. All three neutron radiography beamlines (two at NRAD and one at TREAT) use the indirect radiography methods, where a cassette of different foils is placed in the beamline and activated, then removed and placed next to either film or image plates to render an image [2,4]. The indirect method is not sensitive to gamma rays but is time-consuming. The production rate at NRAD is limited to 14 film radiographs per day, which precludes use of neutron computed tomography (nCT) for routine examination because nCT required hundreds of radiographs to produce a quality reconstruction. Modifications to the existing radiography set-up must be made to capture the number of projections necessary for tomography in a timely manner [5]. Implementing tomography as a routine examination technique thus requires a digital neutron radiography system, but such digital systems have traditionally had problems operating in high-radiation environments. This paper discusses the first efforts to build such digital systems and the initial nCT results that were obtained.

NRAD Facilities

The NRAD reactor is a pool-type, water-moderated, 250 kW$_{th}$ TRIGA reactor. It sits below a hot cell, allowing radioactive materials to be transported directly into either of the two neutron radiography beamlines [3]. Both the East Radiography Station (ERS) and the North Radiography Station (NRS) have radial beamlines directly aligned with the core. Not only does this provide a high neutron flux, but it also creates a high amount of gamma contamination in the beam. The ERS is equipped with an elevator that allows samples to be remotely placed for neutron radiography, as the ERS is directly below the hot cell. With a distance from the beam aperture to the imaging plane of 444.5 cm, the field of view at the imaging plane is 17.8 cm wide by 43.2 cm tall. The beamline is primarily operated at a length-to-diameter ratio (L/D) of 125, yielding a thermal neutron flux of 6.0×10^6 n/cm^{2}s [6]. The NRS utilizes a cask system to transport fuel samples to an elevator which remotely positions the fuel for neutron imaging because the NRS is not straight underneath the hot cell. The NRS has a longer beamline than the ERS, with 1603.8 cm between the beamline aperture and the image plane. The beam is ~60 cm diameter at the image plane, but a beam limiter reduces the field of view to 17.8 cm wide by 43.2 cm tall for traditional neutron radiography. The NRS beam is most often operated at L/D=185 with a neutron flux of $\sim 4.5 \times 10^6$ n/cm^{2}s [3].

Imaging Equipment

The first iteration of the digital imaging system was designed to be low-cost because the high radiation background of the beamline chamber could easily damage the camera. An inexpensive ZWO 178MM Cool CMOS camera was used in conjunction with the SharpCapture 2.9 freeware to capture images. Camera hardware and motor stages were controlled through a combination of a Gertbot and a Raspberry Pi [7,8]. In addition to 10 cm of lead, borated polyethylene plates were also used to shield the camera and Raspberry Pi from the radiation field. The camera in the imaging box, with the assorted wiring to the Raspberry Pi and rotational stage can be seen in Figure 2. The object for these tests was a microwave horn.

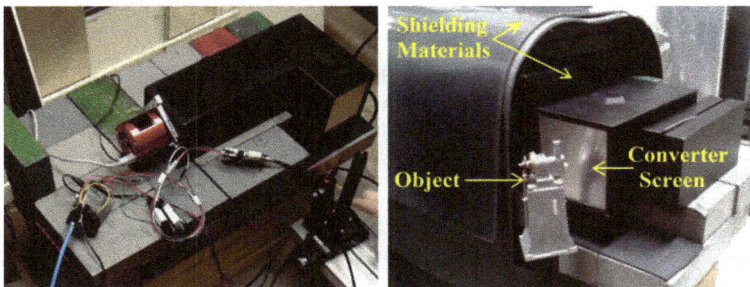

Figure 2. Left, CMOS camera in the imaging apparatus with the connections between the camera, rotating stage and Raspberry Pi controller. Right, Shielding surrounding camera, electronics and imaging box with the microwave horn positioned in front of the converter screen.

A radiograph microwave horn acquired in the ERS is shown in Figure 5a next to a radiograph that was acquired in the NRS with the newer system. A gamma dose rate of 2.0 Sv/hr was measured with an ion chamber dose rate meter placed at the image plane near the scintillator screen. A surrogate prismatic fuel element was also imaged, which was prepared with a gadolinium-dopant liquid penetrant to enhance the visibility of an engineered crack defect inside a coolant channel. The imaging system was able to acquire 32 neutron radiographs in under 12 minutes, which is much more efficient than current daily output of approximately 14 film radiographs. Unfortunately, the Raspberry Pi computer crashed after 44 acquisitions due to the harsh radiation environment. The results obtained before the crash are shown in Figure 3.

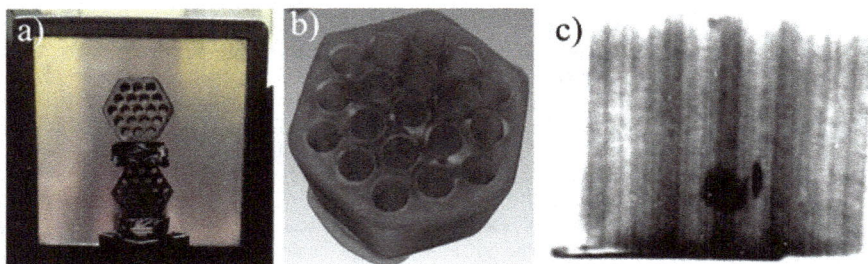

Figure 3. a) Picture of the surrogate fuel element pisitioned in front of the imaging system in the ERS beamline. b) The 3D tomographic reconstruction obtained from only 32 images. c) The 3D rendering emphasizing the gadolinium-penetrant treated crack defect shown as a dark spot in the middle of the image.

The initial success of the first design led to the construction of a second imaging system. The second system was improved by implementing a two-mirror architecture with a longer optical path which helped to reduce the radiation dose delivered to the camera. The camera box was also designed to allow for more robust shielding. The camera was mounted on a translational stage to allow for remote focusing. A simplified version of the ANTARES instrument control at MLZ was used to control the imaging system with the help of a laptop and three Raspberry Pi computers [8]. The field of view was doubled to 20 cm. Vents were also added to the box for cooling the camera the camera. Figure 4 shows the second imaging system design.

A radiograph of the microwave horn were acquired with the newer imaging system in the NRS beamline. Figure 5 shows a side-by-side comparison of radiographs of the microwave horn captured with the first and second imaging systems. A gamma dose rate of 900 mSv/hr was measured at the image plane, less than half the gamma dose rate present for the first measurements in the ERS. The reduced gamma radiation environment, along with the remote focusing and thermal noise reduction, contributed to the capture of an image with significantly improved contrast, signal-to-noise ratio, and resolution. This imaging system allowed for 421 neutron radiographs to be captured in only four hours. Using film and the indirect transfer method, it would take roughly a month of reactor time to acquire the same number of radiographs. This imaging system is able to acquire radiographs more than two orders of magnitude faster than before, which enables nCT at these beamlines despite the high gamma content of the neutron beams.

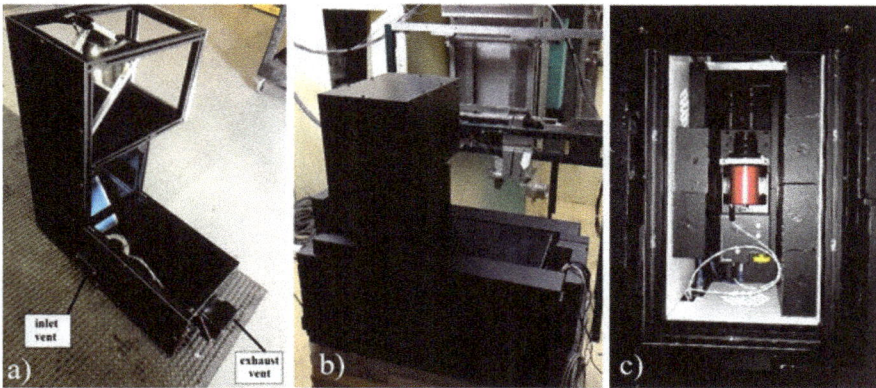

Figure 4. Set-up of the improved imaging system. a) The new box utilizes two mirrors to remove the camera further from the beam's radiation. Vents are also added to reduce thermal noise. b) The box is upgraded to allow for better shielding of the camera and associated electronics. c) The camera is mounted on a linear stage with room for additional shielding inside of the box.

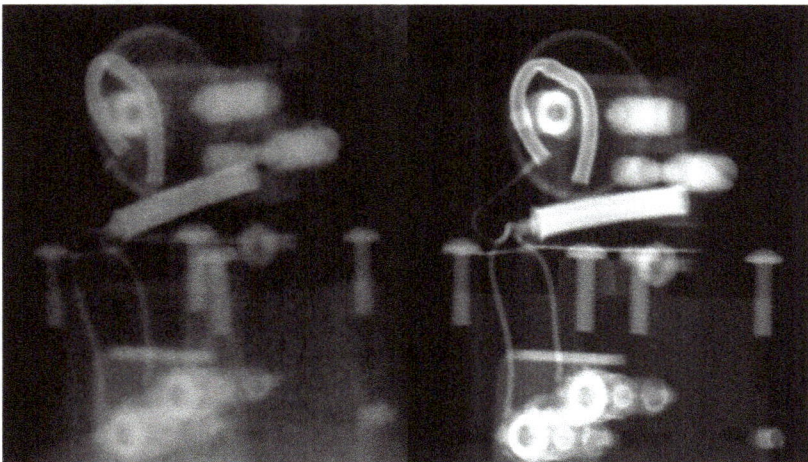

Figure 5. Radiographs of the microwave horn taken with (a) the initial neutron imaging apparatus in the ERS and (b) the improved imaging system in the NRS.

Once 421 the radiographs were obtained, tomographic reconstruction was performed. The resulting reconstruction is shown in Figure 6. The resolution was good enough to distinguish individual internal components of the microwave horn such as screws, wires, coils. The quality of the resulting reconstruction demonstrates that the redesign of the box to accommodate more shielding and the change of controller software and hardware was highly successful in both reducing noise and preventing the controller system from crashing. The improved thermal ventilation and the linear stage to allow for camera focusing both contributed to obtaining a sharper, higher quality images. The ability to run long enough to obtain hundreds of radiographs allows for production of higher quality computed tomography reconstructions than the first system was able to produce.

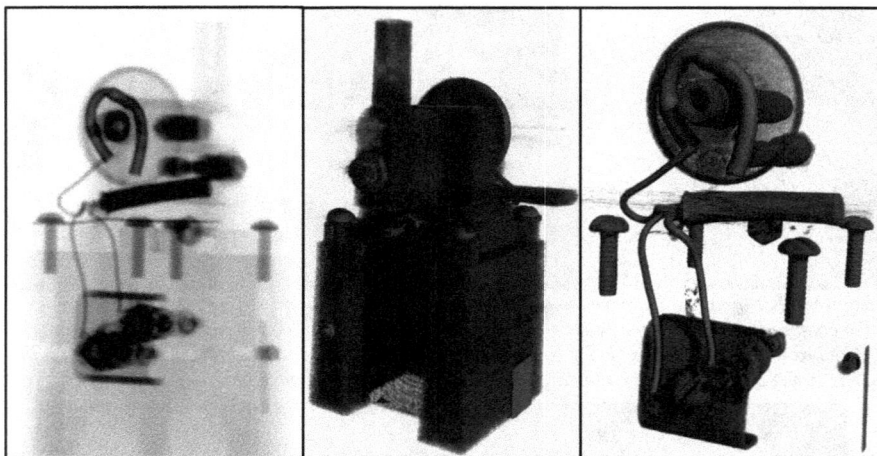

Fig. 6. One of the neutron radiographs (left) used to produce the tomographic reconstruction (middle). Adjusting the visualization threshold levels for the reconstructed object reveals (left) various features such as screws, wires, coils, etc.

Conclusion

The goal of this work was to develop the instrumentation necessary to perform digital nCT. This was accomplished as a collaborative project that built an initial imaging system that led to an improved imaging system that was able to perform quality nCT. Both systems successfully obtained digital neutron radiographs with sufficient image quality to be used in tomographic reconstructions. The improvements made during the construction of the second imaging system allowed stable, quick, and reliable digital neutron radiography. This led to both improved radiographs and higher-quality tomographic reconstructions. Realizing digital neutron radiography capabilities at INL's NRAD facility is an important first step towards implementing digital nCT as a routine examination technique.

References

[1] A.E. Craft and J.D. Barton, "Applications of neutron radiography for the nuclear power industry," Physics Procedia 88 (2017) 73-80. https://doi.org/10.1016/j.phpro.2017.06.009

[2] S.R. Jensen, A.E. Craft, G.C. Papaioannou, W.W. Empie and B.R. Ward, "Reactivation of the Transient Reactor Test (TREAT) Facility neutron radiography program," in this Proceeding.

[3] A.E. Craft, D.M. Wachs, M.A. Okuniewski, D.L. Chichester, W.J. Williams, G.C. Papaioannou, & A.T. Smolinski, "Neutron radiography of irradiated nuclear fuel at Idaho National Laboratory," Physics Procedia, 69 (2015) 483-490. https://doi.org/10.1016/j.phpro.2015.07.068

[4] A.E. Craft, G.C. Papaioannou, D.L. Chichester, and W.J. Williams, "Conversion from film to image plates for transfer method neutron radiography of nuclear fuel," Physics Procedia 88 (2017b) 81-88. https://doi.org/10.1016/j.phpro.2017.06.010

[5] B. Schillinger, E. Lehmann, and P. Vontobel, "3D neutron computed tomography: requirements and applications," Physica B: Condensed Matter 276 (2000) 59-62. https://doi.org/10.1016/S0921-4526(99)01254-5

[6] A.E. Craft, B.A. Hilton, and G.C. Papaioannou, "Characterization of a neutron beam following reconfiguration of the Neutron Radiography Reactor (NRAD) core and addition of new fuel elements," Nuc. Eng. & Tech. 48 (2016) 200-210. https://doi.org/10.1016/j.net.2015.10.006

[7] B. Schillinger and A.E. Craft, "A freeware path to neutron computed tomography," Physics Procedia 88 (2017) 348-353. https://doi.org/10.1016/j.phpro.2017.06.047

[8] B. Schillinger, A. Craft and J. Krüger, "The ANTARES instrument control system for neutron imaging with NICOS/TANGO/LiMA converted to a mobile system used at Idaho National Laboratory," in this Proceeding.

Neutron Radiography - WCNR-11
Materials Research Proceedings 15 (2020) 48-52

Materials Research Forum LLC
https://doi.org/10.21741/9781644900574-8

The ANTARES Instrument Control System for Neutron Imaging with NICOS/TANGO/LiMA Converted to a Mobile System used at Idaho National Laboratory

Burkhard Schillinger[1, a *], Aaron Craft [2,b], Jens Krüger [1,c]

[1] Heinz Maier-Leibnitz Zentrum and Physics E21, Technische Universität München, Lichtenbergstr.1, 85748 Garching, Germany

[2]Advanced Characterization and Post-Irradiation Examination Department Idaho National Laboratory, USA

[a]Burkhard.Schillinger@frm2.tum.de , [b]Aaron Craft@inl.gov, [c]Jens.Krueger@frm2.tum.de

Keywords: Neutron Imaging, Neutron Computed Tomography, Software, NICOS, TANGO, Entangle, LiMA, Python

Abstract. The Neutron Radiography Reactor (NRAD) at Idaho National Laboratory (INL) was designed for thermal and epithermal neutron radiography for examination of highly-radioactive irradiated nuclear fuel elements, exclusively using transfer foils with film or imaging plates. Use of digital detectors was not foreseen in the high radiation environment. Recent collaborative efforts are seeking to introduce digital neutron imaging systems and neutron computed tomography (CT). Two initial camera detector systems were built using a low-price, but high-quality scientific CMOS camera with massive shielding. With no existing electronic infrastructure, it was a challenge to build standalone systems. In first tests in 2017, a Raspberry Pi computer was used as a stepper motor and CT controller inside the East Radiography Station, which crashed in the high radiation field after a few dozen of images even behind 5 cm of lead shielding. In 2018, the network-based and decentralized instrument control system of the ANTARES imaging facility at Heinz Maier-Leibnitz Zentrum (MLZ) of Technische Universität München was scaled down to one laptop and three Raspberry Pi computers mounted outside the radiography bay, which was sufficient to control rotation and translation stages and to control the camera and record tomography measurements. The system is based on **NICOS, TANGO, entangle, and LiMA,** all four free toolkits for building distributed control systems. Details of the setup are described here. The downscaled system can be used standalone at any facility.

Introduction

The 250 kW Neutron Radiography Reactor (NRAD) at Idaho National Laboratory (INL) was designed for thermal and epithermal neutron radiography for examination of highly-radioactive irradiated nuclear fuel elements [1]. Samples are remotely lowered into the radiography stations from shielding flasks or a large hot cell above by elevators, and a rail transfer system remotely positions radiography cassettes into the detector position for indirect radiography. Opening NRAD's two radiography stations, the East and North Radiography Stations (ERS and NRS, respectively), on a regular basis for user operations was not foreseen when the facility was originally designed, but recent collaborative efforts seeking to introduce digital neutron imaging systems have demonstrated the need for simplified access to the radiography stations, which is an extraordinary challenge due to the high radiation environment and associated shielding requirements, and remote control of various imaging hardware inside the radiography stations. For the first working tomography setup, it was a challenge to build a local standalone instrument control system without any existing electronic infrastructure.

Neutron Radiography - WCNR-11
Materials Research Proceedings **15** (2020) 48-52

Materials Research Forum LLC
https://doi.org/10.21741/9781644900574-8

The first minimalist test system

The first system tested in 2017 used cheap, but high quality scientific CMOS camera type 'ASI178 mm cool' [2] with 3096×2080 pixels,14 bit ADC, and two-stage thermoelectric cooler in combination with a $ZnS+^6LiF$ neutron scintillation screen and mirror in a box. This first experiment was meant to test whether the camera would survive in the high radiation field so a minimalist quick-and-dirty solution was employed to record tomography data: A Raspberry Pi computer was used to control a stepper motor on a rotation stage and generate a sequence of images on consecutive angular positions. Since the camera does not have a hardware trigger input, and no dedicated software was available, the free astronomy software SharpCap [3] was used to control the camera on a Windows PC and record a sequence of images. To trigger the camera, the Raspberry Pi computer was wired to the left mouse button of the Windows PC mouse and generated mouse clicks to start new images in SharpCap on the Windows PC. The Raspberry Pi had to be mounted inside the radiography bay because of cable lengths. The gamma dose rate in the neutron beam at the imaging system was ~2 Sv/hr as measured with an ion chamber, and outside the beam in the ERS also has high gamma dose rate and an ambient neutron field of 10^4-10^5 n/cm^2s because of incoherent neutron scattering of the beam in the air. After only 44 images, it crashed in the high radiation field of the unfiltered radial beam of the ERS even behind a 5 cm lead shielding.

The second system

The second detector was built into a larger box with a two-mirror configuration to allow for massive shielding and also remote focusing with a translation stage carrying the camera. For the first working tomography setup, it was a challenge to build a local standalone instrument control system without any existing network infrastructure. For this, the originally distributed and network-oriented instrument control system of the ANTARES imaging facility at Heinz Maier-Leibnitz Zentrum of Technische Universität München was scaled down to one laptop and three Raspberry Pi computers mounted outside the radiography bay, which was sufficient to control rotation and translation stages and to control the camera and perform image acquisition for tomography. Fig. 1a shows the double-mirror camera box mounted in the NRS and Fig. 1b shows the provisional workspace outside the bay with laptop and the Raspberry Pis and controller under the table. The laptop on the right shows a live view of the last radiography image.

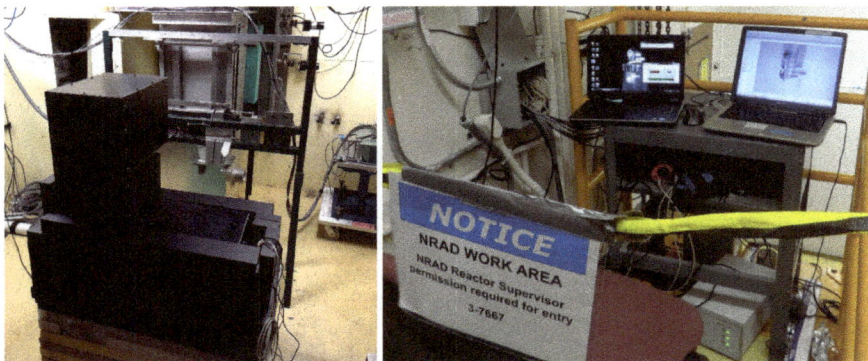

Fig.1a,b: The camera box mounted in the NRS (left) and the provisional work space with laptop and the Raspberry Pis and controller under the table (right).

The ANTARES control system reduced

After the first generation of instruments at the Heinz Maier-Leibnitz Zentrum (MLZ) and FRM II reactor of Technische Universität München was controlled by individually tinkered solutions both in hard- and software, all instrument controls are now being standardized since 2010 using a universal network-based approach.

At the MLZ the NICOS [4] system is used as the instrument control system. It offers the control of an instrument via a command line, graphical user interface, and/or for more complex command sequences Python scripts. Additional features of NICOS are command logging, definitions of data storage position, and displaying of live data (including range selection, gamma spot filtering, and false colors). Fig. 2a,b shows a NICOS graphical interface with history and device list on the left and an open widget on the right.

The access to the hardware in our NICOS setup is done via a network based middleware: TANGO [5], a client-server framework software for instrument control, originally developed at ESRF Grenoble. The TANGO approach allows the definition of software devices which are registered by a unique name in a database of a naming service. Each device is a collection of commands and read/writable attributes, which allows to control the hardware. The translation from the logical layer to real hardware commands is done in servers, running on different machines inside the instrument network. Via the name each device may be accessed and controlled from different clients. One of the clients is the NICOS system.

In our case the LiMA [6] TANGO server is used to control the camera. LiMA is an open source framework, which supports a unified access to a large number of cameras. For all other devices like motors the entangle [7] TANGO server is used. This open source project suppports a unified access to more than 350 TANGO devices.

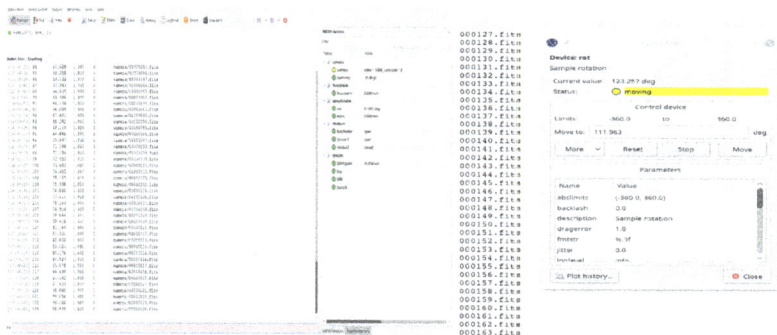

Fig.2a,b: The NICOS graphical interface with history and device list (left) and an open widget (right).

Since the INL's radiography stations were designed exclusively for remote operations outside the bay, not even basic network infrastructure existed inside beyond power outlets. For the first tomography setup, and to maintain maximum flexibility, a small network was built consisting of a network switch, three Raspberry Pi computers, and one Linux laptop (Fig.3). One Raspberry Pi acted as network controller and DNS server, one Raspberry Pi was set up as a motion controller interfacing via USB with a Thorlabs linear stage with its own controller, plus a Thorlabs multi-axis controller for the rotation table. The third Raspberry Pi was used to control the CMOS camera for radiography using the LiMA camera control software [6]. The laptop runs NICOS together with the graphical interface and was used for instrument control and data storage.

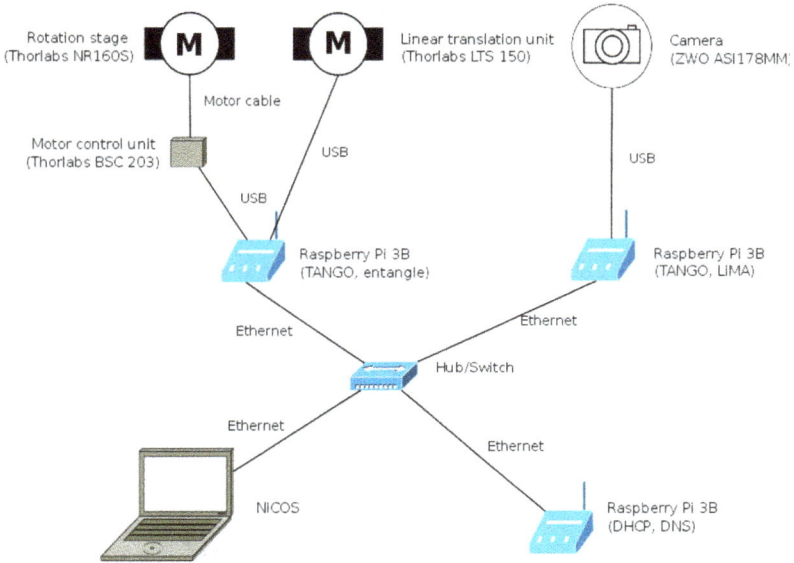

Fig.3: Components and local network of the tomography setup.

First results

The first complete 3D CT recorded at INL shows a microwave horn. Due to impending reactor shutdown, only a reduced data set of 421 images at 30 seconds each was recorded at collimation L/D=185. Fig. 4a-c shows a photo, a radiograph, and a segmented 3D view of the object.

Fig.4a-c: First digital neutron CT at INL: A photo, a radiograph, and a segmented 3D view.

Materials Research Forum LLC
https://doi.org/10.21741/9781644900574-8

Conclusions and Outlook

Using the traditional transfer method, NRAD staff could produce only 14 film radiographs per day, so collecting 421 radiographs in four hours represents a substantial improvement in the production rate by two orders of magnitude. This improvement is directly the result of the digital imaging and control systems.

Following the original TACO/TANGO philosophy, the individual software elements were developed as separate network devices. Since a dedicated instrument network is not yet available at INL, a tiny local network setup was realized to successfully achieve the first computed tomography. All components have now been combined on a single Linux computer for the next phase of introducing user operations, and the first test on-site will be conducted soon. Using the Tango/NICOS systems allows to use the already huge database of instrument servers developed at MLZ for all kinds of components like cameras, motor controllers, data interfaces and other hardware. Electronic shutter control will be added soon.

A smaller camera box with translation and rotation stage using this system was already designed and tested with this control software [8]. The construction will be simplified even more and will in the future be made available for the public for free, to be used as a universal standalone CT system.

References

[1] A.E. Craft, D.M. Wachs, M.A.Okuniewski, et al., Neutron radiography of irradiated nuclear fuel at Idaho National Laboratory, Physics Procedia 69 (2015) 483-490. https://doi.org/10.1016/j.phpro.2015.07.068

[2] https://astronomy-imaging-camera.com/product/asi-178mm-cool/

[3] http://www.sharpcap.co.uk

[4] http://www. nicos-controls.org

[5] http://www. tango-controls.org

[6] http://lima.blissgarden.org/

[7] https://forge.frm2.tum.de/entangle/doc/entangle-master/

[8] B. Schillinger, J. Krüger, A quadruple multi-camera neutron computed tomography system at MLZ, in this issue

Neutron Radiography - WCNR-11   Materials Research Forum LLC
Materials Research Proceedings **15** (2020) 53-57   https://doi.org/10.21741/9781644900574-9

Radiation Degradation of Silicon Crystal Used as Filter for Neutron Radiography

Ladislav Viererbl [a] [*], Jaroslav Šoltés, Miroslav Vinš, Hana Assmann Vratislavská and Alexander Voljanskij

Research Centre Rez Ltd., Hlavní 130, Husinec-Řež, 250 68, Czech Republic

[a] ladislav.viererbl@cvrez.cz

Keywords: Radiation Degradation, Neutron Filter, Silicon Crystal, Neutron Radiography

Abstract. Single crystals of some materials like silicon, sapphire or bismuth are often used as radiation filters for neutron radiography beams. These single crystals are relatively transparent for thermal and cold neutrons and can suppress fast neutrons and gamma radiation. When a single crystal is irradiated the crystal lattice is partially damaged and filter characteristics of the crystal can be deteriorated. This aspect of radiation degradation was studied for silicon single crystals. Experiments for silicon radiation degradation measurements were carried out in the LVR-15 research reactor. A silicon crystal sample (length of 10 cm) was irradiated in the reactor core. Thermal neutron attenuation was measured on the thermal neutron beam used also for neutron radiography facility. The attenuation measurement was made by activation detectors before and after crystal irradiation in the reactor core. A silicon single crystal (length of 100 cm) is used as a neutron filter in the channel used for neutron radiography in the LVR-15 reactor. The results indicate that the radiation degradation of the filter is acceptable low for a few years of facility operation.

Introduction

Horizontal channels of research reactors can be used for neutron radiography. To optimize the neutron beam parameters, filters are inserted into the channel. Single crystals of some materials like silicon, sapphire or bismuth are often used as radiation filters for neutron radiography based on thermal or could neutron beams [1,2]. These crystals are relatively transparent for thermal and cold neutrons and can suppress undesirable fast neutrons and gamma radiation emitting from a reactor core. Cross sections for thermal neutrons of these single crystals are 5 to 10 times lower [3] compared with amorphous or polycrystalline samples of the same materials [4]. When the single crystal is irradiated the crystal lattice is partly damaged and filter characteristics of the crystal can be deteriorated. Here this aspect of radiation degradation was studied for silicon single crystals. Attenuation increasing due to irradiation was measured for thermal neutrons. According to the theory, attenuation parameters for fast neutrons and gamma radiation change insignificantly compare to thermal neutrons. During neutron irradiation of a silicon single crystal, defects in the crystal lattice are induced by two main ways: ^{30}Si nuclide is transmuted to ^{31}P by n-γ reaction and Si nuclei are displaced by fast neutron scattering on the nuclei and produce crystal defects. The second type of induced defects can be partly removed by annealing of the crystal.

Silicon crystal irradiation

Experiments for silicon single crystal radiation degradation were carried out in the LVR-15 research reactor [5], which is a multipurpose facility with nominal thermal power of 10 MW, situated in Řež near Prague. One horizontal channel denoted as HC1 (Fig. 1) is used for neutron radiography facility where cylindrical silicon single crystal with diameter of 78 mm and

Materials Research Forum LLC
https://doi.org/10.21741/9781644900574-9

1000 mm long is inserted into the channel as neutron filter (Fig. 2). The channel is radial type and the facility use multipixel detector placed in the beam axis. Therefore relatively large crystal is needed to significantly suppress undesirable fast neutrons and gamma radiation.

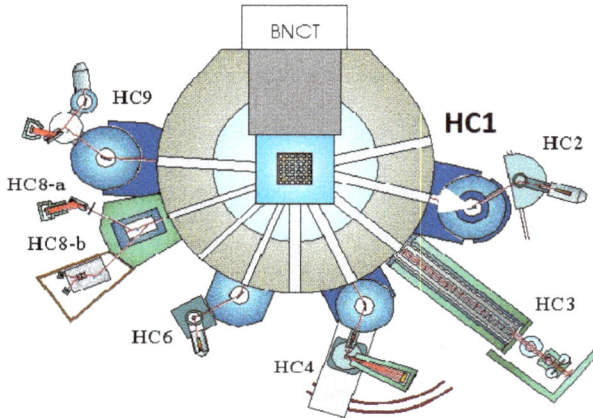

Figure 1. Horizontal channel layout at the LVR-15 research reactor. The channel chosen for the neutron radiography facility is denoted as HC1 [6].

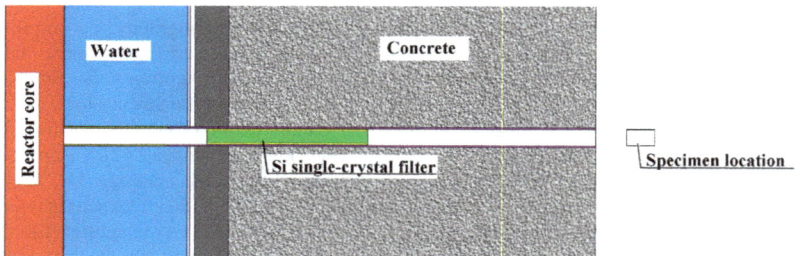

Figure 2. Cross-section of the HC1 horizontal channel with the location of the silicon single crystal used as neutron filter [7].

During reactor operation the silicon crystal used for filter in the horizontal channel HC1 is permanently irradiated from reactor core. In the position of silicon cylinder base faced to reactor core (beginning of the filter), the average neutron fluence rate is estimated to about 8×10^{17} cm^{-2}/year.

For the radiation degradation experiment, a silicon single crystal cylindrical sample (ϕ78 mm $\times$ 100 mm) was used. The sample was irradiated in a vertical channel DONA [5] close to the core of the LVR-15 reactor. During one day irradiation, the sample received total fluence of 7.6×10^{17} cm^{-2}. Then irradiation fluence is roughly equal to the fluence, which the beginning of the filter in the horizontal channel HC1 receives per year.

Thermal neutron attenuation measurement

Thermal neutron attenuation in the silicon single crystal sample was measured before irradiation in the reactor core and after the irradiation. Neutron attenuation was measured on the LVR-15 neutron radiography beam, which is thermal neutron beam on HC1 horizontal channel, the same as discussed in the above chapter. The silicon sample with detectors were irradiated in the beam on "Specimen location" in Fig. 2. The neutron fluence received during this measurement was negligible compare with the fluence in the reactor core described above. Three types (Cu, In, Au) of activation detectors were used. Two sets of detectors were on front face of the silicon sample (Fig. 3 Left) and two sets on the rear face. Activated radionuclides have short half-life then the same detectors were used for both attenuation measurement (before and after irradiation in reactor core) to avoid some uncertainties, e.g. in detector weighting. Silicon cylinder axis was parallel with the beam and the detectors were fixed on both cylinder bases (Fig. 3 Right). Activation detectors were ϕ10 mm × 0.1 mm foils. Materials were pure Cu and 1 % alloy based on aluminium for In and Au.

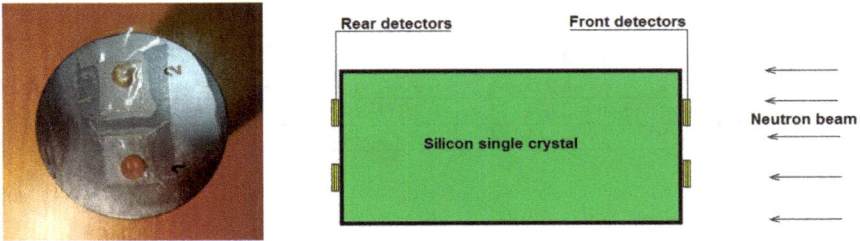

Figure 2. Left - photo of Si crystal cylinder base with fixed activation detectors. Right - arrangement of thermal neutron attenuation measurement.

The silicon sample with the activation detectors was irradiated in the HC1 beam during 24 hours. Then HPGe detector was used for activity measurement [8]. For the attenuation evaluation, neutron capture nuclear reactions were used: ^{63}Cu(n,γ)^{64}Cu for Cu activation detectors, ^{197}Au(n,γ)^{198}Au for Au and ^{115}In(n,γ)^{116m}In for In detectors. From activity values, reaction rates (RR – probability of the reaction per one target atom and per second) were calculated [9] and these were used for attenuation evaluation.

Neutron attenuation C_{ij} of the silicon single crystal was calculated according the formula:

$$C_{ij} = \frac{RR_{Rij}}{RR_{Fij}} + D \,, \tag{1}$$

where RR_{Pij} are reaction rates, index P means front (F) or rear (R) position, index i lower (*1*) or upper (*2*) position (see Fig. 2 Left) and index j represents type of detector (Cu, In, Au). The beam is slightly divergent [7] and D expresses correction on the beam divergency. Then one attenuation measurement gives six C_{ij} values and their mean value can be taken as final value C.

Results

Table 1 shows reaction rates measured before and after silicon sample irradiation in the reactor core. Correction on the beam divergency was evaluated from space distribution of beam fluence rate [7] with result D=0.059. From reaction rates and D, attenuation values C_{ij} were calculated according Eq. 1 and the values are in Table 2. Relative difference between attenuations before and after irradiation in the reactor core are in last column of the Table 2, six values from six pairs of the detectors and the mean value. After irradiation, attenuation of the silicon sample decreased (deteriorated) from 0.802 to 0.785, i.e. relatively by 2.2 %. Uncertainty 1σ of the relative difference was estimated from distribution of individual values with result 0.4 %.

Table 1. Measured reaction rates before and after irradiation in the reactor core for six pairs of activation detectors.

Detector type	Position	RR_{Pij} [1/s]			
		Before irradiation		After irradiation	
		Front	Rear	Front	Rear
In	1	1.14E-14	7.94E-15	1.02E-14	6.96E-15
	2	1.10E-14	8.36E-15	9.89E-15	7.38E-15
Au	1	4.32E-15	3.17E-15	6.28E-15	4.43E-15
	2	4.27E-15	3.32E-15	5.95E-15	4.49E-15
Cu	1	3.07E-14	2.17E-14	2.84E-16	1.98E-16
	2	2.96E-14	2.32E-14	2.66E-16	2.05E-16

Table 2. Attenuation values and relative differences between attenuations measured before and after irradiation in the reactor core.

Detector type	Position	C_{ij}		Relative difference [%]
		Before irradiation	After irradiation	
In	1	0.758	0.741	-2.20
	2	0.816	0.806	-1.29
Au	1	0.793	0.764	-3.63
	2	0.838	0.813	-2.90
Cu	1	0.767	0.754	-1.78
	2	0.842	0.830	-1.36
Mean values		**0.802**	**0.785**	**-2.19**

Conclusion

Silicon single crystal sample was irradiated in the reactor core to neutron fluence roughly equal to the value, which the beginning of the filter in the horizontal channel for neutron radiography receives per year. Thermal neutron attenuation was measured before and after the irradiation. The attenuation of the crystal decreased by 2.2 % ±0.4 %. This change is small but some radiation degradation was proved. In any case, the degradation of the filter in the LVR-15 neutron radiography facility is acceptable low for a few years of facility operation. This result can be also used for similar silicon filter applications.

Acknowledgement

The presented work has been realized within Institutional Support by Ministry of Industry and Trade. Presented results were obtained with the use of infrastructure Reactors LVR-15 and LR-0, which is financially supported by the Ministry of Education, Youth and Sports - project LM2015074.

References

[1] B. M. Rustad, Single-crystal filters for attenuating epithermal neutrons and gamma rays in reactor beams, Review of Scientific Instruments, Vol. 36 (1965) 48-54. https://doi.org/10.1063/1.1719323

[2] R. M. Brugger, A single silicon thermal neutron filter, Nuclear Instruments and Methods, Vol. 135 (1976) 289-291. https://doi.org/10.1016/0029-554X(76)90175-0

[3] A. K. Freund, Cross-sections of materials used as neutron monochromators and filters, Nuclear Instruments and Methods, Vol. 213 (1983) 495-501. https://doi.org/10.1016/0167-5087(83)90447-7

[4] K. Naguib, M. Adib, Attenuation of thermal neutrons by an imperfect single crystal, Journal of Physics D: Applied Physics, Vol. 29 (1996) 1441-1445. https://doi.org/10.1088/0022-3727/29/6/005

[5] M. Koleška, Z. Lahodová, J. Šoltés, L. Viererbl, J. Ernest, M. Vinš, J. Stehno, Capabilities of the LVR-15 research reactor for production of medical and industrial radioisotopes, J. Radioanal. Nucl. Chem., Vol. 305, (2015) 51-59. https://doi.org/10.1007/s10967-015-4025-5

[6] J. Šoltés, L. Viererbl, J. Vacík, I. Tomandl, F. Krejčí, J. Jakůbek, The new facility for neutron radiography at the LVR-15 Reactor, Conference ECNS 2015, Journal of Physics: Conference Series 746 (2016) 012041. https://doi.org/10.1088/1742-6596/746/1/012041

[7] L. Viererbl, J. Šoltés, M. Vinš, Z. Lahodová, V. Klupák, Measurement of thermal neutron beam parameters in the LVR-15 research reactor, Transaction of 15th IGORR Conference, 13-18 October 2013, Deajeon, South Korea (2013).

[8] ASTM E181-10, Standard Test Methods for Detector Calibration and Analysis of Radionuclides (2010).

[9] ASTM E944-08, Standard Guide for Application of Neutron Spectrum Adjustment Methods in Reactor Surveillance (2008).

Neutron Radiography - WCNR-11
Materials Research Proceedings 15 (2020) 58-66

Materials Research Forum LLC
https://doi.org/10.21741/9781644900574-10

Construction of a Quasi-Monoenergetic Neutron Source for Fast-Neutron Imaging

Johnson, Micah S.[1, a *], Anderson, Scott G.[1], Bleuel, Darren L.[1], Caggiano, Joseph A.[1], Fitsos, Peter, J.[1], Gibson, David[1], Gronberg, Jeff[1], Hall, James M.[1], Marsh, Roark[1], and Rusnak, Brian[1]

[1]Lawrence Livermore National Laboratory, 7000 East Ave. Livermore, California, 94551, USA

[a]johnson329@llnl.gov

Keywords: Quasi-Monoenergetic Neutron Source, Fast-Neutron Imaging

Abstract. This paper presents and discusses an approach to fast-neutron imaging that will provide high-resolution detection (*i.e.* ≤ 1 mm) of small features such as inclusions, voids, and variations in density. The application for fast-neutron imaging centers around assessing low-Z materials in high-Z shielded configurations. For this paper we present a simple theoretical argument on the feasibility of fast-neutron imaging and present results from some of our feasibility measurements. Finally, we discuss the requirements and objectives for the fast-neutron imaging system currently under construction at Lawrence Livermore National Laboratory (LLNL).

Introduction

Neutron imaging was first demonstrated by Hartmut Kallmann and Ernst Kuhn ca. 1935 (*c.f.* [1]), shortly after the discovery of the neutron by Chadwick in 1932 [2]. Neutron radiography and x-ray radiography have mutually grown in parallel as technical analogs of each other and are complementary radiographic methods. X-rays and gamma-rays (*i.e.* high-energy x-rays) interact with the electrons of an atom as well as the nucleus of an atom due to the electromagnetic properties of the photon. In general, this implies that photons interact (*i.e.* scatter) more readily with higher-Z materials. Neutrons, however, have no measurable net electric charge and can only interact with the nucleus of the atom via strong-nuclear interactions. The reduced interaction field allows the neutrons to penetrate further than photons, providing additional complementary depth to the probes of radiography. Reaction cross-sections for high-energy neutrons show only slight differences between many nuclei and can be represented by a simple model (see [3]). However, low-energy neutron reaction cross-sections vary considerably as a function of nuclei (*c.f.* [4]). These differences between photon- and neutron-reactions with atoms provide the underpinnings for the complementary nature of their respective radiographies.

Low-energy neutron radiography has been more extensively developed compared to high-energy neutron radiography because of the length scales of interest are on the order of crystalline lengths. (As a matter of reference, sample shapes have lengths approximately $29\text{fm}/\sqrt{T}$, where T is the neutron kinetic energy in units of MeV.) Additionally, it is easier to make low-energy neutron sources, especially with moderated fission reactors. Larger reaction cross-sections with different atomic nuclei also make low-energy neutron imaging particularly enticing. Low-energy neutrons can be used to measure the crystalline structure of solids using Bragg-edge scattering, which was developed by Enrico Fermi in 1947 [5]. Bragg-edges occur at neutron energies whose deBroglie wavelengths are consistent with twice the path-length difference from the lattice spacing as a function of incident angle. For lattice spacings on the order of an Angstrom, the neutron energies are about 1-10 meV. Examples of Bragg-edge scattering with low-energy

Neutron Radiography - WCNR-11   Materials Research Forum LLC
Materials Research Proceedings **15** (2020) 58-66   https://doi.org/10.21741/9781644900574-10

neutrons can be found in [6,7], and references therein. Strain mapping such as tension and torsion can be accomplished with low-energy neutron diffraction. Position-sensitive detectors that surround a sample, irradiated by low-energy neutrons register diffraction peaks (*c.f.* [8]). Stress is then applied to the sample and the crystal dimensions change, resulting in a different diffraction pattern. Examples of strain mapping and other techniques can be found in [9,10], and references therein. Other techniques such as attenuation-based methods are used to test for porosity in materials.

High-energy neutrons ($\geq$ 1 MeV) with ~fm-scale lengths do not have the right wavelengths necessary for applications presented above. Since fast neutrons cannot use the same techniques of diffraction, applications are limited. However, these fast neutrons have penetration power. With few exceptions, the fast-neutron total cross-sections are about 2-5 barns [3]. This cross-section range represents the lowest bounds of the slow-neutron total cross-section. (This is where the hard-sphere cross-section approximation of the nuclear size for low-energy neutrons transitions to neutron-nuclear interactions.) To first order, fast-neutron imaging can resolve mechanical defects of machined parts in heavily-shielded configurations. Fast-neutron imaging

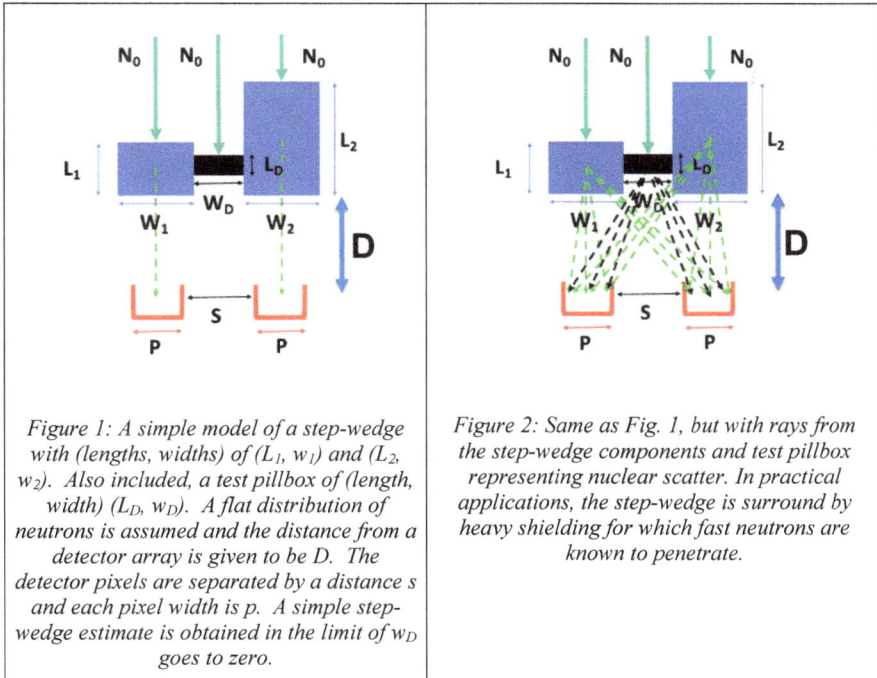

Figure 1: A simple model of a step-wedge with (lengths, widths) of (L_1, w_1) and (L_2, w_2). Also included, a test pillbox of (length, width) (L_D, w_D). A flat distribution of neutrons is assumed and the distance from a detector array is given to be D. The detector pixels are separated by a distance s and each pixel width is p. A simple step-wedge estimate is obtained in the limit of w_D goes to zero.

Figure 2: Same as Fig. 1, but with rays from the step-wedge components and test pillbox representing nuclear scatter. In practical applications, the step-wedge is surround by heavy shielding for which fast neutrons are known to penetrate.

can also resolve voids in a wide variety of materials in heavily-shielded containers. Fast-neutron imaging may have sensitivity of hydride and corrosion features in heavily-shielded scenarios. Regardless of the application objective, the value of fast-neutron imaging is dependent on its resolution. In the next section, we calculate the resolution of fast-neutron imaging with necessary input parameters. We then show results of our feasibility tests and discuss the design requirements of our fast-neutron imaging machine, which is under construction at LLNL.

Model Estimates

Putting aside any tangential benefits from fast-neutron radiography, our first objective is to be able to resolve and contrast a defect of interest against background with an analytical model. To do this, we use a simple step-wedge model in two dimensions with a test pillbox in between the two steps.

The analytical model we derive here is applicable to any imaging technique; the differences are in the input values for cross-sections and detector response. To begin, we define a six-step propagation (see Tab. 1) from the object plane (*i.e.* with the incident flux) to the digitizing plane assuming a lens-coupled CCD system.

Table 1: The (reduced) flux is described by N, with subscripts denoting the n-th step in the propagation series. The variance is denoted by σ^2. Attenuation through the material is denoted by τ, which contains the relevant cross-sections. The other coefficients are efficiencies related to the detector. The values given are with respect to 10-MeV neutrons for scintillator response and nominal values for the CCD and ADC collections and conversions.

- N_1 (object transmission) $= \tau N_0$ $(\tau = f(E)$; binomial) $(N_0$ Poisson)
  - $\sigma_{N_1}^2 = N_0 \sigma_\tau^2 + \sigma_{N_0}^2 \tau^2 = N_0 \tau (1-\tau) + N_0 \tau^2 = \tau N_0 = N_1$ $(\Rightarrow N_1$ Poisson)

- N_2 (scintillator reactions) $= \xi N_1$ $(\xi = f(E)$; binomial)
  - $\sigma_{N_2}^2 = N_1 \sigma_\xi^2 + \sigma_{N_1}^2 \xi^2 = N_1 \xi (1-\xi) + N_1 \xi^2 = \xi N_1 = N_2$ $(\Rightarrow N_2$ Poisson)

- N_3 (scintillator light yield) $= \delta N_2$ $(\delta = f(E)$; Poisson)
  - $\sigma_{N_3}^2 = N_2 \sigma_\delta^2 + \sigma_{N_2}^2 \delta^2 = N_2 \delta + N_2 \delta^2 = \delta N_2 (1+\delta) = (1+\delta) N_3$ $(\Rightarrow N_3$ non-Poisson)

- N_4 (CCD light collection) $= \eta N_3$ $(\eta \sim 2.822\text{E-}05$; binomial)
  - $\sigma_{N_4}^2 = N_3 \sigma_\eta^2 + \sigma_{N_3}^2 \eta^2 = \cdots = \eta N_3 (1+\eta\delta) = (1+\eta\delta) N_4$ $(\Rightarrow N_4$ non-Poisson)

- N_5 (CCD electron yield) $= \varepsilon N_4$ $(\varepsilon \sim 0.900$; binomial)
  - $\sigma_{N_5}^2 = N_4 \sigma_\varepsilon^2 + \sigma_{N_4}^2 \varepsilon^2 = \cdots = \varepsilon N_4 (1+\varepsilon\eta\delta) = (1+\varepsilon\eta\delta) N_5$ $(\Rightarrow N_5$ non-Poisson)

- N_6 (ADC count conversion) $= g N_5$ $(g \sim 0.655$; constant)
  - $\sigma_{N_6}^2 = N_5 \sigma_g^2 + \sigma_{N_5}^2 g^2 = \cdots = g N_5 g (1+\varepsilon\eta\delta) = g(1+\varepsilon\eta\delta) N_6$ $(\Rightarrow N_6$ non-Poisson)

In the context of our six-step model, contrast and fidelity is defined to be:

- $C \equiv \dfrac{\Delta N_6(total)}{\Sigma N_6(total)}$
- $F \equiv \dfrac{\Delta N_6(total)}{\sigma_{N_6(total)}}$,

where the Δ-values are the differences in the final counts measured, the Σ-value is the total final counts, and σ is the final count uncertainty. We parallelize our model into two scenarios where one source is assumed mono-energetic and the other to be a flat uniform spectral distribution, *i.e.* broadband uniform. Without too much difficulty, we derive an expression for contrast to be:

$$C = \tanh \Delta\gamma \left[\frac{1}{1 + 2\frac{W_D}{W}\left(1 + \frac{s}{p}\right)\frac{e^{-\Delta\phi}}{\cosh \Delta\gamma}} \right],$$

where:

- $\Delta\gamma \equiv \sqrt{2}(a_2 - a_1)$;
- $\Delta\phi \equiv \sqrt{2}(a_D - \bar{a})$;
- $a_i \equiv n_i \sigma_i L_i$

Here n is the density of the component (see Figs. 1 and 2), σ is the total cross-section (where the scattering component is distinct from 0-degrees), and L are the component lengths, given in Figs. 1 and 2. One important feature for the contrast is that it is explicitly independent on spectral shape, *i.e.* broadband uniform or mono-energetic. The implicit dependence on spectral shape is in the cross-sections.

The fidelity can be shown to be:

- $F_U = \frac{2E_0\sqrt{\Sigma N}(\sqrt{1+\varepsilon\eta\delta_0}-1)}{\sqrt{g}\varepsilon\eta\delta_0} \tanh \Delta\gamma \left[\frac{1}{1 + 2\frac{W_D}{W}\left(1 + \frac{s}{p}\right)\frac{e^{-\Delta\phi}}{\cosh \Delta\gamma}} \right]$

- $F_M = \frac{2E_0\sqrt{\Sigma N}}{\sqrt{g}\sqrt{1+\varepsilon\eta\delta_0}} \tanh \Delta\gamma \left[\frac{1}{1 + 2\frac{W_D}{W}\left(1 + \frac{s}{p}\right)\frac{e^{-\Delta\phi}}{\cosh \Delta\gamma}} \right]$

Fidelity shows explicit spectral shape dependence. The ratio of these two is easily shown to be:

$$\frac{F_M}{F_U} = \varepsilon\eta\delta_0 \left[\frac{1}{1 - \frac{1}{\sqrt{1+\varepsilon\eta\delta_0}}} \right]$$

Since $\varepsilon\eta\delta_0 > 0$ (*i.e.* positive-definite); this implies that $\frac{F_M}{F_U} > 1$, making mono-energetic sources the better fidelity. Resolution is defined to be: $R = \left|F_{measured} - F_{expected}\right|$. In the case of $W_D \sim W$ and $W_D \lesssim s$, $R = R_G(1 + \tanh \Delta\gamma)$, where R_G is the geometric (*i.e.* line-of-sight) resolution and is a function of step-width, pixel width, and pixel separation. For very small, $W_D \ll W$, $R = R_G\left(1 + \frac{W_D}{W}\left(1 + \frac{s}{p}\right)\frac{\tanh \Delta\gamma \, e^{-\Delta\phi}}{\cosh \Delta\gamma}\right)$, where it neatly reduces to line-of-sight resolution of a step-wedge if the width of the test pillbox is zero. Using the nominal values in Tab. 1 and average cross-sections for fast neutrons, and a signal-to-noise ratio of 5%, and a flux on target of 10^9 mono-energetic 10-MeV neutrons, we estimate that the performance range is around 0.65-0.85 mm. This range is subjective to uncertainties in the cross-sections and detector efficiencies. Moreover, this simplistic model does not consider scattering from shielding materials expected to be present, which is why we are considering fast neutrons. To perform more realistic calculations, we must use Monte-Carlo techniques to consider the scattering channels, especially from shielding.

Our realistic model is shown in Fig. 3. Figure 3 (Left) shows an Image Quality Indicator (IQI) consisting of concentric shells of high-Z and low-Z materials. The direction of the simulated beam is in the +z direction. Our results are based on analysis of a test pillbox (see Fig. 3 top), which is a volume, void of material. The pillboxes are placed at (x,z)= (0,0) and at the y-locations along the dashed lines shown in Fig. 3. The effective areal densities for the beam-path intersection of each pillbox are, 8.43 g/cm^2, 137 g/cm^2, and 191 g/cm^2. The results of the analysis are given in table 2.

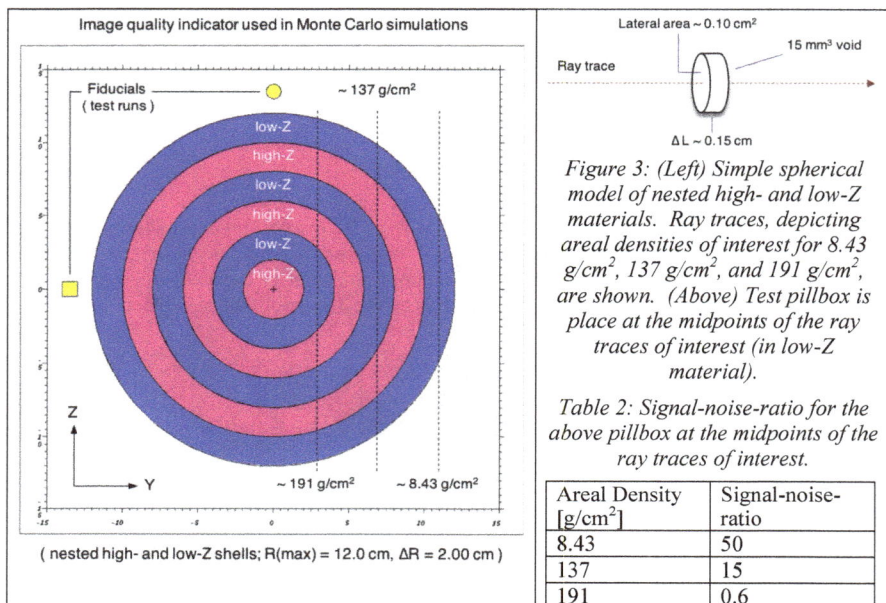

Figure 3: (Left) Simple spherical model of nested high- and low-Z materials. Ray traces, depicting areal densities of interest for 8.43 g/cm^2, 137 g/cm^2, and 191 g/cm^2, are shown. (Above) Test pillbox is place at the midpoints of the ray traces of interest (in low-Z material).

Table 2: Signal-noise-ratio for the above pillbox at the midpoints of the ray traces of interest.

Areal Density [g/cm²]	Signal-noise-ratio
8.43	50
137	15
191	0.6

The results in Tab. 2 indicates good resolution for the areal densities denoted in Fig. 3 for the pillbox with a diameter of ~3.7 mm. It should be noted that the signal-noise-ratio for 137 g/cm^2, is approximately equivalent to x-ray radiography using a 9-MeV bremsstrahlung source. This represents the limit for high-energy x-ray radiography for test pillboxes depicted in Fig. 3.

Feasibility Measurements

Some of our feasibility measurements have been published by a number of us [11-16]. We report on one of our feasibility measurements that was performed with 10-MeV quasi-monoenergetic neutrons using the tandem Van De Graaf accelerator at the Edwards Accelerator Lab at Ohio University.

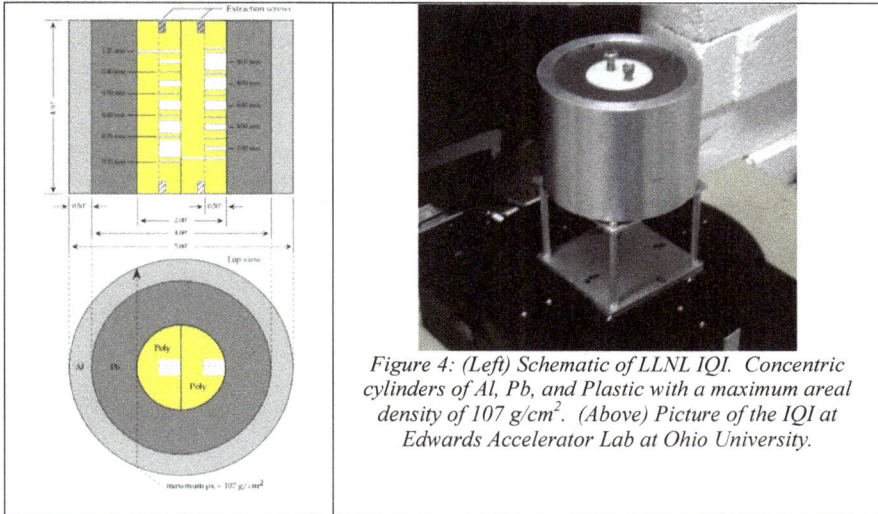

Figure 4: (Left) Schematic of LLNL IQI. Concentric cylinders of Al, Pb, and Plastic with a maximum areal density of 107 g/cm². (Above) Picture of the IQI at Edwards Accelerator Lab at Ohio University.

The IQI used is shown in Fig. 4. It is important to note that the maximum areal density is 107 g/cm^2. Our assessments, made for the above model (*i.e.* Fig. 3), is that the contrast for x-ray should be the same for 9-MeV bremsstrahlung x-rays and 10-MeV neutrons. Our results are shown in Figs. 5 and 6. In both measurements, we used a lens-coupled CCD camera and used CT techniques with the same reconstruction methods.

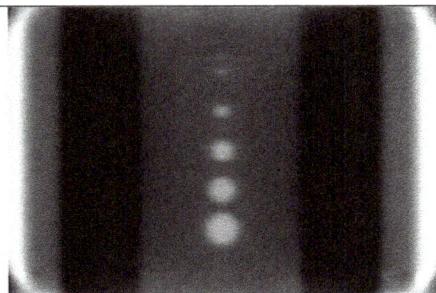

Figure 5: Neutron CT of IQI pictured in Fig. 4. The smallest test pillboxes in the plastic are clearly visible, and the contrast of the interface between the Pb and plastic is excellent.

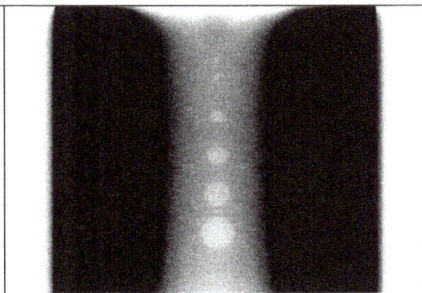

Figure 6: High-energy x-ray CT of same IQI in Figs. 4 and 5. Although the pillboxes are clearly visible, the contrast of the interface between the Pb and plastic is poor.

Although the test pillboxes are clearly visible in both the neutron- and x-ray-CT, the contrast between the low-Z and high-Z interface is poor for the x-ray. Both of these observations are consistent with our calculations and estimates discussed above.

In December 2017 we measured the above IQI, using a flat panel array, at the Los Alamos Neutron Science Center (LANSCE) at Los Alamos National Laboratory (LANL). LANSCE has a spallation source, which will provide a broadband flux. This data is still being analyzed.

Construction of the LLNL source

With the successes of our modelling and validation of our estimates, LLNL has begun construction on a 10-MeV quasi-monoenergetic neutron source at LLNL [17-28]. To meet demands of a high-brightness source, we are using a 300-μA, 100-Hz, deuteron accelerator, which consists of 2 radio frequency quadruples (RFQ) and 1 drift tube linac (DTL). The deuterons are accelerated to 7 MeV and impinge on a pulsed, windowless deuterium gas target system. (The necessity of the windowless gas target system is that the peak power on target is expected to be at 56 kW or 2.1 kW average.) The Q-value for D(d,n) reaction then boosts the neutrons to 10-MeV. The length of the gas target is 4 cm and is maximum on the flight line at nearly 4 atm-gauge. The result is a neutron rate of 10^{11} n/sr/s in the kinematically-focused forward cone of 10-degrees. The object- and image-plane will be adjustable between 1 and 5 meters for different magnifications. The source spot size (lateral area) will be around 10-mm^2, which is necessary for the sub-mm resolution in the object plane, we are striving for.

Outlook

Our plan for December 2018 is to measure the above IQI with filled-in features, using a flat panel array, at the Los Alamos Neutron Science Center (LANSCE) at Los Alamos National Laboratory (LANL). LANSCE has a spallation source, which will provide a broadband flux. We will then remeasure the IQI at LLNL's 10-MeV quasi-monoenergetic source, currently under construction at LLNL. For the latter measurement we will use a flat panel array so that the nearly every aspect of the two measurements (LANL and LLNL) are equivalent with the exception of the source. This will validate and verify our assessments made above for broadband and monoenergetic effects on fidelity and resolution.

Acknowledgements

This work performed under the auspices of the U.S. Department of Energy by Lawrence Livermore National Laboratory under Contract DE-AC52-07NA27344.

References

[1] J. S. Brenzier, "A review of significant advances in neutron imaging from conception to the present", Phys. Proc. **43** pp. 10-20 (2013). https://doi.org/10.1016/j.phpro.2013.03.002

[2] J. Chadwick, F. R. S., "The Existence of a Neutron", Proc. R. Soc. Lond. A **136** pp. 692-708 (1932). https://doi.org/10.1098/rspa.1932.0112

[3] R.W. Bauer, J.D. Anderson, S.M. Grimes, and V.A. Madsen, "Application of simple Ramsauer model to neutron total cross sections", UCRL-JC-127199 (1997)

[4] Hartnig, C. and Manke, I., "Neutron and synchrotron imaging, in-situ for water Visualization", In Encyclopedia of Electrochemical Power Sources, Garche, J., *et al.* (Eds), Elsevier (2009), pp 738. https://doi.org/10.1016/B978-044452745-5.00078-2

[5] E. Fermi, W. J. Sturm, and R. G. Sachs, "The Transmission of Slow Neutrons through Microcrystalline Materials", Phys. Rev. **71** n 9, pp. 589-594 (1947). https://doi.org/10.1103/PhysRev.71.589

[6] W. Kockelmann, G. Frei, E.H. Lehmann, P. Vontobel, and J.R. Santisteban, "Energy-selective neutron transmission imaging at a pulsed source", Nucl. Instrum. And Meth. A **578** pp. 421-434 (2007). https://doi.org/10.1016/j.nima.2007.05.207

[7] Gian Song, *et al.*, "Characterization of Crystallographic Structures Using Bragg-Edge Neutron Imaging at the Spallation Neutron Source", J. Imaging **3**, 65 (2017). https://doi.org/10.3390/jimaging3040065

[8] Winfried Kockelmann, *et al.*, "Status of the neutron imaging and diffraction instrument IMAT", Phys. Proc. **69** pp. 71-78 (2015). https://doi.org/10.1016/j.phpro.2015.07.010

[9] Nikolay Kardjilov, Ingo Manke, André Hilger, Markus Strobl, and John Banhart, "Neutron imaging in materials science", Materials Today Vol. 14 No. 6 (2011). https://doi.org/10.1016/S1369-7021(11)70139-0

[10] Gian Song, *et al.*, "Ferritic Alloys with Extreme Creep Resistance via Coherent Hierarchical Precipitates", Sci. Rep. 5, 16327 (2015). https://doi.org/10.1038/srep16327

[11] F. Dietrich and J. Hall, "Report on measurements at Ohio University to estimate backgrounds for neutron radiography in the 10-14 MeV region", UCRL-ID-127520 (LLNL, 1997). https://doi.org/10.2172/16133

[12] J. Hall, F. Dietrich, C. Logan, and B. Rusnak, "Recent results in the development of fast neutron imaging", UCRL-JC-140345 (LLNL, 2000), Proc. 16th Int. Conf. on the Applications of Accelerators in Research and Industry (Denton, TX, 2000), AIP CP576, 1113 (2001).

[13] J. Hall, "Uncovering hidden defects with neutrons", *Science & Technology Review*, UCRL-52000-01-5 (LLNL, 2001)

[14] J. Hall, F. Dietrich, C. Logan, and B. Rusnak, "High-energy neutron imaging development at LLNL", (LLNL, 2001), Proc. American Nuclear Society (2001).

[15] J. Hall, B. Rusnak, and P. Fitsos, "High-energy neutron imaging development at LLNL", UCRL-CONF-230835 (LLNL, 2007), presented at Proc. 8th World Conference on Neutron Radiography (Gaithersburg, MD, 2007). https://doi.org/10.2172/900879

[16] M. Johnson, S. Anderson, D. Bleuel, P. Fitsos, D. Gibson, J. Hall, R. Marsh, B. Rusnak, and J. Sain, "Development of a high-brightness, quasi-monoenergetic neutron source for neutron imaging", Proc. Conference on the Application of Accelerators in Research and Industry (CAARI 2016). https://doi.org/10.1016/j.phpro.2017.09.018

[17] F. Dietrich and J. Hall, "Detector concept for neutron tomography in the 10 - 15 MeV energy range", UCRL-ID-123490 (LLNL, 1996). https://doi.org/10.2172/226435

[18] F. Dietrich, J. Hall, and C. Logan, "Conceptual design for a neutron imaging system for thick target analysis operating in the 10 - 15 MeV energy range", UCRL-JC-124401 (LLNL, 1996), in Proc. 14th Int. Conf. on the Applications of Accelerators in Research and Industry (Denton, TX, 1996), AIP CP392, 837 (1997). https://doi.org/10.1063/1.52470

[19] J. Hall, F. Dietrich, C. Logan, and G Schmid, "Development of high-energy neutron imaging for use in NDE applications," UCRL-JC-134562 (LLNL, 1999), SPIE 3769, 31 (1999) and AIP 497, 693 (1999) (abridged version).

[20] B. Rusnak and J. Hall, "An accelerator system for neutron radiography", UCRL-JC-139558 (LLNL, 2000), Proc. 16th Int. Conf. on the Applications of Accelerators in Research and Industry (Denton, TX, 2000), AIP CP576, 1105 (2001).

[21] B. Rusnak, J. Hall, and W. Hibbard, "A deuterium accelerator for neutron radiography", UCRL-JC-145234 (LLNL, 2001), Proc. American Nuclear Society (Reno, NV, 2001). https://doi.org/10.1063/1.1395498

[22] B. Rusnak, J. Hall, and S. Shen, "A rotating aperture deuterium gas cell for high brightness neutron production", UCRL-PROC-212293 (LLNL, 2005), Proc. 2005 Particle Accelerator Conference (Knoxville, TN, 2005)

[23] B. Rusnak, J. Hall, P. Fitsos, R. Souza, and M. Jong, "A large-format imaging optics system for fast neutron radiography", UCRL-CONF-232018 (LLNL, 2007), in Proc. 2007 Particle Accelerator Conference (Albuquerque, NM, 2007). https://doi.org/10.1109/PAC.2007.4440657

[24] P. Fitsos, S. Edson, J. Hall, and B. Rusnak, "Design and fabrication of a precision pulsed valve system for neutron imaging", LLNL-CONF-491678 (LLNL, 2011), Proc. 26th Annual Meeting of the American Society for Precision Engineering (ASPE) (Denver, CO, 11/11).

[25] V. Tang, B. Rusnak, S. Falabella, J. McCarrick, H. Wang, J. Hall, and J. Ellsworth, "Fusion-driven gamma and fast neutron radiography test-bed at Lawrence Livermore National Laboratory", Proc. 22nd Int. Conference on Application of Accelerators in Research and Industry (CAARI) (Fort Worth, TX, 08/12).

[26] B. Rusnak, J. Hall, P. Fitsos, M. Johnson, D. Bleuel, A. Weidrick, M. Crank, S. Anderson, R. Marsh, D. Gibson, J. Sain, and R. Souza, "Development of a high-brightness source for fast neutron imaging", Proc. 2013 North American Particle Accelerator Conference (Chicago, IL, 10/16)

[27] B. Rusnak, J. Hall, P. Fitsos, M. Johnson, D. Bleuel, M. Crank, S. Anderson, R. Marsh, D. Gibson, J. Sain, L. Kruse, G. Anderson, S. Fisher, D. Nielsen, K. Lange, D. Jamero, and O. Alford, "Advancement of an accelerator-driven high-brightness source for fast neutron imaging", Proc. of 2017 International Conference on Particle Accelerators (Copenhagen, Denmark, 05/17).

[28] R. Marsh, G. Anderson, S. Anderson, D. Bleuel, M. Crank, P. Fitsos, D. Gibson, J. Hall, M. Johnson, B. Rusnak, J. Sain, R. Souza, and A. Wiedrick, "High average power deuteron beam dynamics", Proc. of 2017 International Conference on Particle Accelerators (Copenhagen, Denmark, 05/17).

Neutron Radiography - WCNR-11
Materials Research Proceedings 15 (2020) 67-73

Materials Research Forum LLC
https://doi.org/10.21741/9781644900574-11

Improvement of Neutron Color Image Intensifier Detector using an Industrial Digital Camera

Takashi Kamiyama[1, a *], Koichi Nittoh[2, b] and Kazuyuki Takada[3, c]

[1] Faculty of Engineering, Hokkaido University, Kita 13 Nishi 8, Kita-ku, Sapporo 060-8628, Hokkaido, Japan

[2] Toshiba Technical Services International Corp., 8 Shinsugita, Isogo-ku, Yokohama 235-8523, Kanagawa, Japan

[3] TAKADA KIKAI Co., Ltd., 5, Kita 3 Higashi 4, Chuo-ku, Sapporo 060-0033, Hokkaido Japan

[a]takashik@eng.hokudai.ac.jp, [b]koichi1.nittoh@glb.toshiba.co.jp, [c]takata-k@sea.plala.or.jp

Keywords: Neutron Radiography, Image Intensifier, CMOS Camera, Peltier Cooling

Abstract. Among various kinds of detectors being developed for neutron transmission imaging, Neutron Image Intensifier (NII) is still a promising option in terms of field of view and sensitivity. Since NII has a position resolution higher than 0.1 mm, it can be expected to obtain a high-sensitivity and high-gradation fine image with a camera to be combined [1]. In this research, we selected CMOS cameras for industrial applications with linearity over a wide sensitivity range, and examined the performance of the camera, especially the cooling effect of the sensor which is important in long-time measurement. Based on this result, we developed a portable imaging system that compactly integrates a cooled camera with NII, and constructed a system that can be used quickly in various neutron facilities.

Introduction

In recent years, with the progress in compact neutron facilities development, an application of high-sensitive Neutron Image Intensifier (NII) is expected for neutron imaging under low flux. NII is a vacuum tube that converts an input neutron image to a visible image [1, 2]. Neutrons react at the NII incident window and produce charged particles, which are converted into fluorescence and electrons at the subsequent phosphor and photoelectric layers, respectively. The emitted electrons are accelerated by the electric field toward the phosphor luminescent window while maintaining the neutron incident image. For the neutron detection by the image intensifier some developments were necessary in the input and output windows with short decay time to apply to pulsed neutron sources [3, 4].

As the imaging system the NII is combined with an optical system which consists of a lens and a camera. NII can adapt a kind of different cameras such as single lens reflexes, high speed cameras, video cameras, etc., but the current whole detector system size with neutron shielding containers becomes large, then it is not easy to carry and setup it for experiments. Also, in case of the NII imaging with a digital camera, we should consider the noise due to increasing the camera amplifier gain besides the radiation origin. Such noise is unique to high-sensitivity digital cameras, and we call it as high-sensitivity noise in this paper.

Against to these issues, we improved the NII imaging system. Here, we introduce the developed system which is portable so that on-site imaging is possible and which enables stable imaging with less noise even for long-time exposure.

Selection of digital camera

In proper condition of the camera combined with our NII is to have a 1:1 aspect ratio to maximize the effective field of view, because the origin of the NII is in the Color Image

Intensifier (CII) for X-ray [5], which has a circular output fluorescent screen. As the NII camera is remotely operated and displayed, USB-3.0 compatible camera which has a transfer rate above 60 fps and more than 2,076,600 pixels equivalent to the current HDTV is preferable, because it can handle video and operation with a single cable at high communication speed. By integrating short exposure time images with a high-speed compact monitoring video camera, it is possible to suppress noise and take a wide luminance range than using a single lens reflex camera with long exposure time. We also created capturing software at 32 bits/pixel to expand the dynamic range of integrated images. We conducted tests and built-in designs using cameras of following two types,

Type-A: CMOSIS CMV-4000, USB-3.0, square 4.19 million pixels view field, up to 90 fps,
Type-B: SONY IMX250, USB-3.0, square 5.01 million pixels view field, up to 75 fps.

We consider three kinds of noise. In case of long-time exposure in digital camera photography, the imaging sensor (CCD or CMOS) generates unnecessary electrons, so called dark current noise, which leads to bright spots on the screen. Also, amplifier noise and heat fog occur from the heat generation of the electronic circuit, and appears on the monitor as pink as the background on the entire screen. High sensitivity noise is caused by raising the ISO sensitivity or gain of the digital camera against dark images. Generally, in digital cameras, the last noise is dominant when imaging a dark subject in a short time, which leads to gritty images. When combined with long-time exposure, dark current noise becomes conspicuous especially. Many cameras are equipped with a noise reduction function that cancels above mentioned noises. Especially, the dark current noise is unique to the camera, but the noise has regularity. Namely, when imaging under the same condition, the noise occurs in the same position. Therefore, it is possible to reduce the dark current noise by imaging with the shutter closed under the same condition and subtracting the noise image (dark current correction), though the imaging time becomes doubled. In the case of repeated imaging with the same setting, efficiency can be improved by reuse of dark current correction data. However, the processing is not appropriate when the imaging conditions changed due to the long time operation. That is, the noise pattern varies by amplifier noise or heat fog and the subtraction is not always valid, so that it is important to keep the temperature of the camera constant.

The above-mentioned noise related issues are similarly examined for astronomical photography camera. In recent digital cameras, high sensitivity noise has been decreasing due to the evolution of the type of image pickup sensor, the circuit configuration and the image processing engine. However, it is necessary to note that the dynamic range is narrowed by these image processing operations.

The camera noise related common issue is the temperature of the camera. Cooled CCD cameras are used as a long-time imaging camera for astronomical photography. They have not yet been introduced for NII because of its high price and rapid advancement of camera sensors and difficulty in dealing with new models. In this system, we evaluated the characteristics of general surveillance cameras with added cooling function to them for improving the performance of stable operation, and the change in characteristics before and after cooling.

The relationship between neutron flux and imaging time is as follows. For example, if neutron transmission images can be continuously taken with a moving average of 1 second at 30 fps under the flux of 3×10^6 n/cm^2/sec as typical small reactors, it would take as much as 1000 seconds (17 minutes) under flux 3×10^3 n/cm^2/sec to get the similar brightness single frame image, and 10000 seconds (2.8 hours) under flux 3×10^2 n/cm^2/sec as neutron generators, respectively. Cameras installed in the previous NII systems were video cameras, high-definition digital single-lens reflex cameras or high-speed cameras [3, 4]. In case of exposing with 17 minutes or 2.8 hours in the single lens reflex camera, bulb shooting will be performed during that

Neutron Radiography - WCNR-11   Materials Research Forum LLC
Materials Research Proceedings **15** (2020) 67-73   https://doi.org/10.21741/9781644900574-11

time. In high-speed cameras, assuming that the maximum exposure time is 15 seconds, it is necessary to accumulate 67 images in 17 minutes or accumulate 667 images in 2.8 hours. Again, keeping the temperature of the camera stable during these measurements is essential for noise processing.

Relationship between camera temperature and dark current image
Transition characteristics of the camera temperature and image brightness after turning on the camera power were measured for the Type-A camera. Each camera image was accumulated during the elapsed time with external light blocked by the lens cap. Under the fixed condition of 15 seconds of camera exposure time, the initial image under ambient temperature of about 14°C was just black, but after longer elapsed time it became gray as the result of the camera temperature increasing. The camera temperature rose sharply to near 35°C in about 30 minutes after turning on the camera power. We repeated background imaging in every 30 minutes with accumulating 1800 images at a high camera gain and 1 second exposure time. The result is as shown in Fig. 1, the background RGB component of average brightness for elapsed time is increasing due to the increase of the dark current noise.

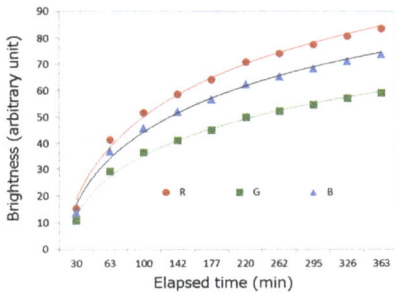

Fig.1. Elapsed time dependence of dark current image of Type-A camera [8].

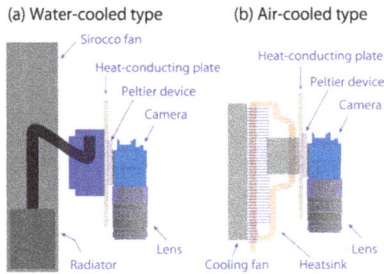

Fig.2. Camera cooling systems.

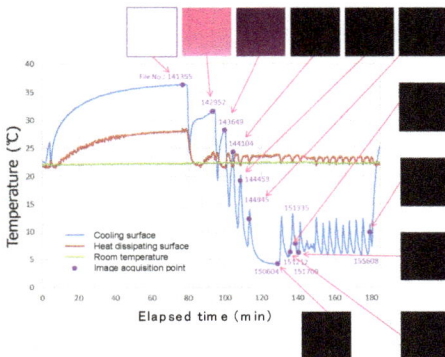

Fig.3. Dark current images and temperature dependence of Type-A camera [8].

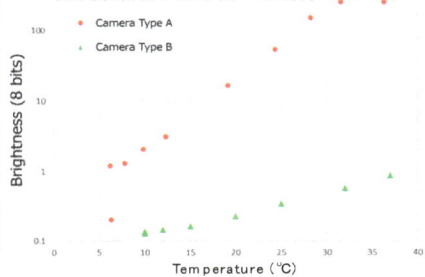

Fig.4. Temperature characteristics of the Type-specific cameras [8].

Neutron Radiography - WCNR-11   Materials Research Forum LLC
Materials Research Proceedings 15 (2020) 67-73   https://doi.org/10.21741/9781644900574-11

In order to maintain the camera body at a low temperature, a Peltier device was applied. The heat-conducting plate which receives heat from the Peltier device is cooled either by using a water-cooling type (Fig. 2(a)) or an air-cooling type (Fig. 2(b)). Using these simple cooling systems, the cooling performance was measured without turn-on the camera. Both Peltier devices used were 6A type and the measurement was carried out under fixed conditions of an applied voltage of 5V and a steady state current of 1.5A. Measurements were carried out in the ambient temperature of about 24°C in both cases. However, the air-cooling type was confirmed that the temperature on both sides on the heat-conducting plate of the Peltier device and the heat radiator changed largely against slight change in room temperature. Namely, among these two types of coolers, it was shown that the water-cooling type is stable and the cooling capacity is higher. The relationship between camera temperature and dark current image was measured with the camera power on and the Peltier device operated at 5V (1.5A) together with the water-cooling unit active. The measured results are shown in Fig. 3. The operation was started with the cooling unit stopped. When the camera temperature exceeded 35°C, the Peltier cooling unit was repeated manually turn on and off, and the dark current images of 15 second exposure time were acquired at the points marked closed circle dropped. The noise decreased from the white image with more than 250 in 8bit brightness range at 37°C to gray then black image with decreasing temperature.

About the new Type-B camera, images of the dark current noise against the camera temperature were measured similarly with the Type-A. Although the noise of the Type-B increases as the camera temperature rises, the noise component is drastically smaller than the Type-A. Figure 4 shows the temperature dependence of the image brightness taken by the Type-B together with that by the Type-A. Under Peltier conditions cooled to 15°C or less, it was confirmed that the dark current noise of the Type-A was 1% or less than that at room temperature. On the other hand, by the Type-B it was suppressed to 0.1% or less even at 20°C. Then we adopted the Type-B camera.

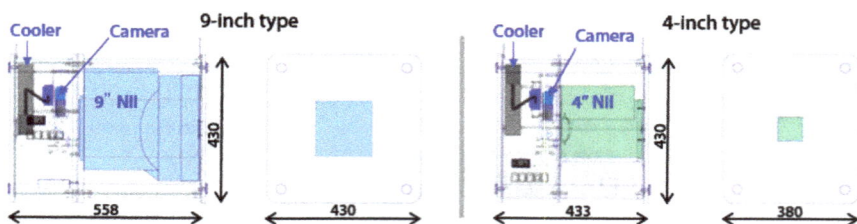

Fig.5. Design drawings of 9 and 4-inch NII systems.

Incorporation of cooled camera into NII
An imaging system combined the NII with the cooled camera has to be newly designed and fabricated in consideration of radiation shielding. In order to make the system as compact as possible, it is desirable to reduce the distance from the NII output phosphor screen to the camera. In the actual optical design, equipment placements were decided considering the size of a lens, a mirror, the camera and the cooling unit, as well as their radiation shielding arrangement. In particular, the camera was placed right angle bent by the mirror to prevent direct hits from the incident neutron irradiation. Figure 5 shows the camera built-in configuration of 4-inch and 9-inch NII system, respectively. The container of the system is covered with neutron shielding material of 50% boron carbide in epoxy resin.

Neutron Radiography - WCNR-11   Materials Research Forum LLC
Materials Research Proceedings **15** (2020) 67-73   https://doi.org/10.21741/9781644900574-11

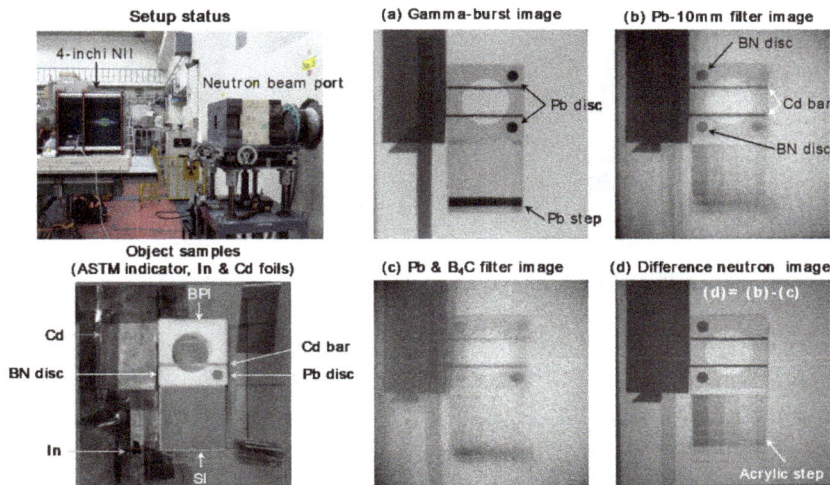

Fig.6. Neutron imaging results at HUNS.

Figure 6 shows the results of imaging using the Hokkaido University neutron source (HUNS), which is a compact accelerator pulsed source. In the measurement, we used BPI and SI indicators according to ASTM standard as shown in the lower left of Fig. 6. Because HUNS is the pulsed neutron source using an electron linac, intense burst X-rays are initially produced besides neutrons. In the case of missing lead filter as a shield for X-rays in front of the NII input window, an image by the burst X-rays is dominant as shown in Fig. 6 (a) where the upper two lead discs and the lower horizontal lead step can recognize. Inserting a 10 mm lead filter results as Fig. 6 (b). The image of the lead discs becomes paler, and two discs of boron nitride turn visible on the left side. Further, an image in which a boron filter was inserted is shown in Fig. 6(c), and from here it can be confirmed the weakened irradiations of both X-ray and neutron. By subtracting Fig. 6 (c) from Fig. 6 (b), the image of Fig. 6 (d) is obtained of a neutron image. In this last image, there were no shadows of the lead parts, but an image of the boron nitride discs, the step of cadmium and polyethylene were confirmed.

Figure 7 shows the variation of brightness by changing the camera temperature from 10 to 39°C with the same setup of Fig. 6. We set an ROI on the lead step wedge of the SI indicator, then the step-like brightness changes were determined depending on the temperature. The right side of Fig. 7 shows the ROI brightness normalized to 1 at an outside place of the samples. When the camera temperature is 15°C or less, the brightness deviation is less than 1%, but with increasing its temperature the deviation rises such as 1.3% at 25°C and 3.8% at 39°C. This result includes the fluctuation of the neutron source itself as well as the fluctuation due to the temperature control of the camera.

71

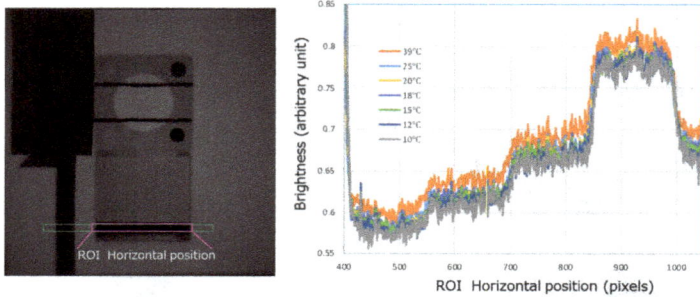

Fig.7. Change the status of the brightness values of the image with a change in temperature of the camera.

Figure 8 shows the results of the brightness change for 50 minute accumulation in 16 bits using the 9-inch NII under 12°C control. It is confirmed that the amounts of the fluctuation are ±195 with respect to the average brightness of 30302.5, so it means the stable measurement is performed within the 1.3% fluctuation, which included the neutron source instability.

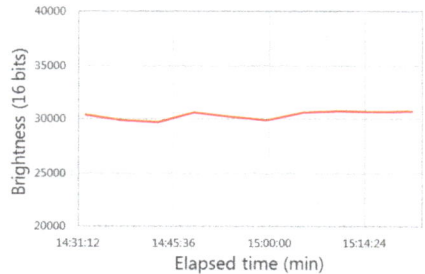

Fig.8. Stability check of 9-inch NII camera temperature setting.

Conclusion

We have developed a compact NII system aiming to stable operation for a long-time while reducing noise. In particular, with a low neutron flux or a subtle change imaging, it is necessary to perform long-time exposure imaging and multiple image accumulation since the variation in the number of neutrons is small. For the purpose, we selected industrial CMOS cameras with linearity over a wide sensitivity range, and examined the performance of the cameras, especially the cooling effect which is important in the long-time measurement. Through the experiments, it became clear that there was a difference in temperature dependence on the types of imaging CMOS device in the camera, and it was also understood that the temperature characteristics differ depending on the cooling method. Based on the above results, we have developed a portable system with the compact neutron shielding. As a result, we succeeded in developing a prototype NII system that operates stably and is not susceptible to temperature changes.

Acknowledgements

This research is partially supported by "AS272I001c, Adaptable and Seamless Technology Transfer Program through Target-driven R&D, JST", by "the Development of Non-Destructive Methods Adapted for Integrity test of Next generation nuclear fuels project by the Ministry of Education, Culture, Sports, and Technology (MEXT), Japan", and by "JSPS KAKENHI Grant Number 17H03515".

References

[1] K. Nittoh, Neutron Color Image Intensifier, Hamon, 22 (2012) 322-328 [in Japanese]. https://doi.org/10.5611/hamon.22.4_322

[2] K. Nittoh, C. Konagai, T. Noji, K. Miyabe, New feature of the neutron color image intensifier, Nucl. Instr. and Meth. A, 605 (2009) 107-110. https://doi.org/10.1016/j.nima.2009.01.136

[3] K. Nittoh, C. Konagai, M. Yahagi, Y. Kiyanagi, T. Kamiyama, Development of Neutron Color Image Intensifier for Pulsed Neutron Source, Physics Procedia, 69 (2015) 177-184. https://doi.org/10.1016/j.phpro.2015.07.025

[4] K. Mochiki, K. Ishizuka, K. Morikawa, T. Kamiyama, Y. Kiyanagi, Development of a New High-Frame-Rate Camera for Pulsed Neutron Transmission Spectroscopic Radiography, Physics Procedia, 69 (2015) 143-151. https://doi.org/10.1016/j.phpro.2015.07.021

[5] K. Nittoh, C. Konagai, T. Noji, Development of multi-color scintillator based X-ray image intensifier, Nucl. Instr. And Meth. A, 535 (2009) 686-691. https://doi.org/10.1016/S0168-9002(04)01658-4

[6] ASTM Designation, Standard Method for Determining Image Quality in Direct Neutron Radiographic Examination, E545-91 (1991)

[7] Riso National Laboratory, Standardization Activities of the Euratom Neutron Radiography Working Group, RISO-M-2356 (1982)

[8] T. Kamiyama, K. Nittoh, Consideration of the neutron image intensifier camera, Hamon, 28 (2018) 77-83 [in Japanese]

Gamma Discriminating Scintillation Screens for Digital Transfer Method Neutron Imaging

Aaron Craft[1, a*], Christian Grünzweig[2, b], Manuel Morgano[2, c] William Chuirazzi[1, d], and Eberhard Lehmann[2, e]

[1]Idaho National Laboratory, PO Box 1625, MS 2211, Idaho Falls, ID 83415, USA

[2]Paul Scherrer Institute, Laboratory for Neutron Scattering and Imaging, CH-5232 Villigen PSI, Switzerland

[a]aaron.craft@inl.gov, [b]christian.gruenzweig@psi.ch, [c]manuel.morgano@psi.ch, [d]william.chuirazzi@inl.gov, [e]eberhard.lehmann@psi.ch

Keywords: Neutron Imaging, Scintillator Screen, Indirect Transfer Method, Digital Imaging, Post Irradiation Examination, Nondestructive Evaluation

Abstract. A collaborative project between Idaho National Laboratory (INL) and Paul-Scherrer Institute (PSI) is investigating a new type of scintillation screen that uses ZnS scintillator material with a dysprosium neutron converter instead of traditional prompt converters such as ^{6}Li. Such a screen exposed to a neutron beam creates a latent image by neutron activation of the dysprosium in the scintillator screen. The activated screen is transported into a camera box allowing the camera to read a digital image from the photons emitted by the activated scintillation screen. Such an imaging system combines modern camera-based system architecture with the approach of traditional indirect transfer method radiography. The results show for the first time that the combination of dysprosium with a scintillation material like ZnS can produce light which is measurable under common camera-based detection conditions and that neutron radiographic images of reasonable quality can be produced. The resolution was poorer than expected at ~ 300 µm, but is on the order of the desired resolution of 100 µm. Potential improvements and additional converter materials may be investigated in the future that could increase the light output and improve spatial resolution.

Introduction

As the demand for power increases worldwide, an emphasis has been placed on nuclear energy to meet this need. New nuclear reactor designs utilize a variety of fuels, which must be properly studied and characterized before they can be commercially manufactured. Neutron radiography provides more comprehensive information about the internal geometry of irradiated nuclear fuels and components than any other nondestructive examination technique to date [1]. Transfer method neutron radiography can be used to study irradiated materials despite the very high gamma radiation fields emitted from them [2,3,4]. In the transfer method, a cassette of converter foils is placed in a neutron beam behind an object and exposed to the neutron beam. The converter foils become activated in the pattern of the neutron beam, forming a latent image as a pattern of activation on the foil. The activated foils are coupled to an imaging medium (e.g. film [2], image plates [3]), which are exposed by the decay radiation as the foil decays. Another technique is to use image plates containing dysprosium [4]. The main advantage of this method is its complete insensitivity to gamma rays. However, it is very time consuming and labor intensive.

Current neutron radiography work is trending towards the use of digital systems due to their higher detection efficiency, position stability, flexibility in the field-of-view, high spatial resolution, and faster image acquisition. Most digital systems rely on ^{6}LiF/ZnS:X scintillator screens placed directly in the neutron beam to produce an image. These screens are sensitive to the gamma rays, which has precluded use of camera-based imaging systems for examination of irradiated materials because the object itself is a strong source of gamma radiation.

This paper discusses a project to develop a new type of scintillator screen that combines some of the benefits of modern digital imaging systems with the indirect transfer method by replacing the promptly-decaying neutron converter material with dysprosium which decays over time. Different configurations of scintillator material and dysprosium converter were fabricated. These novel screens were tested to determine their light output during exposure and subsequent decay, and to measure the effective spatial resolution.

Proposed Imaging Methodology for Using Dysprosium-Based Screens

The proposed imaging process for dysprosium-based scintillator screens is a hybrid between the indirect transfer method and the common direct imaging method with camera-based systems. An illustration of the proposed process can be found in Figure 1. The scintillator screen is first placed behind the imaging object in the neutron beam and exposed directly in the neutron beam. The scintillator screen becomes activated in the neutron beam, creating a latent image in the pattern of activated dysprosium on the scintillator screen. This first step is analogous to the indirect transfer method. After some time, either the neutron beam is turned off (shutter closed) or the activated scintillator screen is transported out of the beam into a light-tight camera box, the latter of which is depicted in Figure 1. The decay products from the activated dysprosium interact with the scintillator material, emitting visible light that is read by the digital camera. In the present tests, the beam shutter was closed to turn off the neutron beam instead of physically moving the screens.

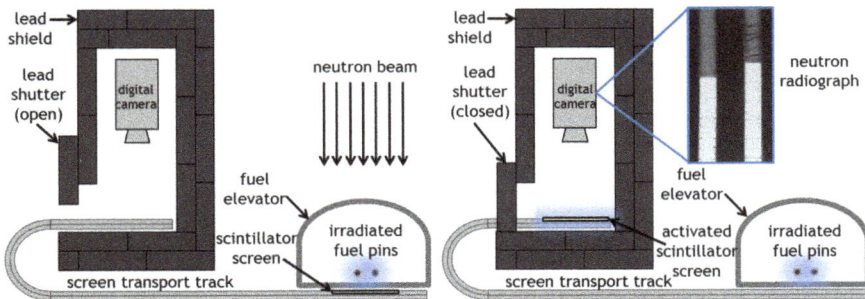

Figure 1. Illustration of the digitial indirect neutron imaging with scintillator screen concept. (Left) The screen is placed in the beamline and exposed behind the imaging object. (Right) The screen is then transported away from the fuel into the camera box and the digital camera captures the latent image on the scintillator screen.

Scintillator Screens

Scintillator screens were manufactured with a variety of compositions and thicknesses. The ZnS:Cu scintillator was mixed with Dy_2O_3 for some screens and was deposited directly onto a dysprosium foil for others. Natural dysprosium consists of seven stable isotopes, but ^{164}Dy, which comprises 28.18% of natural dysprosium, is the isotope useful in imaging. When ^{164}Dy absorbs a neutron, it can decay in one of two paths, shown in Equations 1 and 2. Dysprosium was chosen because its shorter decay path has a quick half-life (1.26 minutes) and large absorption cross-section (~1000 b), making it suitable for quickly producing neutron radiographs [5]. This is especially useful in an application when many shots must be taken, such as computed tomography.

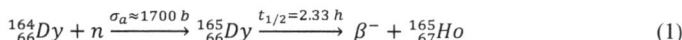

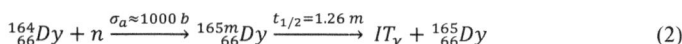

$$^{164}_{66}Dy + n \xrightarrow{\sigma_a \approx 1700\ b} {}^{165}_{66}Dy \xrightarrow{t_{1/2}=2.33\ h} \beta^- + {}^{165}_{67}Ho \tag{1}$$

$$^{164}_{66}Dy + n \xrightarrow{\sigma_a \approx 1000\ b} {}^{165m}_{66}Dy \xrightarrow{t_{1/2}=1.26\ m} IT_\gamma + {}^{165}_{66}Dy \tag{2}$$

A total of 23 screens, each 25 mm × 25 mm square, were fabricated with varying substrate, converter material, phosphor grain size, material ratio, and scintillator thickness. A list of all 23 screens and their properties is detailed in Table 1. Regarding the ZnS:Cu grain size, three sizes were used: fine (d50=3.05 μm), medium (d50=4.67 μm) and large (d50=9.67 μm).

Measurements

The scintillator screens were tested at the NEUTRA beamline at Paul Scherrer Institute (PSI) [6]. The light output of each screen was measured with individual images taken every 2 seconds for total exposure times of 1-20 minutes. Dark-field and open-beam images were acquired and used to correct the images per standard procedures. Light output was analyzed for both the exposure and decay phases of the experiments.

Table 1. List of scintillator properties

Number	Substrate	Absorber Material	Phosphor (grain size)	Mixture	Thickness [um]
1	Al-plate	Dy_2O_3	ZnS:Cu (large)	1:2	30
2	Al-plate	Dy_2O_3	ZnS:Cu (large)	1:2	50
3	Al-plate	Dy_2O_3	ZnS:Cu (large)	1:2	100
4	Al-plate	Dy_2O_3	ZnS:Cu (large)	1:2	200
5	Al-plate	Dy_2O_3	ZnS:Cu (large)	1:1	75
6	Al-plate	Dy_2O_3	ZnS:Cu (large)	1:3	100
7	Al-plate	$Dy(O_2CH)_3$	ZnS:Cu (medium)	1:2	60
8	Al-plate	Dy_2O_3	ZnS:Cu (large)	1:2	90
9	Al-plate	Dy_2O_3	ZnS:Cu (medium)	1:2	100
10	Al-plate	Dy_2O_3	ZnS:Cu (fine)	1:2	100
11	Dy-Foil	Dy_2O_3	ZnS:Cu (large)	1:2	35
12	Dy-Foil	Dy_2O_3	ZnS:Cu (large)	1:2	50
13	Dy-Foil	Dy_2O_3	ZnS:Cu (large)	1:2	100
14	Dy-Foil	Dy_2O_3	ZnS:Cu (large)	1:2	225
15	Dy-Foil	Dy_2O_3	ZnS:Cu (large)	1:1	85
16	Dy-Foil	Dy_2O_3	ZnS:Cu (large)	1:3	100
17	Dy-Foil	Dy_2O_3	ZnS:Cu (large)	1:2	100
18	Dy-Foil	Dy_2O_3	ZnS:Cu (medium)	1:2	100
19	Dy-Foil	Dy_2O_3	ZnS:Cu (fine)	1:2	100
20	Dy-Foil	none	ZnS:Cu (large)	pure	60
21	Dy-Foil	none	ZnS:Cu (large)	pure	100
22	Dy-Foil	none	ZnS:Cu (large)	pure	200
23	Dy-Foil	none	ZnS:Cu (fine)	pure	100

Since the light output of a scintillator is proportional to its activity, light production increases while the screen is exposed to the neutron beam, then decays once the neutron exposure ceases. The screens were exposed for periods of 1, 5 and 10 minutes then allowed to decay with the beam turned off. Images were acquired read every 2 seconds. The light output over time was then integrated to determine the total light output during the 5-minute decay phase. The results of the 25 mm × 25 mm screens are shown in Figure 2. Experimental results revealed that samples #3 and #4, both consisting of an absorber material of Dy_2O_3 in a 1:2 wt. ratio with large grain ZnS:Cu, produced the greatest light output for the screens on an aluminum substrate. Some improvement in light output was seen for similar screens on a dysprosium foil substrate (i.e. compare #3 and #4 to #13 and #14). Figure 3 shows the light output per pixel measured every

Materials Research Forum LLC
https://doi.org/10.21741/9781644900574-12

two seconds during a 5-minute exposure and subsequent decay phases for the seven screens that exhibited the highest light output (#'s 3-4, 13-14 & 20-22).

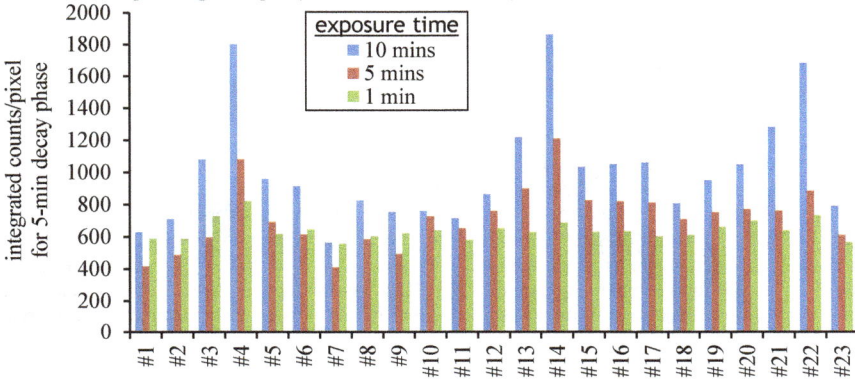

Figure 2. The integrated light output of scintillator screens measured in counts per pixel during the 5-minute decay phase.

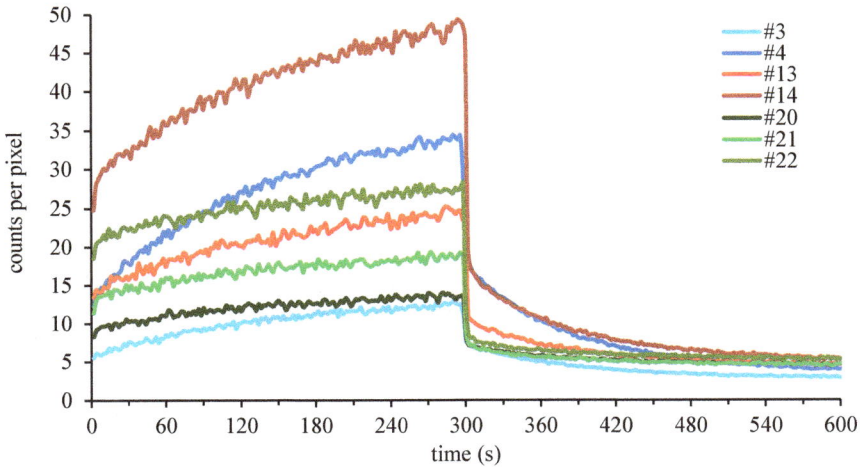

Figure 3. The light output per pixel of the seven screens that exhibited the highest light output. Individual images were taken every 2 s during 5-minute exposure and 5-minute decay phases. The light output builds up as the screens are exposed to the neutron beam for a longer amount of time. The beam is shut off after 300 seconds and the screens' light output decays away with time.

The brightest seven screens were subjected to a 20-minute neutron exposure, and the results are shown in Figure 4. Due to limited beam time, some screens were not allowed to decay completely (takes ~24 hours for ten half-lives) and residual decay may offset the light output values in Figure 4 compared to Figure 3, but the buildup and decay behaviors can still be compared. Recall from Table 1 that screens #3 and #4 mix converter with scintillator, #13 and #14 have scintillator material deposited on a converter substrate, and screen #20, #21 and #22 include both mixed converter and a converter substrate. As expected, screens with both mixed converter and a converter substrate (screens #13 and #14) provided the highest light output. Mixing the scintillator with Dy_2O_3 (screens #3 and #4) provided higher light output than scintillator deposited on a dysprosium foil (screens #21 and #22) for the same scintillator thickness. Thus, mixing converter material with the scintillator was more effective in producing high light output than depositing scintillator on a converter substrate. Also, the difference in light output was much less for the thinner, 100 μm thick screens (comparing screens #3 to #21, and #4 to #22), suggesting that the beta particle from activated dysprosium produces the majority of the light in the first 100 μm of scintillator. Furthermore, for screens with scintillator material on a converter substrate, there is only marginal increase in light output by increasing the screen thickness to 200 μm (compare screens #21 and #22). The screens provided sufficient integrated light output to produce usable images, though there is much room for improvement to increase the light output.

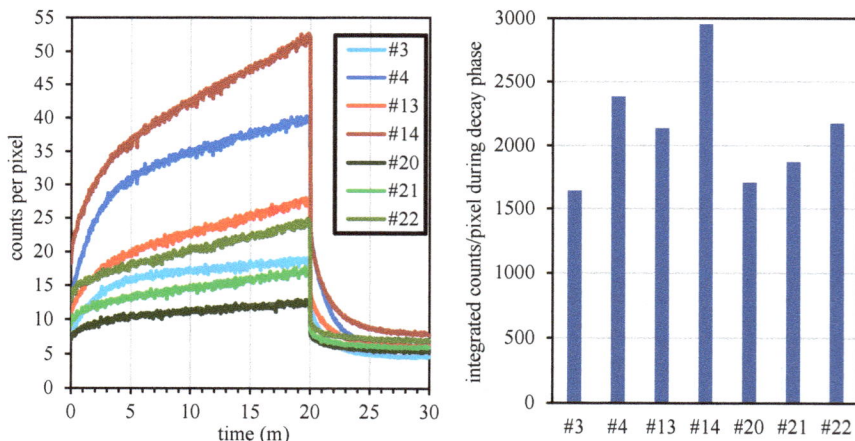

Figure 4. (Left) Light output measured every 2 seconds during the 20-minute exposure and 10-minute decay. (Right) Integrated light output for the 10-minute decay phase.

In addition to light output, the spatial resolution of each scintillator was tested. This was accomplished by imaging a Siemens star resolution test pattern with each scintillator screen. The exposure time for the resolution testing was 20 minutes. Upon obtaining the images, the images were gamma-spot filtered, dark-field corrected, and open-beam normalized. Multiple dark-field and open-beam images were acquired, and the median values used for corrections. The test patterns were then analyzed to see the smallest discernable line-pairs. Figure 5 shows images taken with the seven brightest scintillator screens. The resolution was measured by taking the tangential line profile of a circle centered on the Siemens star and extending the radius out to the location where the star's spokes first become visible. A standard 200 μm thick ^{6}LiF/ZnS screen

Neutron Radiography - WCNR-11
Materials Research Proceedings 15 (2020) 74-79

Materials Research Forum LLC
https://doi.org/10.21741/9781644900574-12

was used as a comparison. The best spatial resolution from the screens in this experiment was ~300 µm, which was not as good as the ~200 µm given by the standard screen.

Conclusions

The purpose of this study was to develop and test a dysprosium-based scintillator screen that could be used in a novel, hybrid neutron imaging method combining the indirect transfer method with digital camera-based neutron radiography. The results show for the first time that dysprosium can be used with a standard scintillator phosphor to create a scintillator screen that can produce a useful amount of light after exposure to a neutron beam. Although the resolution was not as high as desired, the resulting radiographs were of reasonable quality. With further improvements (e.g. using enriched ^{164}Dy), dysprosium-based screens could be usable for imaging highly-radioactive objects. Additionally, other converter materials may be investigated in future work with the goal of improving light output and spatial resolution.

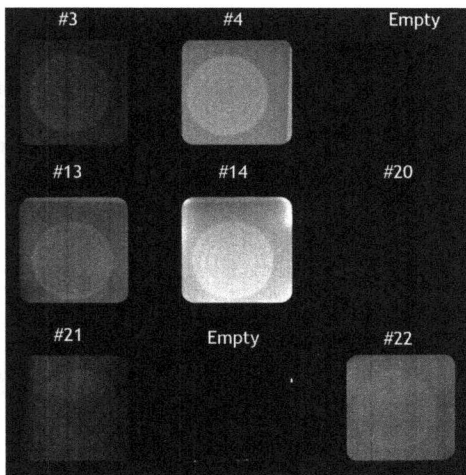

Figure 5. Radiographs of a Seimens star acquired with the brightest seven scintillator screens.

References

[1] A.E. Craft and J.D. Barton, "Applications of neutron radiography for the nuclear power industry," Physics Procedia 88 (2017) 73-80. https://doi.org/10.1016/j.phpro.2017.06.009

[2] A.E. Craft, D.M. Wachs, M.A. Okuniewski, D.L. Chichester, W.J. Williams, G.C. Papaioannou, & A.T. Smolinski, "Neutron radiography of irradiated nuclear fuel at Idaho National Laboratory." Physics Procedia, 69 (2015) 483-490. https://doi.org/10.1016/j.phpro.2015.07.068

[3] A.E. Craft, G.C. Papaioannou, D.L. Chichester, & W.J. Williams, "Conversion from film to image plates for transfer method neutron radiography of nuclear fuel." Physics Procedia, 88 (2017) 81-88. https://doi.org/10.1016/j.phpro.2017.06.010

[4] P. Vontobel, M. Tamaki, N. Mori, T. Ashida, L. Zanini, E.H. Lehmann, & M. Jaggi, "Post-irradiation analysis of SINQ target rods by thermal neutron radiography." Journal of nuclear materials, 356(1-3) (2006) 162-167. https://doi.org/10.1016/j.jnucmat.2006.05.033

[5] E.M. Baum, H.D. Knox, and T.R. Miller, "Chart of the Nuclides," 16th ed., Knolls Atomic Power Laboratory (2002).

[6] E.H. Lehmann, P. Vontobel, and L. Wiezel, "Properties of the Radiography Facility NEUTRA at SINQ and its Potential for Use as European Reference Facility." Nondestructive Testing & Eval 16 (2001) 191-202. https://doi.org/10.1080/10589750108953075

Neutron Radiography - WCNR-11   Materials Research Forum LLC
Materials Research Proceedings **15** (2020) 80-85   https://doi.org/10.21741/9781644900574-13

Imaging Based Detector with Efficient Scintillators for Neutron Diffraction Measurements

Matt W. Seals[a], Stephen B. Puplampu[a*], Dayakar Penumadu[a], Richard A. Riedel[b], Jeff R. Bunn[b], Christopher M. Fancher[b]

[a] Department of Civil and Environmental Engineering, University of Tennessee, Knoxville, TN 37996, USA

[b] Oak Ridge National Lab, Oak Ridge, TN 37892, USA

email: dpenumad@utk.edu

(*) corresponding author

Abstract. The Anger Camera developed by the detector group at the Oak Ridge National Laboratory was utilized for the present work for its unique advantage of employing multiple modules to obtain large active measurement area for detecting diffracted/scattered thermal neutrons. Considering the relatively small flux associated with diffracted/scattered neutrons, suitable efficiency with high spatial resolution is a requirement for utilizing two-dimensional imaging detectors. The potential to implement pulse shape (in addition to pulse height) discrimination-based scintillators further enhances the ability to detect diffracted neutrons with improved signal to noise ratio. In this paper, initial results associated with 6Li glass-based scintillator will be presented. The authors explored the feasibility of using this system to detect and quantify diffraction peaks and peak shifts at the Neutron Residual Stress Facility (NRSF2), High Flux Isotope Reactor (HFIR) in the Oak Ridge National Laboratory (ORNL). Suitable camera mounting and shielding had to be developed. Reference measurements using polycrystalline powders with known atomic planar spacing will be discussed along with measurement settings associated with expected resolution for peak shift measurements. Initial results are promising and demonstrate that a suitable scintillation-based neutron detecting system is viable for residual stress-based diffraction measurements. Small area detectors are also feasible with suitable consideration to scattering volume and distance to detector.

User Acknowledgments

This research [or, A portion of this research] used resources at the High Flux Isotope Reactor [and/or Spallation Neutron Source, as appropriate], a DOE Office of Science User Facility operated by the Oak Ridge National Laboratory.

Introduction

Anger camera is a variant within the CCD imaging family that allows for fast read-out of large multi-anode photo-multiplier tubes [1-3]. Instead of approaching the issue of readout by establishing a channel for each anode, Anger camera reduces the entire output of the PMT down to four channels allowing for fast data acquisition over the entire photo-sensitive area in real-time. The method by which this is performed is by injecting the output signal of the PMT onto a resistive network when a particle is incident on the scintillator of the PMT as depicted in Figure 1.

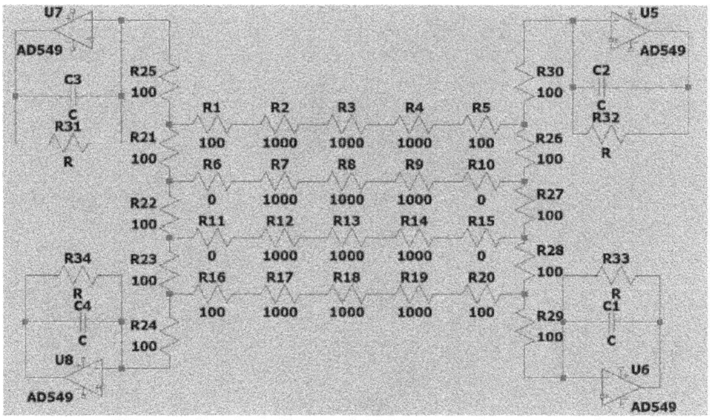

Neutron Radiography - WCNR-11   Materials Research Forum LLC
Materials Research Proceedings **15** (2020) 80-85   https://doi.org/10.21741/9781644900574-13

Figure 1: 4x4 Anger logic resistive network

The resistive network is what performs the Anger logic; the position of incidence of a particle on the scintillator can be obtained by the output of the PMT onto the network. Any response from the PMT results in a response at the four outputs of the network in the form of a voltage. These four voltages, $V_a, V_b, V_d, and\ V_c$ can be used to determine the location of incidence as a coordinate pair (X, Y) via the formulae Eq. 1 and Eq. 2:

$X_{position} = \dfrac{(V_a + V_b) - (V_c + V_d)}{V_a + V_b + V_d + V_c}$	*(Eq. 1)*
$Y_{position} = \dfrac{(V_a + V_d) - (V_c + V_b)}{V_a + V_b + V_d + V_c}$	*(Eq. 2)*

The scintillating material used in this study is a 6Li glass scintillator which has found extensive use in the field of neutron detection [4-6]. Other neutron detection methods lack spatial resolution, and many are by default one dimensional only looking at the number of neutrons that interact with the detector about the vertical channels since the study of neutron diffraction is primarily concerned with the angle at which the bulk of neutrons are diffracting. Typically, the detector type used is the helium-3 tube which is large with a facing size of up to an inch or more in diameter per anode. This results in quite low spatial resolution and a large form-factor.

The Anger camera allows for quite high spatial resolution since it incorporates as many as 1024 anodes into a 116 mm by116 mm area. In this study, authors seek to implement an Anger camera in diffraction mode for residual stress measurements. Steps towards this implementation require ability to achieve precise, repeatable angular motion and positioning. Furthermore, pulse shape discrimination provides significant assistance in noise reduction however appropriate shielding is still necessary for diffraction. Instrument calibration is also needed to correctly process acquired data. Typically, calibration requires alignment scans in the diffraction plane and acquisition of diffraction peaks from well-known standard powders; additional considerations have to account for the geometry of the detector being a flat surface moving along an arc.

Experimental Method

To set up and operate the Anger camera at the NRSF2 instrument, additional components had to be put together to ensure accuracy and repeatability of measurements. A stage for the anger camera was secured to the side of the NRSF2 detector; a rotational degree of freedom on the mounted stage, allows the camera to be precisely pointed at the diffracting sample. In addition, the height was carefully adjusted to make sure the detector lied in the horizontal plane of the

Neutron Radiography - WCNR-11   Materials Research Forum LLC
Materials Research Proceedings 15 (2020) 80-85   https://doi.org/10.21741/9781644900574-13

incident beam. Shielding was another significant component; its purpose is to prevent background radiation from striking the detector and ensuring the Anger camera only "sees" the sample on the NRSF2 stage. Shielding schematics are shown in Figure 2:

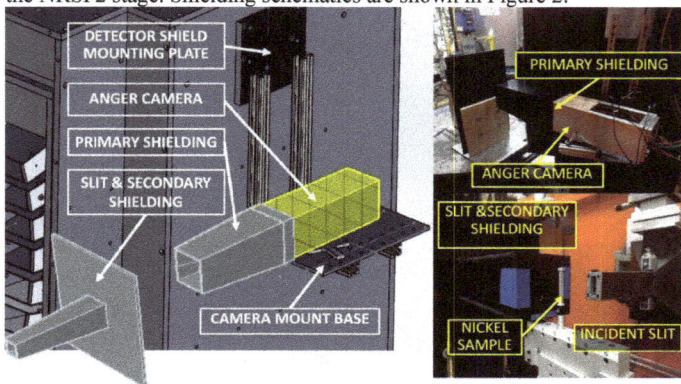

Figure 2: Schematic representation and actual pictures of shielding components and camera mounted on NRSF2 detector

Due to special constrains and the operating distance of the camera from the diffracting sample, shielding could not be implemented as a single snout going from the detector to the sample; as a solution, shielding was split into two parts such that the overall effect reproduces the effect of a single snout. The Anger camera was connected to and integrated with the NRSF2 control workstation. This enabled us to control both instruments from one computer. Reference data is obtained from the main instrument then, having a set angular offset between the Anger camera and the NRSF2 detector, the main detector is driven to a position such that the Anger camera can record the same peak.

Results

Imaging data collected using Anger camera comes in the form of raw binary packets consisting of 262144 or 65536 16-bit integer values, depending on the number of PMT's utilized. Using a full 4-array of PMT's results in a 512x512 image as shown in Figure 3(a).

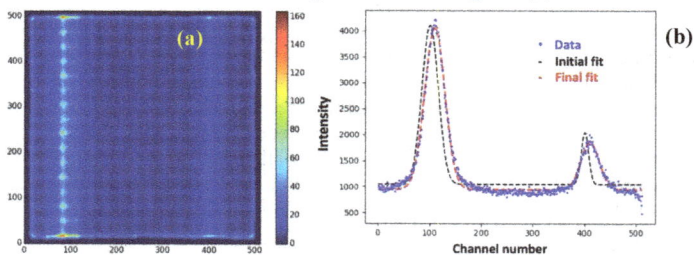

Figure 3: (a) example heat-map image from Anger camera depicting two peaks; colormap shows counts per pixel over image acquisition time (b) Fitted vertical integration

The processing of the data involves converting the raw pixel array into a 16-bit image using the Python Imaging Library. The data is then integrated vertically such that if the image pixels are expressed in row-column format as $I(M, N)$ then the value for each channel according to Eq. 3 will be:

$$Y_N = \sum_{i=0}^{N} I(m, i) \; where \; m \in M \qquad (Eq. 3)$$

Where Y_N indicates the number of neutrons counted on a channel as a result of the vertical sum of the pixels on that channel. Figure 3(b) was processed to obtain the plot shown in Figure 3(b) where a Gaussian peak fit is implemented. Peak information such as Signal to Noise Ratio (SNR) and full-width-at-half-max (FWHM) can be used to describe the performance of the camera in combination with different scintillators while looking at neutron scattering from various reference samples. All acquired images were normalized to remove noise due to background and detector intensity variations. Dark current (recording data with the shutter closed) and open beam images (recording data with open shutter and vanadium incoherent scatterer in place of the sample) were recorded and used for the normalization process. Images were normalized according to Eq. 4 where I_{RAW} is the intensity of the recorded image, I_{DC} is the intensity of the dark field image and I_{OPEN} is the intensity of the open beam image.

$$I_{RAW} = \frac{I_{RAW} - I_{DC}}{I_{OPEN} - I_{DC}} \qquad (Eq. 4)$$

Initial measurements consisted of standard powder samples diffraction peaks. These samples are used the NRSF2 instrument calibration at the beginning of each cycle. Single peaks were acquired for Body Centered Cubic Iron, Germanium, Nickel and Inconel samples. Count time was 60 seconds for each measurement. Example normalized images are shown in Figure 4 with superimposed intensity vs. channel plot obtained from vertical image integration.

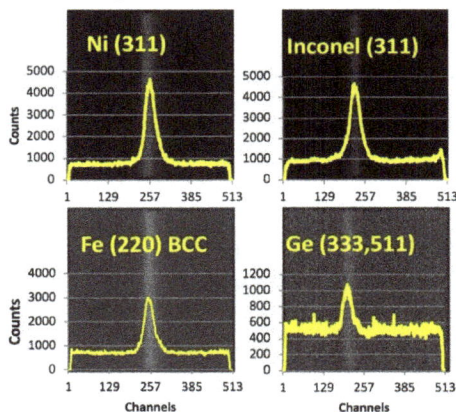

Figure 4: Normalized two-dimensional imaging of diffraction peaks with superimposed intensity plot

Discussion

Diffraction peaks were selected for 2θ values as close to $90°$ as possible; for quantification of these measurements, curvature was neglected and, subsequent to normalization, raw intensity was obtained by vertical image integration. To determine detector resolution, a 2θ sweep using an Inconel powder sample was performed. During this set of scans, two peaks swept across the detector. All obtained images were integrated and peak positions within the field of view of the camera were determined by the channel number. The channel number peak positions are then plotted versus angular 2θ position as shown in Figure 5 and the slope yields the channel per degree value giving an idea of what the spatial resolution of the detector is for the given operating sample to detector distance. For the operating distance of 750mm, the Anger camera was found to have an approximate resolution of 57 pixels per degree.

Inconel (311) and (222) 2θ sweep:

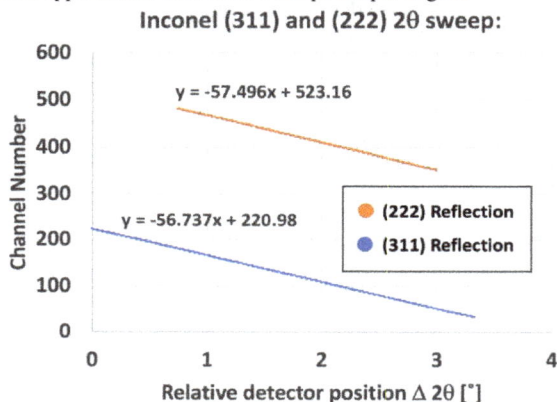

$y = -57.496x + 523.16$

$y = -56.737x + 220.98$

- (222) Reflection
- (311) Reflection

Channel Number

Relative detector position $\Delta 2\theta$ [°]

Figure 5: Peak position (channel) vs. 2θ position

Conclusions

Integration of the Anger camera in a diffraction instrument was successful. Development and implementation of shielding was part of the study and performed as intended (system performance without shielding was not investigated). For portable applications, a smaller footprint would be desirable however shielding dimensions are geometrically determined by sample to detector distance. Overall the Anger camera proved to be a viable instrument for neutron diffraction measurements. The results show a high degree of 2-D spatial resolution and for the working distance of 750 mm the Anger camera was found to cover a 2θ range of approximately $8°$. Future work will be aimed at evaluating the possibility of using the Anger camera for residual strain measurements; it will be critical to determine the magnitude of the strain change that can be determined by processing the two-dimensional images recorded by the Anger camera and successfully resolving shifts in peak position.

References

[1] R. A. Riedel, C. Donahue, T. Visscher, and C. Montcalm, "Design and performance of a large area neutron sensitive anger camera," *Nuclear Instruments and Methods in Physics Research Section A: Accelerators, Spectrometers, Detectors and Associated Equipment*, vol. 794, pp. 224-233, 2015. https://doi.org/10.1016/j.nima.2015.05.026

[2] P. D. Olcott, J. A. Talcott, C. S. Levin, F. Habte, and A. M. K. Foudray, "Compact readout electronics for position sensitive photomultiplier tubes," *IEEE Transactions on Nuclear Science,* vol. 52, no. 1, pp. 21-27, 2005. https://doi.org/10.1109/TNS.2004.843134

[3] S. Siegel, R. W. Silverman, S. Yiping, and S. R. Cherry, "Simple charge division readouts for imaging scintillator arrays using a multi-channel PMT," *IEEE Transactions on Nuclear Science,* vol. 43, no. 3, pp. 1634-1641, 1996. https://doi.org/10.1109/23.507162

[4] L. M. Bollinger, G. E. Thomas, and R. G. Ginther, "Glass Scintillators for Neutron Detection," *Review of Scientific Instruments,* vol. 30, no. 12, pp. 1135-1136, 1959. https://doi.org/10.1063/1.1716471

[5] J. M. Neill, D. Huffman, C. A. Preskitt, and J. C. Young, "Calibration and Use of a 5-Inch Diameter Lithium Glass Detector," (in English), *Nuclear Instruments & Methods,* vol. 82, pp. 162-&, 1970. https://doi.org/10.1016/0029-554X(70)90343-5

[6] C. Coceva, "Pulse-shape discrimination with a glass scintillator," *Nuclear Instruments and Methods,* vol. 21, pp. 93-96, 1963. https://doi.org/10.1016/0029-554X(63)90092-2

Neutron Radiography - WCNR-11
Materials Research Proceedings 15 (2020) 86-91

Materials Research Forum LLC
https://doi.org/10.21741/9781644900574-14

Commissioning of the NDDL-40 Micro-Channel Plate Neutron Detector System at Oregon State University

Nicholas M. Boulton[1,a*], Steven R. Reese[2,b], and Aaron E. Craft[1,c]

[1]Idaho National Laboratory, PO Box 1625, MS 2211, Idaho Falls, ID 83415, USA

[2]Radiation center, 3451 Jefferson Way, Corvallis, OR 97330, USA

[a]Nicholas.Boulton@inl.gov, [b]Steven.Reese@oregonstate.edu, [c]Aaron.Craft@inl.gov

Keywords: Neutron Radiography, Neutron Tomography, Micro-Channel Plate

Abstract. The Neutron Radiography Facility (NRF) at Oregon State University (OSU) has been modified to begin working on the non-destructive evaluation of concrete materials to study the early stages of shrinkage, cracking, and water transport of concrete during the curing process. The objective of this work was to investigate the efficiency and spatial resolution of the NDDL 40 micro-channel plate (MCP) detector for the use of neutron radiography and tomography to determine its applicability for examining concrete. Working in collaboration with the School of Civil and Construction Engineering, the NRF at OSU has added a NDDL-40 vacuum-sealed neutron imaging detection system with a delay line system readout developed by NOVA Scientific. This study found that the system installed at the NRF was capable of a maximum spatial resolution of ~250 μm with a neutron detection efficiency of 5.49%. Significant artifacts from the detector system and image noise degraded the quality of the tomographic reconstruction to such an extent that this neutron imaging system could not be used to visualize the desired phenomena in concrete.

Introduction

Oregon State University acquired a NDDL-40 detector system from NOVA Scientific. for examination of water transport through concrete. The detector was evaluated through a series of measurements at the Oregon State TRIGA Reactor (OSTR) to determine special resolution and detector efficiency. Similar testing has been performed previously using an earlier MCP system [1]. The detector was advertised to provide spatial resolution less than 50 μm. The vacuum-sealed detector contained two neutron sensitive micro-channel plates with a delay line anode output. The MCP's are 1 mm thick glass plates with millions of 4-11 μm diameter channels in a hexagonal array through each plate [2]. These micro-channel plates are doped with boron, because of its large thermal neutron absorption cross-section. A neutron absorbed by the ^{10}B undergoes a ^{10}B(n,α)Li^7 interaction, producing daughter products that interact with the glass, creating secondary electrons. A voltage potential across the plates allows the secondary electrons to create an electron cascade with a net signal gain of approximately 10^7. This creates a detectable pulse from the anode which is registered by the detector as a count.

MCP detector systems have been used extensively for neutron imaging of dynamic phenomena due to their exceptionally fast dynamic response [3]. The detector response of some MCP neutron imaging systems is so fast that it can be used with pulsed neutron sources for time-of-flight (TOF) measurements, enabling a wide range of scientific applications. An MCP system was used to perform TOF resonance absorption measurements at LANSCE to visualize the distribution of ^{235}U and ^{238}U in nuclear fuel pellets [4]. Another benefit of MCP's is their lower gamma-ray sensitivity relative to many other neutron imaging technologies. MCP have been demonstrated to be able to directly examine highly radioactive irradiated nuclear fuel [5,6].

Reference Standard Test Objects

ASTM E545 [7], describes a method for determining image quality of a thermal neutron beam that is specifically used for radiography. The standard describes the use two different test objects. The first standard test object is called the Beam Purity Indicator (BPI), which is constructed as a polytetrafluoroethylene block containing two boron nitride disks, two lead disks, and two cadmium strips [7]. This standard provides a method for determining the thermal neutron content and gamma content measures image sharpness and overall quality of direct neutron radiography with gadolinium conversion screens with film and is not directly applicable to digital neutron radiography systems. The second indicator used is the Sensitivity Indicator (SI). This indicator qualitatively determines the sensitivity of the neutron radiograph by observing holes and gaps of known dimensions between each shim. While not directly applicable to digital neutron radiography systems, the standard [7] can be applied to the extent applicable to determine the category of the facility. Another image quality indicator used was a cadmium strip provided by NOVA Scientific was imaged to obtain additional data on the detectors image resolution and sharpness. The strip was a 0.050 cm thick cadmium strip with a series of holes along the center line. The smallest holes are 250 μm diameter with a separation of 250 μm between holes.

Measurement of the Neutron Beam Flux

Before determining the detector efficiency, the thermal neutron flux needed to be measured. This was performed using gold activation foils. The gold foil is activated and decays concurrently, and the activity can be determined at any point in time using the differential equation, $dN/dt=R-\lambda N$, where dN/dt, the rate of change in radioactive nuclei over time, is the difference between the rate of activation, R, and the activity of the sample at a point in time, $A(t)=\lambda N(t)$, which is the product of the decay constant, λ, and the number of radioactive nuclei, $N(t)$. Thus, the activity at a point in time, is given by:

$$A(t) = N(t)\lambda = R\left(1 - e^{-\lambda t_0}\right). \#(1)$$

After a 7-hour beam exposure time with the OSTR power at 1 MW, the activity of each of the two foils was measured using a calibrated HPGe detector. Because the foils are continually decaying after irradiation, the time between each step was accounted for through the following equation:

$$A = \frac{\lambda(C - B)}{\varepsilon(1 - e^{-\lambda t_0})(e^{-\lambda t_1} - e^{-\lambda t_2})}, \#(2)$$

where C is the counts measured from the detector, B is the number of background counts expected in the counting time, t_0 is the time of irradiation, t_1 is the time of the foils removal from the neutron flux, t_2 is the time the foils are counted by the HPGe detector, and ε is the counting efficiency of the detector. Each gold foil was of the same purity, size and thickness. For the measurement, one bare foil was placed in front of the beam while the other foil was placed in a Cd cover in the beam. The difference in radioactivity between the two foils can be used to determine the flux from thermal and higher-energy neutrons. By subtracting the amount of activation of the Cd covered gold foil from the bare gold foil the activity from only thermal neutrons can be calculated.

Gold has an average thermal absorption cross-section of 98.65 barns. After neutron absorption, the ^{197}Au (100% natural abundance) becomes ^{198}Au which has a half-life of 2.7 days and decays via beta emission to become ^{198m}Hg that promptly decays into stable ^{198}Hg via emission of a 411 keV gamma-ray. Thermal neutron flux was obtained from the activities measured with bare and Cd-covered foils given material constants, decay time and irradiation time, as shown in Eq. 3.

$$A = \left(\phi_{th}\sigma_{th} + \phi_{epi}\sigma_{epi}\right) * N\left(1 - e^{-\lambda t}\right). \#(3)$$

Solving for ϕ_{th} gives,

$$\phi_{th} = \left(\frac{A_{Au}}{N_{Au}(1 - e^{-\lambda t})} - \left(\phi_{epi} * \sigma_{epi}\right)\right)\bigg/\left(\sqrt{\frac{\pi}{2}} * \sigma_{th}\right). \#(4)$$

Table 1 shows values used to calculate the thermal neutron flux of OSTR's Beam Port 3, which was determined to be $9.42\times10^5 \pm 1.55\times10^4$ $n/cm^2/s$ with the reactor power at 1 MW.

Table 1. Measurements used for Gold foil experiment.

Variable	Value
Irradiation time [hr]	6.96
λ_{Au-198} [hr^{-1}]	6.17
N_{Au-197}, bare gold [atoms]	2.65×10^{20}
N_{Au-197}, Cd-covered [atoms]	2.63×10^{20}
Thermal cross section [cm^2]	9.87×10^{-23}
Epithermal cross-section (cm^2)	1.55×10^{-21}
Activity, bare [Bq]	3.17×10^3
Error of activity, bare [Bq]	5.20×10^1
Activity, Cd-covered [Bq]	9.34×10^2
Error of activity, Cd-covered [Bq]	1.53×10^1

Neutron Detection Efficiency

The rate of neutrons incident on the detector must be calculated from the neutron flux and the number of neutrons registered must be measured by the detector. The neutron flux (9.4242×10^5 $n/cm^2/s$) multiplied by the area of the 2 cm diameter detector gives the rate of neutrons incident on the detector as 1.184×10^7 n/s. The number of neutron counts registered was counted by the detector software during exposure, which was 6.5×10^5 n/s. The total efficiency is then the ratio of the two, giving an approximate detector efficiency of 5.49%.

Image Processing of Neutron Radiographs

After image acquisition, image post-processing optimizes sharpness and contrast to clean up each image. Issues that can reduce the quality of the image include noise from the detector, foreign objects in the field of view, ghosting, or non-uniform image acquisition parameters. Images were processed after each exposure using ImageJ image processing software [8]. For most corrections, a flat-field image (i.e. background and/or open beam image), is taken before each radiograph or series of radiographs with the object of interest positioned outside the field of view. During post processing the flat-field is divided into the raw image containing the desired object to be viewed. Flat-field normalization corrects for variations of the detector response and beam non-uniformity. The software uses the following equation to normalize the intensities across the image:

$$I_1 = \left(\frac{i_1}{i_2}\right) * k_1 + k_2, \#(5)$$

where I_1 is the new image after processing, i_1 is the image with the object, i_2 is the flat-field image, k_1 is the average intensity in i_1, and k_2 is a minimum intensity which was always 0.

Multiplying by the k_1 scales the grayscale values to more completely fill the histogram per the available bit-depth of the final image. The image may need to be further processed by adjusting the contrast or removing noise to better visualize features of interest. Fig. 1 shows the process at which the radiographs are processed and corrected. Figs. 1a and 1b show the flat field and uncorrected radiograph of the BPI, respectively. Flat field correction was applied to the image of the BPI using Eq. 5, resulting in the image shown in Fig. 1c. The result shows the removal of detector noise and high intensities that are in the radiograph that are not part of what is being radiographed. The resulting image was then processed by normalizing the histogram and enhancing contrast, which yields the image of the BPI shown in Fig. 1d.

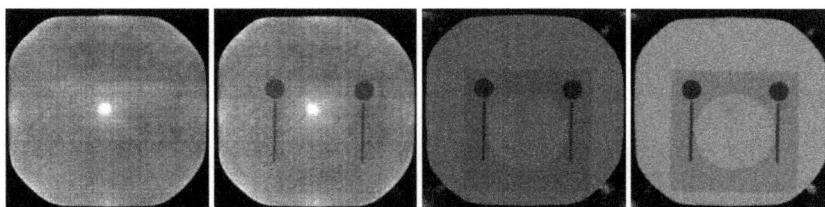

Fig. 1. Resulting radiographs and processed images of a BPI. a) Flat-field image of the beam. b) Radiograph of the BPI. c) Processed radiograph. d) Processed radiograph with enhanced contrast.

Neutron Tomography

For tomography reconstruction, the post processing is similar to radiography image processing, but the process is applied to multiple images. This is done by placing the desired object on a motorized stage to precisely rotate the object. A 500 mm linear travel (IMS500PP) and 360° rotational (URS75BPP) stage were acquired from the Newport Corporation. Newport's proprietary software allows stages to be controlled remotely from outside the NRF. This allows the desired object to be moved horizontally and rotated while keeping the shutter for the beam open. For adequate tomography, radiography images should be acquired over a minimum range of 180°. A set of 180 radiographs were taken using the maximum field of view of 4 cm and L/D of 100.

The minimum distance between the center of rotation and the detector based on the dimensions of the BPI is 13.125 mm (the minimum distance from the center to the outside edges of the 25×25×8 mm BPI) plus the ~12.5 mm distance from the detector to the vacuum housing window, for a total minimum object-to-detector distance of 25.625 mm. Thus, the minimum geometric unsharpness for L/D of 100 would be 256.25 μm when performing neutron tomography. The distance from the vacuum housing window to the detector would add to the object-detector distance, further increasing geometric unsharpness.

A macro was written in ImageJ to process the images automatically, applying a flat-field correction and enhancing contrast for each image. After processing projections with ImageJ, Octopus reconstruction software was used to perform tomographic reconstruction of the images [9]. Using Octopus, the processed images from ImageJ were first cropped to the field of view of the object. Even though the setup for the experiment could be assumed to be a parallel beam, the mode set in Octopus was set to cone beam with the appropriate source-to-object distance and source-to-detector distance for the experiment. The software can adjust filters to remove more noise from each projection. A bi-linear interpolation filter characterized each pixel and voxel to 1 μm. Sinograms are produced from the filtered projections using filtered back-projection.

Results & Discussion

Fig. 2a and 2b show the processed radiographs of the SI and BPI, respectively. For radiographs, the objects were placed directly on the detector window to reduce geometric unsharpness. Considering the radiograph of the SI in Fig. 2a, there are no visible holes (i.e. does not meet any category requirements) but five visible shims (category IV for this metric). The metrics derived from the BPI in Fig. 2b are not applicable for digital neutron radiography systems, so they cannot be used in determining the ASTM facility category. Examination of Fig. 2c shows the radiograph of a cadmium strip used to help quantify the spatial resolution. The effective spatial resolution of the detection system was found to be in the range of ~250 μm, which is the diameter of the smallest holes in the cadmium strip with a known spacing of 250 μm between each hole. The samples being placed on the window of the vacuum housing would give an object-to-detector distance of ~12.5 mm, so the resulting geometric unsharpness with an L/D of 100 would be 125 μm, which may have limited the spatial resolution obtained in these measurements.

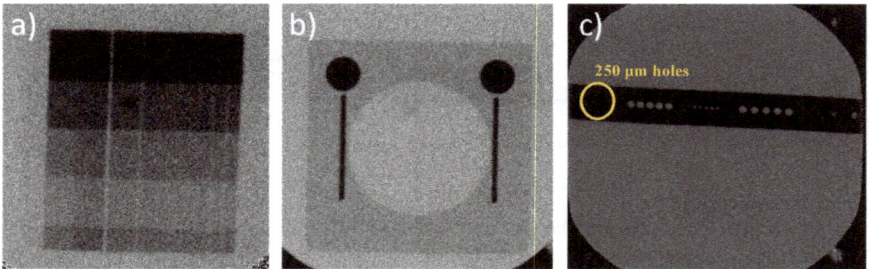

Fig. 2. a) Neutron radiographs of the SI and b) BPI (right). c) Image of a cadmium strip with various size holes.

Fig. 3 shows a representative radiograph, sinograms from multiple projections, and the resulting tomographic reconstruction. While the resulting tomographic reconstruction visibly shows the major features of the BPI (e.g. cadmium wires, boron nitride discs), artifacts from inconsistent detector performance and the significant amount of noise in each image yielded poor overall quality of the resulting reconstruction. While the boron discs and cadmium wires rendered very well, the polyethylene block is hardly visible.

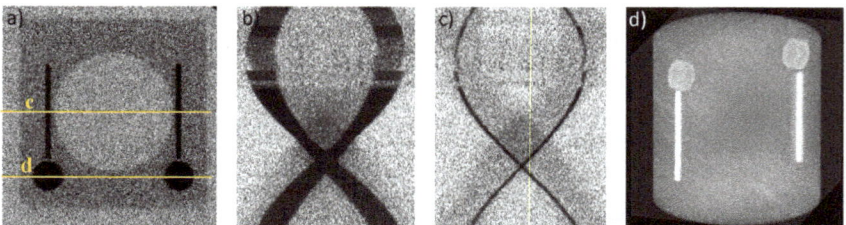

Fig. 3. a) Radiograph of a BPI. b&c) Resulting sinograms. d) Isomeric view of the resulting tomographic reconstruction.

Neutron Radiography - WCNR-11   Materials Research Forum LLC
Materials Research Proceedings **15** (2020) 86-91   https://doi.org/10.21741/9781644900574-14

Tomographic reconstruction was done using the Octopus software. Fig. 3 shows a pair of sinograms that were rendered using radiographs of the BPI. The left sinogram follows the boron disc of the BPI while the sinogram on the right shows the path the cadmium wires within the BPI. Inconsistencies are visible in the sinograms that can be attributed to periodic, temporary, and unexplained degradation of contrast signal to noise ratio of the NDDL40 detector system.

Conclusion

The objective of this work was to investigate efficiency and spatial resolution of the NDDL40 MCP vacuum sealed detector from NOVA Scientific for the use of neutron radiography and tomography. The detector is based on using borated micro channel glass plates to detect incoming thermal neutrons through the $^{10}B(n,\alpha)^{7}Li$ interaction. The NDDL 40 MCP detector has been shown to be capable of producing neutron radiographs and tomography. Radiographs exhibited inconsistent image quality due to the varying background, low signal to noise ratio, and low detector efficiency. This led to issues in rendering quality tomographic reconstructions. More than half of the 190 projections taken for tomography had to be reacquired. The NDDL-40 MCP detector installed at the OSTR Beam Port 3 demonstrated a neutron detection efficiency of 5.49% with a spatial resolution of approximately 250 μm.

References

[1] W.J. Williams, "Neutron Radiography and Tomography: Determining and Optimizing Resolution of Neutron Sensitive Multi Channel Plate Detectors," Corvallis, 2013.

[2] O. H. Siegmund, J. V. Vallerga, A.S. Tremsin, J. Mcphate and B. Feller, "High Spatial Resolution Neutron Sensing Mincochannel Plate Detectors," Nuclear Instruments and Methods in Physics Research A 576, 178-182, 2007. https://doi.org/10.1016/j.nima.2007.01.148

[3] C.D. Ertley, O.H.W. Siegmund, J. Hull, A. Tremsin, A. O'Mahony, C.A. Craven, and M.J. Minot, "Microchannel Plate Imaging Detectors for High Dynamic Range Applications," IEEE Trans. Nuc. Sci. 64(7) 1774-1780, 2017. https://doi.org/10.1109/TNS.2017.2652222

[4] A.S. Tremsin, S.C. Vogel, M. Mocko, M A M Bourke, V. Yuan, R.O. Nelson, D.W. Brown, and W.B. Feller, "Non-destructive studies of fuel pellets by neutron resonance absorption radiography and thermal neutron radiography," Nuc. Mat. 440, 633-646, 2013. https://doi.org/10.1016/j.jnucmat.2013.06.007

[5] Tremsin, A.S., Craft, A.E., G.C. Papaioannou, et al., "On the possibility to investigate irradiated fuel pins nondestructively by digital neutron radiography with a neutron-sensitive microchannel plate detector with Timepix readout," Nucl. Instr. Meth. in Physics Research A, 927, 109-118, 2019. https://doi.org/10.1016/j.nima.2019.02.012

[6] Tremsin, A.S., Craft, A.E., A.M.M. Bourke, et al., 2018. Digital neutron and gamma-ray radiography in high radiation environments with an MCP/Timepix detector. Nucl. Instr. Meth. in Physics Research A 902, 110-116, 2018. https://doi.org/10.1016/j.nima.2018.05.069

[7] ASTM E545-15, "Standard Test Method for Determining Image quality in Direct Thermal Neutron Radiographic Examination," ASTM International, West Conshohocken, PA, 2014.

[8] W. Rasband, "ImageJ," 1997. [Online]. Available: http://imagej.nih.gov.

[9] Octopus Imaging, Octopus Reconstruction User Manual, Ghent, 2016.

Neutron Radiography - WCNR-11
Materials Research Proceedings 15 (2020) 92-96

Materials Research Forum LLC
https://doi.org/10.21741/9781644900574-15

A Quadruple Multi-Camera Neutron Computed Tomography System at MLZ

Burkhard Schillinger[1, a *], Jens Krüger [1,b]

[1] Heinz Maier-Leibnitz Zentrum and Physics E21, Technische Universität München, Lichtenbergstr.1, 85748 Garching, Germany

[a]Burkhard.Schillinger@frm2.tum.de , [b]Jens.Krueger@frm2.tum.de

Keywords: Detector, Multi-detector, Neutron Imaging, Neutron Computed Tomography

Abstract. Most neutron imaging systems can accommodate large samples, but recent interest is more focused on small cm-sized samples. With a small field of view for a camera-based detection system, the neutron flux per pixel decreases, and measurement time increases. An earlier approach split a large field of view into smaller fields for individual CT measurements using several rotation axes in the field of view of the camera, but reduces the available amount of camera pixels per tomography field, while many applications require the highest possible resolution even or especially for very small samples. A new approach at MLZ uses a multiple camera system with multiple rotation stages to make better use of the full size of the original neutron beam. With four cameras stacked in a 2x2 matrix, only two rotation stages are required where samples are stacked in an aluminum tube with cutouts above each other. A small, but high quality cooled CMOS camera is employed; joint shielding is built up on the outside of the boxes with lead bricks and PE plates. The first prototype is already working; four more camera boxes are currently in production and will be completed soon. First high-resolution results are shown.

Introduction

Most neutron imaging systems can accommodate large samples of 15 -30 cm size, but recent interest is more focused on small cm-sized samples. In cone-beam X-ray tomography, the beam is emitted from the focal spot of the X-ray tube and illuminates the detector in fixed distance, the flux per pixel remains constant, while the projection ratio and effective resolution can be varied by moving the sample closer to or farther from the focal spot. Since neutron imaging uses a quasi-parallel beam, the neutron flux per area remains constant, and higher resolution can only be achieved with smaller detector pixels. The neutron flux per pixel of the detector decreases with decreasing pixel size, and measurement time increases. Moreover, looking only at small samples with a small detector field of view leaves most of the large neutron beam unused.

There were approaches to split a large field of view into smaller fields for individual CT measurements using a cogwheel-based adapter for the rotation stage [1] or using individual micro rotation stages [2], but this leaves a smaller amount of camera pixels per tomography field, while many applications require the highest possible resolution even or especially for very small samples.Using a super-high resolution camera with 50 megapixel or so would require expensive non-standard optics, increase readout time and data rates, and would be inflexible due to mechanical constraints for setting up samples, and often leave large parts of the field of view unused.

An alternative approach is followed at MLZ, using a multiple camera system with multiple rotation stages to make better use of the full size of the original neutron beam. With four cameras, only two rotation stages are required where samples are stacked in an aluminum tube with cutouts above each other. Cameras are stacked with two on top of each other, and two stacks beside each other.

Neutron Radiography - WCNR-11   Materials Research Forum LLC
Materials Research Proceedings **15** (2020) 92-96   https://doi.org/10.21741/9781644900574-15

A compact high-resolution camera detector

Most facilities use a high-end CCD or CMOS camera mounted in a large box with a mirror and scintillation screen plus massive lead shielding around the camera against stray gamma radiation and scattered neutrons, often permitting a variation of the field of view by moving the camera within the box. Using such a massive geometry is very inflexible to accommodate more than one camera. Our new approach is to use multiple small camera boxes as compact as possible, with the shielding mounted separately on the outside of the box to allow for flexible installation. The box only contains two lead shields in front and back of the camera, the first surrounding the lens, sideways shielding is only mounted externally. In addition, the part containing the scintillation screen and mirror is designed as a separate item so different sizes can be adapted to the camera. For the compact setup, a small, but high quality cooled CMOS camera type 'ASI178 mm cool' [3] with 3096x2080 pixels and 14 bit ADC is employed.

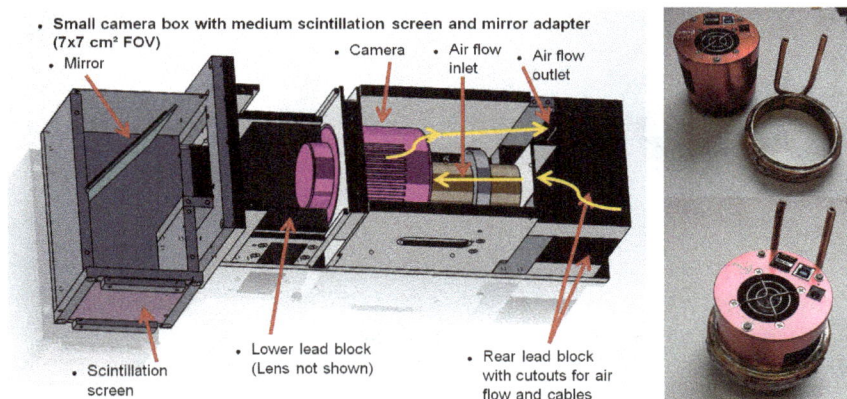

Fig. 1a,b: Schematic drawing of the camera box and photo of the water cooler ring

Fig. 1a shows the schematic drawing of the camera box. A lead block with cutouts for air flow and cables is situated behind the camera, a two-piece lead block surrounds the lens on the front side of the camera. The mirror box is connected by a flange and can be rotated by 180° to face the other way, or it can be replaced by a mirror box of different size. The selection of the focal length of the C-mount lens allows for a wide variation of the field of view. Once the camera box was mounted within external shielding, the air flow was insufficient for the Peltier cooler built into the camera, so a simple water cooler ring (Fig. 1b) was designed to remove the heat.

The camera stack

With four boxes stacked in a 2x2 matrix, only two rotation stages are required if samples are mounted in aluminum tubes with cutouts and shelves as displayed in Fig.2 The thin aluminum is nearly transparent for neutrons and does not hinder the tomography of the sample. This kind of sample holder is routinely employed for sequential multi-CT by vertical sample movement at the ANTARES facility. Fig. 2b shows two micro rotation stages as employed in [2].

Software

The ANTARES neutron CT facility uses a distributed TANGO/Entangle server system with a graphical client system named NICOS, which is described in more detail in [4] in this issue.

Each physical device is driven by a respective software device. Using multiple rotation devices within one tomography scan has already been implemented for the setup described in [2]. Using more than one detector proved more complicated, since multiple data paths have to be set and used for the multiple detectors. The software generates one path for the measurement with four (or multiple) subdirectories for the respective cameras. For now, only four synchronous identical tomography measurements can be carried out, i.e. with identical angular positions and exposure times.

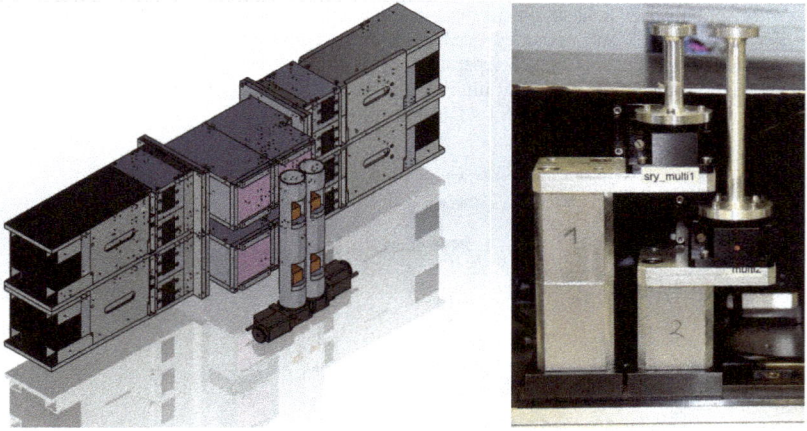

Fig. 2a,b: Schematic drawing of the camera stack and photo of micro rotation stages

High-resolution measurements with the first prototype
Problems with an external manufacturer who produced the lead blocks far outside specified measures have delayed the completion of the setup, but successful measurements were performed with the first prototype. The quality will be the same for the multiple setup. Fig. 3a shows the prototype mounted on the detector table of the ANTARES facility, with shielding added in Fig. 3b. The standard rotation stage of ANTARES was employed.

Fig. 3a,b: The prototype detector mounted in the ANTARES facility, bare (Fig.3a) and with added lead shielding (Fig.3b)

Neutron Radiography - WCNR-11   Materials Research Forum LLC
Materials Research Proceedings **15** (2020) 92-96   https://doi.org/10.21741/9781644900574-15

Results

First measurements were performed with a 50 μm LiF+ZnS scintillation screen and a 20 μm Gadox screen. For each CT, 1278 projections were measured over 360°, with 6 seconds exposure time each, and 9 μm effective pixel size. The Gadox screen produced much less light output and thus a weaker signal, but longer exposure times were not possible within the available beam time. Fig. 4 shows reconstructions of a lady's watch. The graininess is attributed to the coarseness of the scintillation screen.

Fig. 4: CT reconstruction of a lady's wrist watch with ZnS+LiF screen at 9 μm pixel size

Fig. 5a,b shows a photo and reconstruction of a dried hornet, recorded with the same settings. Fig. 6a shows the innards of the hornet recorded with the LiF+ZnS screen, Fig. 6b was recorded with the Gadox screen and shows even higher resolution, but worse statistics.

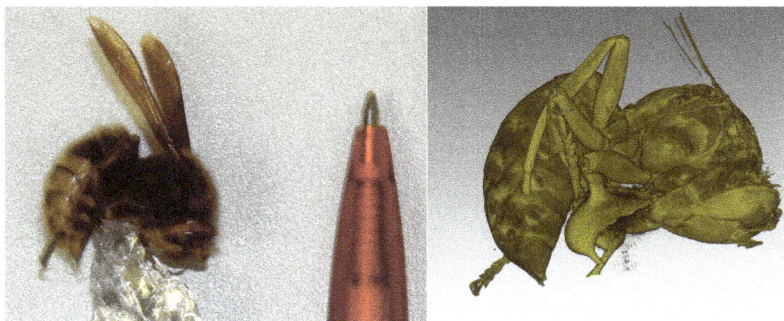

Fig. 5a,b: Photo and CT reconstruction of a hornet with ZnS+LiF screen at 9 μm pixel size

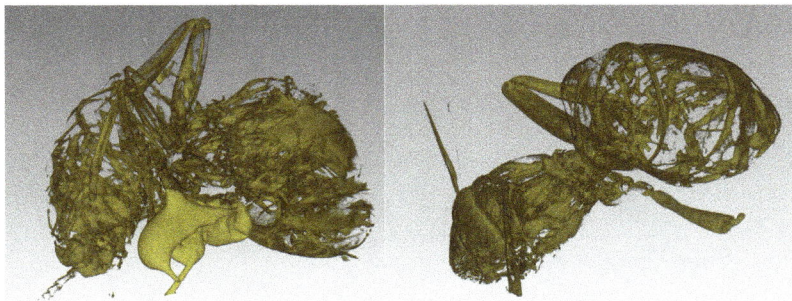

Fig. 6: Photo and CT reconstruction of a hornet with ZnS+LiF screen (left) and Gadox (right). The Gadox screen gives even higher resolution.

Conclusions and Outlook

A 2x2 camera stack is used for multiple synchronous tomography measurements of small samples in a large neutron beam. A compact, but high-quality CMOS detector system was built to be flexibly stacked with external shielding and to be connected to different mirror boxes for different fields of view. The construction of the camera box is currently being worked over for simplification, and the construction drawings and the software described in [4] will be made available to the public in the future. Since the intensity of current neutron sources is limited and cannot be significantly increased, better utilization of available neutron beams can be achieved by synchronous multiple measurements.

References

[1] P. Trtik, F. Geiger, J. Hovind, U. Lang, E. Lehmann, P. Vontobel, S. Peetermans, Rotation axis demultiplexer enabling simultaneous computed tomography of multiple samples, published online 2016 Apr 18. https://doi.org/10.1016/j.mex.2016.04.005

[2] B. Schillinger, D. Bausenwein, Quadruple axis neutron computed tomography, Physics Procedia 88 (2017) 196 – 199. https://doi.org/10.1016/j.phpro.2017.06.027

[3] https:// astronomy-imaging-camera.com/product/asi-178mm-cool/

[4] B. Schillinger, A. Craft, J. Krüger, The ANTARES instrument control system for neutron imaging with NICOS/TANGO/LiMA converted to a mobile system used at Idaho National Laboratory, in this issue

Neutron Radiography - WCNR-11
Materials Research Proceedings **15** (2020) 102-107

Materials Research Forum LLC
https://doi.org/10.21741/9781644900574-17

Development of Event-Type Neutron Imaging Detectors at the Energy-Resolved Neutron Imaging System RADEN at J-PARC

Joseph Don Parker[1,a*], Masahide Harada[2,b], Hirotoshi Hayashida[1,c], Kosuke Hiroi[2,d], Tetsuya Kai[2,e], Yoshihiro Matsumoto[1,f], Takeshi Nakatani[2,g], Kenichi Oikawa[2,h], Mariko Segawa[2,i], Takenao Shinohara[2,j], Yuhua Su[2,k], Atsushi Takada[3,l], Toru Tanimori[3,m] and Yoshiaki Kiyanagi[4,n]

[1]Neutron Science and Technology Center, CROSS, Tokai, Ibaraki 319-1106 Japan

[2]J-PARC Center, Japan Atomic Energy Agency, Tokai, Ibaraki 319-1195 Japan

[3]Graduate School of Science, Kyoto University, Kyoto 606-8502 Japan

[4]Graduate School of Engineering, Nagoya University, Nagoya, Aichi 464-8603 Japan

[a]j_parker@cross.or.jp, [b]harada.masahide@jaea.go.jp, [c]h_hayashida@cross.or.jp, [d]kosuke.hiroi@j-parc.jp, [e]tetsuya.kai@j-parc.jp, [f]y_matsumoto@cross.or.jp, [g]takeshi.nakatani@j-parc.jp, [h]kenichi.oikawa@j-parc.jp, [i]segawa@post.j-parc.jp, [j]takenao.shinohara@j-parc.jp, [k]yuhua.su@j-parc.jp, [l]takada@cr.scphys.kyoto-u.ac.jp, [m]tanimori@cr.scphys.kyoto-u.ac.jp, [n]kiyanagi@phi.phys.nagoya-u.ac.jp

Keywords: Energy-Resolved Neutron Imaging, Neutron Imaging Detectors, Micropattern Detectors, Lithium Glass Scintillators

Abstract. At the RADEN beam line of the Materials and Life Science Experimental Facility at the Japan Proton Accelerator Research Complex, we combine cutting-edge, event-type imaging detectors with a high-intensity, pulsed neutron beam to perform *energy-resolved neutron imaging*. In particular, the μNID (Micropixel-chamber-based Neutron Imaging Detector), with its unique combination of a large 10 cm × 10 cm field-of-view, 100 μm spatial resolution, and recent improvements in the rate performance to over 1 Mcps, is quickly becoming the main event-type detector for such measurements at RADEN. To improve the ease-of-use of the μNID system, we have recently redesigned the control hardware and software to allow full integration into the RADEN experiment control system and developed a web-based user interface for data processing. Further development efforts for the μNID, including a new reduced-pitch readout for improved spatial resolution and a boron-converter based μNID that achieves a count rate of over 20 Mcps, are also ongoing. In addition, we are studying super resolution methods to improve the spatial resolution of a lithium-glass scintillator pixel detector.

Introduction

At the RADEN instrument [1], located at beam port 22 of the high-intensity, pulsed neutron source of the Materials and Life Science Experimental Facility (MLF) at the Japan Proton Accelerator Research Complex (J-PARC), we take advantage of the accurate measurement of neutron energy by time-of-flight to perform *energy-resolved neutron imaging*. By analyzing the two-dimensionally resolved, energy-dependent neutron transmission, these techniques can image macroscopic distributions of microscopic properties of bulk materials *in situ*, including crystallographic structure and residual strain (Bragg-edge transmission [2]), nuclide-specific density and temperature distributions (resonance absorption [3]), and internal/external magnetic fields (polarized neutron imaging [4]). At RADEN, we use advanced, event-type neutron imaging detectors, based on micropattern detectors or fast lithium-glass scintillators with fast, all-digital data acquisition systems for high count rate and sub-μs time resolution, to measure the

Neutron Radiography - WCNR-11
Materials Research Proceedings 15 (2020) 102-107

Materials Research Forum LLC
https://doi.org/10.21741/9781644900574-17

energy-dependent neutron transmission at all points over a sample in a single measurement. The quantitative nature and potentially short measurement times make these techniques very attractive for both scientific and industrial applications.

The event-type detectors currently available at RADEN include two micropattern detectors, the µNID (Micropixel-chamber-based Neutron Imaging Detector) [5,6] and nGEM (boron-coated Gas Electron Multiplier) [7], and a lithium glass scintillator pixel detector, the LiTA12 (^{6}Li Time Analyzer, model 2012) [8]. The performance of these detectors, as measured at RADEN, has recently been presented in Ref. [9], and the results are summarized in Table 1.

Table 1: Performance of event-type detectors at RADEN. 'Peak count-rate capacity' and 'Effective peak count-rate' refer to the global instantaneous peak rates (i.e., peak rates over the entire detector) at the limit of the hardware and with less than 2% event loss, respectively.

Detector	µNID	nGEM	LiTA12
Type	Micropattern	Micropattern	Scintillator
Neutron converter	^{3}He	^{10}B	^{6}Li
Area [cm^2]	10×10	10×10	4.9×4.9
Time resolution [ns]	250	15	40
Spatial resolution [mm]	0.1	1	3
Efficiency @25.3 meV [%]	26	10	23
Peak count-rate capacity [Mcps]	8	4.6	8
Effective peak count-rate [Mcps]	1	0.18	6

We are continually working to improve our event-type neutron imaging detectors for better spatial resolution and shorter measurement times and, as a user facility, to improve the ease-of-use of their control and analysis software. In particular, we have recently redesigned the µNID control software to allow full integration into the experiment control system at RADEN, and we have developed an easy-to-use web-based user interface for data processing. We are also developing a new 215-µm pitch readout for the µNID for improved spatial resolution and a µNID with boron-based converter for increased count rate via a much-reduced event size. In addition, we have tested super resolution techniques to improve the spatial resolution of the LiTA12 detector. These development efforts are described below.

µNID Development
Integration into the RADEN Control System. RADEN features an experiment device control system based on the IROHA2 software framework developed at the MLF [10]. As its most visible feature, IROHA2 provides a user-friendly web interface, making the system accessible from any device with a web browser. The IROHA2 framework includes *device control servers* for controlling and monitoring beamline devices, experimental equipment, etc. It also allows the control of detector systems via the DAQ-Middleware framework developed at KEK, which consists of customizable, modular data acquisition software components with a unified interface [11]. Finally, an *instrument management server* provides overall control of the instrument components and measurement process, and a *sequence server* provides for automated, multi-step measurements via simple Python-based scripting.

In order to integrate the µNID into the RADEN control system, we have developed a new combined DAQ control and power hardware unit, along with the corresponding device control module for IROHA2, and we have rewritten the data acquisition software using the DAQ-Middleware framework. (The manufacture of the DAQ control and power unit and coding of the device module and data acquisition software were done by Bee Beans Technologies, Inc.) The

Neutron Radiography - WCNR-11   Materials Research Forum LLC
Materials Research Proceedings **15** (2020) 102-107   https://doi.org/10.21741/9781644900574-17

new DAQ control unit and device control module allow full control of the µNID, including detector power, setting of all detector parameters, and real-time monitoring of detector status, all from within the IROHA2 system. Additionally, the IROHA2 system is able to remotely control data acquisition via the new DAQ-Middleware-based data acquisition software, allowing the µNID to be used for automated measurements.

As a demonstration of the integrated RADEN/µNID control system, we have carried out a computed tomography measurement as shown in Fig. 1. The sample was an iron step-wedge of size 5 cm × 2.5 cm × 1 cm, and projections from 0 to 180° in 2° steps, with an exposure time of 8 minutes per step, were taken with an automated measurement sequence. At the time of the measurement, the MLF beam power was 500 kW, and the observed peak count rate at the detector was 4.4 Mcps. As seen in Fig. 1, we were able to successfully reconstruct a three-dimensional image using this data. (The reconstruction and visualization were performed with the commercial software, VGSTUDIO MAX.) In the present case, we have used the full neutron energy spectrum in the reconstruction. Due to the nature of the event-type detectors, however, we can return to the original event data and easily select any neutron energy range for the tomographic reconstruction, adding significant quantitative power and flexibility to computed tomography measurements at RADEN.

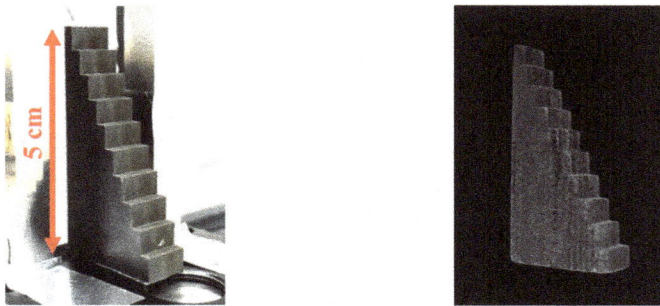

Figure 1. Automated computed tomography measurement with the µNID. (Left) Photograph of Fe step-wedge sample. (Right) Three-dimensional image reconstructed from 91 projections (0 to 180°, 2° steps).

Data Processing Software. The µNID system produces copious amounts of raw hit data that is currently saved and processed off-line using custom software. This data processing software, which is based on C++ and runs on Linux or macOS, decodes the binary hit data, clusters the hits into neutron events, and calculates the neutron position (using a *template* fit), time-of-flight, and energy deposition as described in Refs. [5,6]. Over the last several years, we have completely re-written the data processing software in order to optimize the performance from the ground up, including optimization of the position calculation and implementation of a new clustering algorithm for an improved spatial resolution of 0.1 mm and an increased rate performance to more than 1 Mcps, respectively, as described in Ref. [9].

To ease the task of data processing, we have also developed a new, web-based graphical user interface (GUI). Modeled loosely after the IROHA2 device control system and written in Python with JavaScript and HTML/CSS, the µNID data processing GUI runs as a server-client system, allowing data processing to be performed from any device with a web browser, independent of

Neutron Radiography - WCNR-11   Materials Research Forum LLC
Materials Research Proceedings 15 (2020) 102-107   https://doi.org/10.21741/9781644900574-17

the client operating system. The server-side runs on a Linux or macOS system, where the actual data processing takes place, and acts as the interface between the browser-based GUI and the underlying data processing software. The functionality of the GUI includes user/data management, a guided step-by-step data calibration procedure, data processing job queue, data visualization with interactive plots, and conversion of event data to multipage TIFF. This new GUI system is now in use at RADEN since last year, and we are continuing to make improvements in functionality and performance. The μNID data processing software and GUI use only open-source libraries.

New μNID Development. We are developing a new readout element with reduced pitch for improved spatial resolution and a μNID with a ^{10}B-based neutron converter for increased count rate. These are described in Ref. [9] in detail and are summarized below.

The current readout element of the μNID, referred to as the μPIC (Micro-Pixel Chamber), is a micropattern readout developed at Kyoto University and manufactured by DaiNippon Printing Company, Ltd. The μPIC consists of orthogonal anode and cathode strips with a 400-μm pitch on a polyimide substrate and provides gas gain via its unique geometry (maximum gain factor: 6000) [12]. The new readout element is made using MEMS-based (Micro-Electro-Mechanical Systems) manufacturing processes to achieve a reduction in strip pitch from the current 400 μm down to 215 μm, with a corresponding increase in spatial resolution expected after optimization of the gas mixture. Initial testing of a MEMS μPIC with a silicon substrate confirmed the basic operation of the new readout. However, instability in the gain arising from the silicon substrate was observed, and we are now investigating an alternate substrate material.

The μNID with thin-film boron converter achieves an increase in peak count-rate capacity to over 20 Mcps via a reduced event size as compared to the μNID with ^{3}He (i.e., the alpha and Li nucleus produced by the n-^{10}B reaction travel a much shorter distance in the gas of the detector compared to the proton and triton of the ^{3}He case). A prototype detector has been tested at RADEN, and a 22 Mcps peak count-rate capacity and 0.45 mm spatial resolution were confirmed, matching expectations. The current boron converter, consisting of a 1.2-μm layer of ^{10}B deposited on the drift cathode, provides a neutron detection efficiency of only 3 to 5 %, and we are now considering ways to increase this detection efficiency.

LiTA12 Development
The LiTA12 detector, developed at KEK, consists of a 16 × 16 array of Li-glass scintillator pixels (type: GS20, size: 2.1 mm × 2.1 mm × 1 mm) matched to a Hamamatsu H9500 multi-anode photomultiplier tube with a 3-mm anode pitch and total area of 4.9 cm × 4.9 cm. Due to the very fast decay time of the Li-glass scintillator, this detector has the potential for very high neutron count rates up to 100 Mcps. The front-end electronics of the current system, however, limit the count rate to 6 Mcps, but this can be increased in the future. This high count-rate capacity and the possibility to easily increase the efficiency by increasing scintillator thickness make the LiTA12 an attractive candidate detector for neutron resonance absorption techniques using epithermal neutrons [13].

The main drawback of the current LiTA12 detector, however, is the poor spatial resolution, which is limited by the 3-mm anode pitch. To address this, we have been investigating so-called *super resolution techniques* for improved spatial resolution, including charge centroiding and multi-image composition. Charge centroiding is achieved by replacing the scintillator pixels with a single scintillator plate, which allows the light from a single neutron event to spread over multiple anodes. A refined position is then calculated by finding the center-of-gravity of this light distribution. As reported in Ref. [13], tests of such a detector were carried out at BL10 of the J-PARC MLF, and an improvement in spatial resolution to around 0.7 mm was confirmed. We are also investigating an image compositing technique that uses multiple images taken at sub-pixel shifts of the detector to reconstruct an image with improved spatial resolution [14]. Fig. 2 shows the results of a test measurement carried out at RADEN. A total of 36 images of a gadolinium test target [15] were taken by scanning the LiTA12 in 0.5-mm steps in the horizontal and vertical directions. The combined image, constructed very simply by dividing each 3-mm pixel into a 6 × 6 grid and placing the pixel values from each image at the corresponding grid point, indicates that the reconstruction of a higher spatial resolution image may be possible from this data. Based on these results, we are now considering a more sophisticated image reconstruction technique.

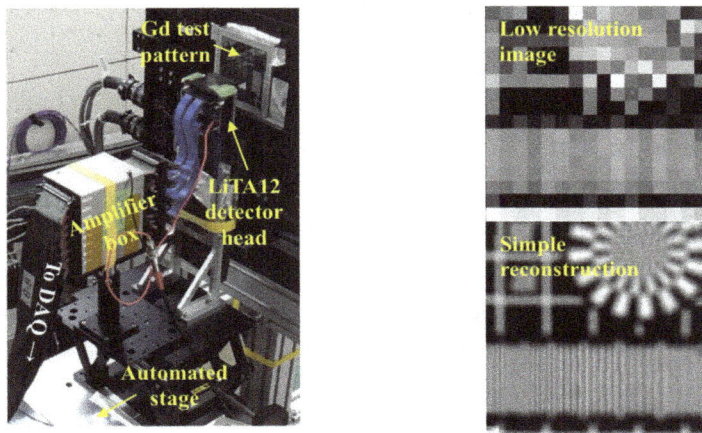

Figure 2. Multi-image compositing with the LiTA12. (Left) Photograph of the LiTA12 mounted on a remote-controlled stage behind the Gd test pattern. (Right, top) Single image taken by the LiTA12 with a 4.9 cm × 4.9 cm field of view and 3-mm pixel size. (Right, bottom) Composited image combining 36 separate measurements taken at 0.5 mm steps in the horizontal and vertical directions.

Summary

At RADEN, we are continuing development of our event-type neutron imaging detectors to better meet the requirements of energy-resolved neutron imaging techniques. For the µNID, we have integrated the detector into the RADEN experiment device control system and redesigned the data processing software for an overall improvement in performance and ease-of-use. We are also working to provide improved spatial resolution and increased peak count-rate capacity up to 22 Mcps for the µNID by developing a new readout element with reduced pitch and a µNID with boron-based converter for reduced event size, respectively. Additionally, we are investigating super resolution techniques to improve the spatial resolution of the LiTA12 detector from the current 3 mm down to 0.7 mm or less.

Acknowledgements

This work was partially supported by the Momose Quantum Beam Phase Imaging Project, ERATO, JST (Grant No. JPMJER1403). Detector testing at RADEN was carried out under Instrument Group Proposal Nos. 2017I0022 and 2018I0022 and CROSS Development Proposal Nos. 2017C0004 and 2018C0002.

References

[1] T. Shinohara et al., Final design of the energy-resolved neutron imaging system "RADEN" at J-PARC, J. Phys.: Conf. Series 746 (2016) 012007. https://doi.org/10.1088/1742-6596/746/1/012007

[2] H. Sato, O. Takada, K. Iwase, T. Kamiyama, and Y. Kiyanagi, Imaging of a spatial distribution of preferred orientation of crystallites by pulsed neutron Bragg edge transmission, J. Phys.: Conf. Series 251 (2010) 012070. https://doi.org/10.1088/1742-6596/251/1/012070

[3] H. Sato, T. Kamiyama, and Y. Kiyanagi, Pulsed neutron imaging using resonance transmission spectroscopy, Nucl. Instr. and Meth. A 605 (2009) 36-39. https://doi.org/10.1016/j.nima.2009.01.124

[4] T. Shinohara et al., Quantitative magnetic field imaging by polarized pulsed neutrons at J-PARC, Nucl. Instr. and Meth. A 651 (2011) 121-125. https://doi.org/10.1016/j.nima.2011.01.099

[5] J.D. Parker et al., Neutron imaging detector based on the μPIC micro-pixel chamber, Nucl. Instr. and Meth. A 697 (2013) 23-31. https://doi.org/10.1016/j.nima.2012.08.036

[6] J.D. Parker et al., Spatial resolution of a μPIC-based neutron imaging detector, Nucl. Instr. and Meth. A 726 (2013) 155-161. https://doi.org/10.1016/j.nima.2013.06.001

[7] S. Uno, T. Uchida, M. Sekimoto, T. Murakami, K. Miyama, M. Shoji, E. Nakano, and T. Koike, Development of a two-dimensional gaseous detector for energy-selective neutron radiography, Phys. Proc. 37 (2012) 600-605. https://doi.org/10.1016/j.phpro.2012.01.035

[8] S. Satoh, Development of a new exclusive function for a 2012 model ^{6}Li time analyzer neutron detector system, JPS Conf. Proc. 8 (2015) 051001. https://doi.org/10.7566/JPSCP.8.051001

[9] J.D. Parker et al., Development of energy-resolved neutron imaging detectors at RADEN, JPS Conf. Proc. 22 (2018) 011022. https://doi.org/10.7566/JPSCP.22.011022

[10] T. Nakatani, Y. Inamura, T. Ito, and T. Otomo, The control software framework of the web base, JPS Conf. Proc. 8 (2015) 036013. https://doi.org/10.7566/JPSCP.8.036013

[11] H. Maeda, Y. Nagasaka, H. Sendai, E. Inoue, E. Hamada, T. Kotoku, N. Ando, S. Ajimura, and M. Wada, Control functionality of DAQ-Middleware, J. Phys.: Conf. Series 513 (2014) 012020. https://doi.org/10.1088/1742-6596/513/1/012020

[12] A. Ochi, T. Nagayoshi, S. Koishi, T. Tanimori, T. Nagae, and M. Nakamura, A new design of the gaseous imaging detector: Micro Pixel Chamber, Nucl. Instr. and Meth. A 471 (2001) 264-267. https://doi.org/10.1016/S0168-9002(01)00996-2

[13] T. Kai et al., Characteristics of the 2012 model lithium-6 time-analyzer neutron detector (LiTA12) system as a high efficiency detector for resonance absorption imaging, Phys. B, in press. (DOI:10.1016/j.physb.2017.11.086)

[14] S. Farsiu, M.D. Robinson, M. Elad, and P. Milanfar, Fast and robust multiframe super resolution, IEEE Trans. Image Proc. 13 (2004) 1327-1344. https://doi.org/10.1109/TIP.2004.834669

[15] M. Segawa et al., Spatial resolution test targets made of gadolinium and gold for conventional and resonance neutron imaging, JPS Conf. Proc. 22 (2018) 011028. https://doi.org/10.7566/JPSCP.22.011028

Neutron Radiography - WCNR-11
Materials Research Proceedings **15** (2020) 97-101

Materials Research Forum LLC
https://doi.org/10.21741/9781644900574-16

High-resolution Detector for Neutron Diffraction and Quantification of Subsurface Residual Stress

Stuart R. Miller[1,a*], Matthew S.J.Marshall[1,b], Megan Wart[1,c],
Pijush Bhattacharya[1,d], Stephen Puplampu[2,e], Matthew Seals[2,f],
Dayakar Penumadu[2,g], Rick Riedel[3,h], and Vivek V. Nagarkar[1,i]

[1]Radiation Monitoring Devices, Inc., Watertown, MA 02472, USA

[2]University of Tennessee, Knoxville, TN 37996, USA

[3]Neutron Scattering Science Division, Oak Ridge National Laboratory, PO Box 2008, Building 7962, Oak Ridge, TN 37831-6393, USA

[a*]smiller@rmdinc.com, [b]mmarshall@rmdinc.com, [c]mwart@rmdinc.com,
[d]pbhattacharya@rmdinc.com, [e]spuplamp@vols.utk.edu, [f]mseals2@vols.utk.edu,
[g]dpenumad@utk.edu , [h]riedelra@ornl.gov, [i]vnagarkar@rmdinc.com

Keywords: Neutron Scattering, Neutron Diffraction, Anger Camera, Residual Stress Measurement, Lithium Sodium Iodide

Abstract. We have developed a new high-resolution large-area detector for neutron diffraction imaging, specifically for the measurement of subsurface residual stress in engine components. Neutron diffraction typically requires monochromatic thermal or cold neutrons, neutron flux at the detector is orders of magnitude lower than for standard neutron radiography. Therefore the detection efficiency for incident neutrons must be maximized in order to reduce acquisition times. Here we use the high absorption cross section of LNI (^{6}Li$_x$Na$_{1-x}$I:Eu,Tl) scintillators coupled to an Anger camera consisting of an array of silicon photomultiplier (SiPM) detectors. The LNI scintillator developed at RMD is derived from the well-known NaI scintillator and comes in two formats for this application, a vapor-deposited film and a crystal sliced into 1-2 mm thick layers for imaging. In either case, 95% enriched ^{6}Li was utilized. LNI crystal slices have demonstrated high light yield, for example a 1 mm thick sample with 10% Li produced a 40,900 photons/MeV gamma response and a neutron response with gamma equivalent energy (GEE) of 3.76 MeV, indicating 153,800 photons/neutron. In general higher Li content decreases the light yield but increases the GEE up to 4.8 MeV, which is approximately the theoretical maximum. Use of spatially resolved detectors for diffraction signals from single and polycrystalline materials is gaining strong interest from the neutron community, particularly from advanced light source users who benefit from plenty of photons to work with. Since diffraction signals from polycrystalline materials are inherently weak, the use of this approach at neutron facilities, which typically provide a low flux of neutrons in the thermal or cold energy ranges, poses unique problems. This paper presents aspects of utilizing high-resolution and high efficiency neutron detectors for obtaining important engineering measurements such as residual stress considering the target materials of interest. We present for the first time measurements associated with diffraction signals captured using an Anger camera with high spatial resolution for an example scattering polycrystalline powder appropriate for the incident mono-energetic cold neutrons at the CG1 (Cold Guide Hall) beamline of the High Flux Isotope Reactor at the Oak Ridge National Laboratory.

Introduction

Neutron diffraction is an important tool for crystallography to provide information about the structure of materials. While these measurements are typically done at a high flux beamline from a reactor or spallation source, the scattered neutrons incident at the detector plane are typically quite low in quantity due to the fact that they are scattered from a target and also monochromatized. Thus the detector requirements for neutron diffraction imaging place a premium on detection efficiency in order to capture as many of the incident neutrons as possible.

The Anger Camera, developed by the detector group at the Oak Ridge National Laboratory, coupled with a highly efficient scintillator, provides the required detection efficiency [1,2]. In addition it has the advantage of a large active measurement area and high spatial resolution for detecting diffracted/scattered thermal neutrons. Implementation of pulse shape (in addition to pulse height) discrimination further enhances the ability to detect diffracted neutrons with enhanced signal to noise ratio.

Here we are developing a neutron imaging detector specifically for the quantification of residual stress in metallic engine components. The goal is to make this possible with a neutron generator in a laboratory environment. This makes detector requirements even more critical due to a lower flux of neutrons incident on the detector.

A key component of our detector development is the scintillator that captures neutrons and converts them efficiently to optical photons for detection by the Anger camera. RMD is developing the LNI ($^6Li_xNa_{1-x}I$:Eu,Tl) scintillator that combines the high absorption of ^6Li with the well-known NaI scintillator; LNI technology is patented under US 9,417,343 [3].

Anger Camera

The anger camera is the detector of choice for this application due to its high detection efficiency as well as the capability to use PSD and PHD for neutron/gamma discrimination. The Anger camera is a variant within the CCD imaging family that allows for fast read-out of large multi-anode photo-multiplier tubes. Instead of approaching the issue of readout by establishing a channel for each anode; the Anger camera reduces the entire output of the PMT down to four channels allowing for fast data acquisition over the entire photo-sensitive area in real-time [4,5]. The output from an array of PMT's is injected into a resistive network where Anger logic is used to determine the location of incident particles on the scintillator.

The Anger camera used here has an active area of 15 x 15 cm^2 and consists of a 16 x 16 array of SiPM's.

LNI Scintillator Development

The development of the LNI scintillator has two formats, including vapor-deposited films and crystal slices. The advantage of the film format is the columnar structure that produces enhanced spatial resolution for a given thickness of scintillator. **Fig. 1** shows the SEM images of an LNI film showing the columnar structure that allows the scintillation light to be channeled to the detector, allowing thicker films to be used while maintaining high

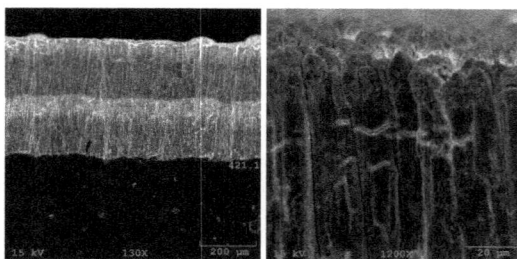

Figure 1: (Left) SEM image of a 420 µm thick LNI film. (Right) Close up of the columns at the top of the film. Note that some degradation of the film structure has occurred due to exposure to room atmosphere.

Materials Research Forum LLC
https://doi.org/10.21741/9781644900574-16

spatial resolution. When coupled to the Anger camera a 400 µm thick film has demonstrated the very high spatial resolution of 350 µm, as shown in **Fig. 2**. This is the highest observed with the Anger camera.

Another approach being explored is to grow the LNI in crystal format and then cut the crystal into thin slices that are 1-4 mm thick, as shown in **Fig. 3**. While this approach can't provide the high resolution observed from films, it allows thicker layers of scintillators to be used to provide the needed stopping power for neutron scatter imaging. Due to the Anger logic method of single photon detection, monolithic crystal slices provide a good combination of efficiency and spatial resolution. The challenge with this approach is to grow crystals large enough for imaging, and cutting thin slices without cracking. Currently we are developing the methods to achieve this.

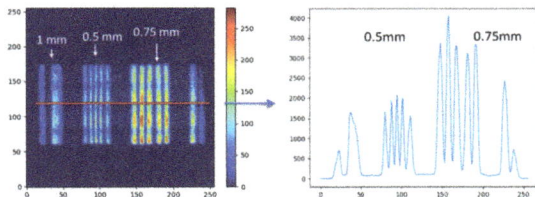

Figure 4: (Left) Resolution mask image acquired using a 4"x4", 500 µm thick, LNI film to an SiPM Anger camera. (Right) Line profile demonstrating a high degree of modulation even for a 0.5 mm pattern. Estimated resolution is ~350 um.

Figure 2: (Left) An LNI crystal with 20% ⁶Li enriched to 95%. (Right) Several crystal slices cut from the crystal at left, with 1, 3 and 2 mm thicknesses, viewed with UV excitation.

Neutron Response

The neutron response was measured by coupling the LNI crystal slices to a 3" PMT (Hamamatsu R6233-100 SBA). The 20% LNI provided a very high gamma-equivalent energy (GEE) of 4.8 MeV, as shown in **Fig. 4**, which is on par with the theoretical Q-value for ⁶Li conversion. This sample produced 15,000 photons/MeV gamma light yield and 72,200 photons/neutron. A sample with 10% ⁶Li however produced 40,900 photons/MeV gamma response and a neutron response with GEE of 3.76 MeV, indicating 153,800 photons/neutron. This is on par with and slightly exceeding previously reported results [6].

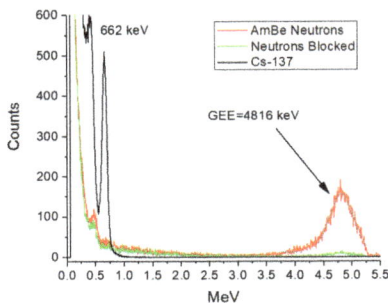

Figure 3: Neutron response of a 2 mm thick LNI crystal sample with 20% ⁶Li. Gamma equivalent energy is 4.8 MeV which is the theoretical maximum.

The neutron response of LNI films has been previously reported [7] and a GEE of 4.3 MeV was observed from a 375 µm thick film.

Both crystals and films provide excellent gamma-neutron discrimination [1,2], which is particularly important for neutron scattering measurements and the Anger camera has the capability to take advantage of this to remove gamma events and increase signal-noise ratio in the images.

Neutron Absorption Efficiency
The neutron absorption efficiencies were measured for LNI films of various thicknesses. The measurements were taken with the beam on and aperture open and beam on and aperture closed and compared to a ^{3}He detector standard. The measured efficiencies as a function of LNI thickness are shown in **5**, and the results are commensurate with the theoretical values with 50% ^{6}Li films.

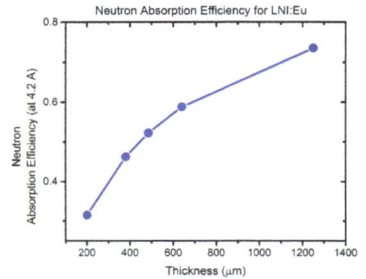

Figure 5: Neutron absorption efficiency as a function of thickness of LNI films for 4.2 Å neutrons.

Diffraction Imaging
Diffraction measurements were demonstrated by coupling a 450 µm thick LNI film to the Anger camera. The diffraction line from a germanium rod was imaged with a 300 s acquisition time with the detector 65 mm from the sample, as shown in **Fig. 6**. Imaging was performed at the HFIR CG-1A beamline. This shows that an LNI film with

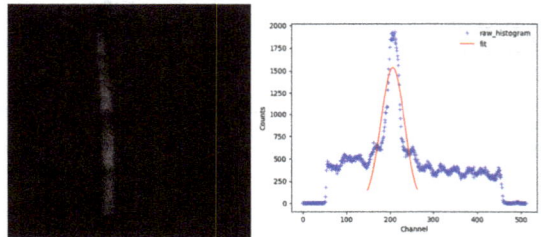

Figure 6: (Left) Diffraction image from a germanium rod acquired with a 450 µm thick LNI:Eu film coupled to the Anger camera. Gamma rejection was done with pulse-height discrimination. (Right) The line profile through the image at left.

about 50% efficiency can produce high qualiy diffraction images, however the image acquisition times are still long. We are developing thicker LNI films that are also brighter, and expect to improve on this result.

Summary
We have developed a new large-area detector for neutron diffraction imaging based on an Anger camera coupled to an LNI scintillator with the following attributes:
- High sensitivity, up to 70%
- High spatial resolution, 350 µm, 1.4 lp/mm
- Neutron/Gamma discrimination with PHD
- The LNI scintillator can be in film or crystal format.

The LNI scintillator coupled to the Anger camera provides the highest possible combination of detection efficiency and spatial resolution. This detector is currently under development for use in the imaging of diffraction peaks from various materials. By mapping diffraction peak changes it is then possible to determine the stress state of the objects of interest. Future developments are planned to optimize the detector to measure and map the residual stress in

Materials Research Forum LLC
https://doi.org/10.21741/9781644900574-16

engine components and to make it possible to perform these measurements in a laboratory setting.

Acknowledgements

This work is currently being supported by Department of Defense Grant N6833517C0250.

References

[1] H.O. Anger, Review of Scientific Instruments, 29(1) (1959) 27. https://doi.org/10.1063/1.1715998

[2] R.A.Riedel, et. al., Design and performance of a large area neutron sensitive anger camera, NIM in Physics Research A 794 (2015) 224–233. https://doi.org/10.1016/j.nima.2015.05.026

[3] V.Nagarkar, H. Bhandari, and O. Ovechkina, United States Patent "Neutron Detector and Fabrication Method Thereof, US 9,417,343 B1, August 16, 2016.

[4] P. D. Olcott, J. A. Talcott, C. S. Levin, F. Habte, and A. M. K. Foudray, "Compact readout electronics for position sensitive photomultiplier tubes," *IEEE Transactions on Nuclear Science,* vol. 52, no. 1, pp. 21-27, 2005. https://doi.org/10.1109/TNS.2004.843134

[5] S. Siegel, R. W. Silverman, S. Yiping, and S. R. Cherry, "Simple charge division readouts for imaging scintillator arrays using a multi-channel PMT," *IEEE Transactions on Nuclear Science,* vol. 43, no. 3, pp. 1634-1641, 1996. https://doi.org/10.1109/23.507162

[6] Nagarkar, V., Ovechkina, E., Bhandari, H., Miller, S., Marton, Zs., Glodo, J., Soundara-Pandian, L., Mengesha, W., Gerling, M., Brubaker, E., 2013. Lithium alkali halides – new thermal neutron detectors with n-γ discrimination, Nuclear Science Symposium and Medical Imaging Conference (NSS/MIC), 2013 IEEE.

[7] Marshall, M.S.J., M.J. More, H.B. Bhandari, R.A. Riedel, S. Waterman, J. Crespi, P. Nickerson, S. Miller, V.V. Nagarkar, "Novel Neutron Detector Material: Microcolumnar LixNa1–xI:Eu," in IEEE Transactions on Nuclear Science, vol. 64, no. 11, pp. 2878-2882, Nov. 2017. doi: 10.1109/TNS.2017.2762859. https://doi.org/10.1109/TNS.2017.2762859

Materials Research Forum LLC
https://doi.org/10.21741/9781644900574-18

One Inch CCD Cameras for Neutron and X-ray Imaging

Alan Hewat

NeutronOptics and Institut Laue-Langevin, Grenoble, France

alan.hewat@neutronoptics.com

Keywords: CCD Cameras, Neutron Imaging, X-ray Imaging, Laue Diffraction

Abstract. Most normal applications of neutron and x-ray imaging and diffraction can be satisfied with inexpensive commercial CCD cameras. Here we have chosen the largest CCD made by Sony, the 1-inch ICX694ALG, and compared it with much more expensive CCD and CMOS cameras offered by specialized companies such as Andor and PCO. We give examples of the use of this CCD for large area (250x200mm) imaging with optical resolution of 90 um, for very high resolution 1:1 macro imaging, where the optical resolution and area is equal to the CCD chip, and for backscattered Laue diffraction where high efficiency is required.

Introduction

Until recently, neutron imaging has been the domain of a few big laboratories in Europe and the USA using expensive equipment. However, neutron imaging is one of the applications that can be conducted on low-flux neutron sources in small laboratories, and indeed it has been a priority for the IAEA in encouraging peaceful uses of nuclear techniques in developing countries. Sufficient neutron intensity can be provided by a 1MW TRIGA reactor, of which there are many in US Universities and in the national laboratories of smaller countries. The TRIGA reactor is very safe, uses relatively low technology, and can be switched on and off as required. Neutron generators using "table top" Deuterium-Deuterium (D-D) or Deuterium-Tritium (D-T) generators are being developed to provide alternative low-flux sources suitable for neutron imaging. A low-cost camera suitable for such low-flux sources, can be used with these new neutron and x-ray generators. Our objective is to satisfy the demand for the wider application of neutron and x-ray imaging. But there is no point in doing "second class" science, so we will show that these low cost cameras can in many cases compete with the more expensive equipment in big laboratories.

The origins of Neutron Cameras, and the Advantages of Simplification

Neutron cameras are almost as old as neutron diffraction itself, and are simply a variation of the even older photographic techniques used from the discovery of x-rays. Indeed neutron cameras are just x-ray cameras with a component (usually LiF) to convert neutrons into ionising particles and x-rays, which are then converted into light using an x-ray scintillator (usually ZnS). A neutron Polaroid film camera was used with a scintillator from the beginning of ILL in the early 1970's, and even earlier elsewhere.

The idea of using a video camera instead of film goes back to Arndt & Ambrose (1968) and was proposed for neutron diffraction by Arndt & Gilmore (1975), with experiments in the 1970's at ILL Grenoble [1]. CCD cameras were used for the ILL NEUTROGRAPH in 2002 [2]. ILL was the world's highest flux neutron source, with intensities of more than an order of magnitude greater than other facilities, but the NEUTROGRAPH had relatively low resolution and dynamic range. ILL eventually stopped neutron imaging but FRM-2 and HMI Germany with PSI Switzerland in particular, continued to develop the technique. Only recently has ILL built a new world class neutron imaging station, on a high resolution and high flux cold guide [3].

ILL did however, continue to develop CCD cameras for sample alignment and diffraction. In 2005, a simple neutron CCD camera [4] was designed to replace the Polaroid neutron camera on most ILL neutron instruments. CCD detectors were also developed for neutron Laue diffraction [5, 6, 7], and proved a great success, partly replacing neutron image plate detectors.

Our current CCD cameras have a number of technical advantages:

- We use high volume commercial CCD units, which are inexpensive to repair & replace.
- Neutron scintillators, front-surfaced mirrors and lenses are identical to those used in leading neutron imaging laboratories.
- We use fixed geometry, with a variable Field-of-View (FOV) in our large cameras obtained by simply exchanging the scintillator front end to change the optical path length.
- This fixed geometry requires no translation components that can fail or malfunction.
- More importantly, fixed geometry means that the CCD unit can attached to the exterior of the camera box, can be easily shielded, and can use simple air and Peltier cooling.
- If the CCD unit was on a motorised rail, it would have to be inside the box or external bellows, and the camera would be more complicated, perhaps requiring water cooling.
- Ordinary 12V power supplies are used, together with standard USB-2 cables with amplified 10m extensions up to 30m total.
- A wide choice of third party software is available for free, as well as an SDK.

Fig 1. Simple neutron alignment camera

Simple Neutron and X-ray Beam Alignment camera

Our current slim camera (fig.1) is designed for checking sample centering and beam homogeneity. It is only 44mm thick for a sensitive area of 100x50mm using a 1/2" or 2/3" Sony CCD, so can fit into the small space between the sample environment and the beam stop. A single 10-20m amplified USB cable is used for power, control and image acquisition.

These simple cameras can be made as large as 200x150mm for guide tube beam homogeneity measurements. With such a large CCD they are very efficient, and weak 10^4 n.cm^{-2}.s^{-1} neutron beams require only a few seconds exposure. Thin but strong carbon fibre windows with appropriate scintillators are used for the similar x-ray cameras.

Fig 2. mini-iCam x-ray alignment camera

Our smallest mini-iCam (fig.2) neutron or x-ray camera is only ~190mm long and has 580 or 1040 pixels over an area of 30mm diameter, so with its efficient f/1.0 lens it is also very bright. The scintillator and carbon fibre window can be exchanged, depending on whether x-rays or neutrons are to be imaged

High Resolution Neutron & X-ray Imaging Cameras using a 1-inch CCD

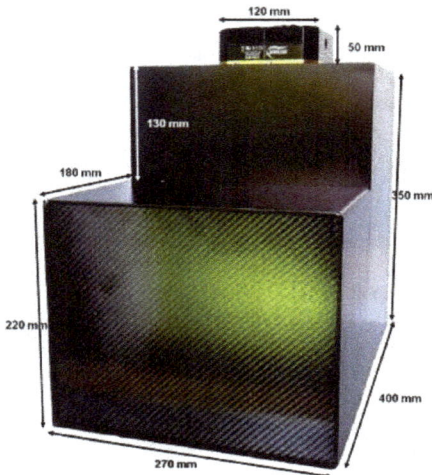

Our imaging camera (fig.3) compares well with more expensive cameras, with low noise, high resolution and fast readout. The scintillator holder on the front section can be changed in-situ to select different neutron and x-ray scintillators (figs.4). For tomography, both software and hardware triggering (via a GPIO socket) is provided.

- Sensor: 1" Sony EXview HAD CCD II
- Optics: High resolution f/1.4 1" lens
- Resolution: 2750 x 2200 pixels
- High sensitivity: (QE~75%), low smear
- Dark current: 0.002 e/pix/s @-10 °C
- Cooling: Regulated Peltier ΔT = -35°C
- Digital Output: 16-bit 65536 levels
- Readout Speed: 6-12 MPixels/s
- Binning and Region-of-Interest
- External Trigger: GPIO synch.

Fig 3. High resolution imaging camera

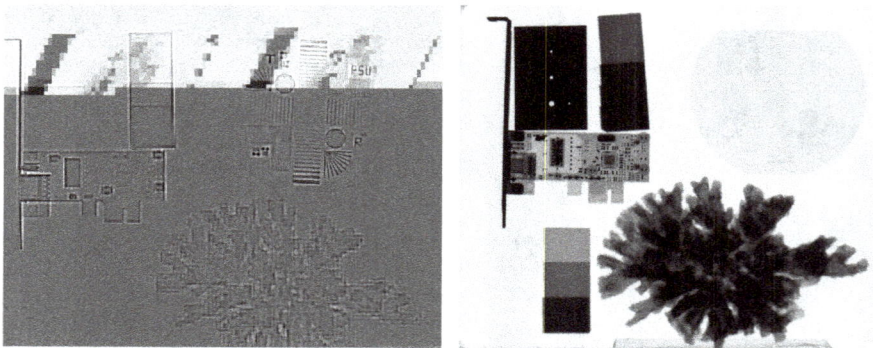

Fig 4a) Images on 100 kW Triga reactor (60s). Fig. 4b) on 60kV/3ma. x-ray source (15s)(R. Zboray PSU)

Comparison of the Sony ICX694ALG 1-inch CCD & Expensive CCD/CMOS Detectors

The *Well Depth* is an important measure of the quality of a detector, since the *Dynamic Range* is the Well Depth divided by the Total Noise. So at first sight the Andor iKon has a big advantage (table 1 below). However, the *Dark Current* noise is exceptionally low for the Sony CCD, so the advantage is not so great in practice. In fact the well depth of the Sony CCD is equal to that of the PCO sCMOS chip, which is an excellent detector. Low dark current can only be achieved with the iKon using extreme cooling, which is certainly not an advantage, especially in an enclosure where air circulation would be limited. The external Sony camera achieves lower dark current with more modest Peltier-air cooling.

The iKon *Read Noise* is also higher unless very slow readout is used. The iKon publicity claims better numbers for individual parameters, but they cannot all be achieved simultaneously eg for readout at 1 frame/sec, the same as the Sony camera, the read noise is much higher at 31.5 ! The iKon is a good camera, but not significantly better for many users, especially considering that it is an order of magnitude more expensive, and also more expensive to maintain, repair and replace.

Table 1. Comparison of Sony ICX694ALG 1-inch CCD & Expensive CCD/CMOS Detectors

CCD	PCO.edge gold 4.2	NOptics VS60	NOptics 11002	iKon L-936
Type	Scientific sCMOS CIS2020	Sony Interline ICX694ALG	"Kodak" Interline KAI-11002	e2V Full Frame CCD42-40
Resolution pixel	2048 x 2048	2759 x 2200	4008 x 2672	2048x2048
Image diag. mm	18.8 (4/3")	16 (1")	43.3 (35mm)	38 (35mm)
Image area mm	13.3x13.3	12.53x9.99	37.25 x 25.70	27.6 x 27.6
Pixel size μm*	6.5 x 6.5	7.4 x7.4	9.0 x 9.0	13.5 x 13.5
Quantum Effic*	>70%	75%	50%	90%
Fullwell e- **	~30,000	~30,000	~60,000	~100,000
Read noise e- **	1	6	13	12
Dark c. e-/pix/s	<0.02@-30°C	0.002@-10 °C	0.03@-20 °C	0.01@-50 °C
Peltier Cooling	Δ -30 °C	Δ -35 °C	Δ -38 °C	Δ -80 °C
Read time (s)***	0.01 to 0.02	1	12 to 22	2 to 10
A/D Readout**	16-bits	16-bits	16-bits	16-bits
Interface	USB 3.0	USB 2.0	USB 2.0	USB 2.0
Relative Cost	16	4	6	50

Very High Resolution Macro Neutron and X-ray Imaging

Following the proposal of Kardjilov et al. [8] we have constructed an inexpensive 1:1 macro imaging camera designed to use the latest Gd_2O_2S:Tb high resolution neutron/x-ray scintillators from RC-TriTec. The photo shows our macro camera optics; without the CCD unit is 290mm high. With a large commercial f=100mm f/2.8 F-mount macro lens, the FOV is equal to the size of the CCD, and the optical resolution approaches the size of the pixels, though the resolution depends of course on the scintillator and collimation.

The camera is shown with a carbon fibre x-ray window, which unscrews to change the scintillator. The C-mount adapter can be unscrewed to take Nikon F-mount cameras up to 35mm full frame. Fine manual focus locking is provided by a thumbscrew mechanism; optional remote focusing can be provided by a motorised drive, controlled by a remote control box or computer.

A compact version using a shorter focal length lens gives larger FOVs with smaller CCDs. A 2x super resolution version can be made to order with an additional 50mm Rodenstock lens following Williams et al.[9].

The optical resolution of such a 1:1 neutron macroscope is comparable to the dimensions of the CCD pixels. Fig.6 shows an optical image of a 50μm grid imaged with 3.9μm pixels (checkered pattern) on our 1:1 macro imaging camera. The 25μm wires are clearly resolved to better than 10μm (2 pixels).

Neutron Radiography - WCNR-11   Materials Research Forum LLC
Materials Research Proceedings **15** (2020) 108-114   https://doi.org/10.21741/9781644900574-18

The real resolution depends of course on the thickness of the scintillator, and for neutrons will be significantly less than the optical resolution. Although even higher optical magnification may be justified for x-rays, there seems little point in that for neutrons; it is more important to maximise the signal/noise from the thin neutron scintillators [10].

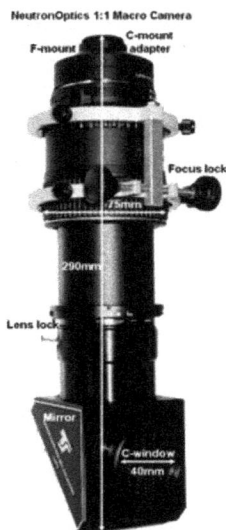

Fig 5. Neutron or X-ray Macro camera

Fig.6 Optical image of 25μm wire grid

Backscattering Laue Camera using a 1-inch Sony CCD

CCD cameras have continued to be important for neutron and x-ray diffraction, and again simplification has been pursued, using only a single 1-inch CCD instead of multiple CCDs. Again the idea was to replace the x-ray Laue camera, or image plate cameras used for crystal alignment in many labs. X-ray intensities are very low in Laue backscattering, so the camera (fig.7) has been optimised to be as close to the source as possible, with the 1-inch CCD permitting a relatively large 120x100mm window and collection times of 2-3 minutes on normal x-ray generators. (Si and Sm2Fe17 micro-crystal mages figs.8 below courtesy of Dean Hudek (Brown University) and Wolfgang Donner & Léopold Diop (TU Darmstad).

Fig.7 Laue x-ray camera

Fig.8a Si (top) and Sm2Fe17 (bottom) imges

Conclusion and Summary

We have shown that the largest Sony CCD, the 1" ICX694ALG is a good choice for a wide variety of neutron and x-ray imaging cameras. It is much less expensive to purchase and maintain than specialised cameras used in big laboratories, yet competes well with such cameras, even on low flux sources. We have given examples of an efficient 250x200mm camera with an optical resolution of 90μm, a 1:1 macro camera with an optical resolution of better than 10μm, and a single-CCD Laue camera that can provide rapid alignment of single crystals on an ordinary x-ray generator.

References

[1] Arndt, U.W. & Ambrose, B.K. (1968) in Cambridge (UK) "An Image Intensifier – Television System for the Direct Recording of X-ray Diffraction Patterns", IEEE. Trans. Nucl. Sci. NS-15, 92-94. https://doi.org/10.1109/TNS.1968.4324920

[2] Brunner, J., A. Hillenbach, E. Lehmann, and B. Schillinger. (2002) "Dynamic neutron radiography of a combustion engine." In Proc. 7th World Conf. on Neutron Radiography, Rome, p. 439.

[3] Tengattini, A., D. Atkins, B. Giroud, E. Andò, J. Beaucour, and G. Viggiani.(2017) "NeXT-Grenoble, a novel facility for Neutron and X-ray Tomography in Grenoble." In Proc. 3rd International Conference on Tomography of Materials and Structures (Lund, Sweden, 26–30 June 2017). https://next-grenoble.fr/

[4] A. Hewat & P. Falus (2007) ILL Annual Report pp. 85-86 "An_Inexpensive_CCD_Neutron_Alignment_Camera"

[5] Ouladdiaf, B., Archer, J., McIntyre, G. J., Hewat, A. W., Brau, D. & York, S. (2006). "OrientExpress: A new system for Laue neutron diffraction" Physica B, 385–386, 1052–1055. https://doi.org/10.1016/j.physb.2006.05.337

[6] A.W. Hewat, B. Ouladdiaf, G.J.McIntyre, M-H.Lemee-Cailleau, D. Brau, S. York (2006) "CYCLOPS, A proposed high flux CCD neutron diffractometer" ILL Millennium Programme Symposium 2006

[7] Bachir Ouladdiaf, John Archer, John R. Allibon, Philippe Decarpentrie, Marie-Helene Lemee-Cailleau, Juan Rodrıguez-Carvajal, Alan W. Hewat, Scott York, Daniel Brau and Garry J. McIntyre (2011) "CYCLOPS – a reciprocal-space explorer based on CCD neutron detectors" J. Appl. Cryst. 44, 392–397. https://doi.org/10.1107/S0021889811006765

[8] Kardjilov, N., M. Dawson, A. Hilger, I. Manke, M. Strobl, D. Penumadu, F. H. Kim, F. Garcia-Moreno, and J. Banhart. "A highly adaptive detector system for high resolution neutron imaging." Nuclear Instruments and Methods in Physics Research Section A: Accelerators, Spectrometers, Detectors and Associated Equipment 651, no. 1 (2011): 95-99. https://doi.org/10.1016/j.nima.2011.02.084

[9] Williams, S. H., A. Hilger, N. Kardjilov, I. Manke, M. Strobl, P. A. Douissard, T. Martin, Heinrich Riesemeier, and J. Banhart. "Detection system for microimaging with neutrons." Journal of Instrumentation 7, no. 02 (2012): P02014. https://doi.org/10.1088/1748-0221/7/02/P02014

[10] Walfort, B., Grünzweig, Ch., Trtik, P., Morgano, M. and Strobl, M. (2018) "Novel scintillation screen with significantly improved radiation hardness and very high light output" WCNR-11 Sydney proceedings.

Methods

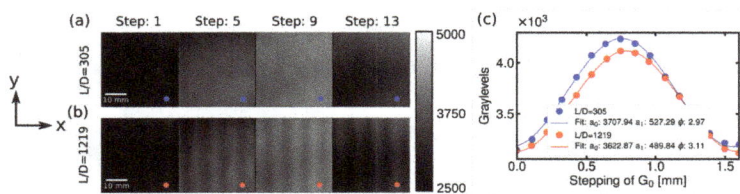

Neutron Radiography - WCNR-11   Materials Research Forum LLC
Materials Research Proceedings **15** (2020) 117-128   https://doi.org/10.21741/9781644900574-19

Various Aspects of the Contrast Modalities of Modulated Beam Imaging

Markus Strobl[1,2,a] *

[1]Paul Scherrer Institut (PSI), Laboratory for Neutron Scattering and Imaging, 5232 Villigen, Switzerland

[2]University of Copenhagen, Niels Bohr Institute, Denmark

[a]markus.strobl@psi.ch

* corresponding author

Keywords: Imaging, Grating Interferometry, Attenuation, Differential Phase Contrast, Dark-Field Contrast

Abstract. Since the introduction of grating interferometers to imaging, in addition to attenuation contrast, differential phase and later also dark-field contrast imaging have been explored intensely. However, in particular with dark-field contrast imaging, imaging entered into a new domain, i.e. the scattering from sub-image-resolution structures. This has led to the need to expand the horizon of considered interactions into the reciprocal space domain of small angle scattering, not necessarily familiar in real space imaging. Correspondingly, description and interpretation and finally quantitative analyses lacked somewhat behind of on the other hand qualitatively invaluable results. Against this background all modalities measured in modulated beam imaging experiments, namely attenuation, differential phase and dark-field contrast, shall be given some additional attention. It will be undertaken to draw a clear picture of analogies, of contrast formation and consequences to interpretation and information content.

Introduction

In imaging with a spatially modulated beam, the beam modulation is superimposed to the image of the sample. Assuming a one-dimensional modulation, it can in general be described by a simple sinusoidal intensity modulation function with a period p superimposed to the conventional radiographic projection $a(x)$ as

$$I(x)=a(x)+b(x)\sin(2\pi x/p+\Delta\phi(x)) \tag{1}$$

Generally, three images of different contrast modalities are/can be deduced from such projection data. Attenuation contrast, manifested as the local average count over one period, or in other words the offset of the modulation $a(x)$; differential phase contrast, the local relative phase of the modulation compared to the undisturbed phase of an empty beam measurement $\Delta\phi(x)$; finally, dark field contrast, the relative local amplitude of the modulation $b(x)/a(x)$. For simplicity it is considered here, that the modulation period is smaller than the spatial resolution of the detector, in which case it is conventionally measured by a scanning procedure and an analyser grating matching the modulation period. This provides a separate modulation in each pixel of an image recorded. Therefore, we can consider the principal spatial resolution limit for all three signals at best twice the pixel size. Note that therefore henceforth x shall be the transversal parameter on the length scale of the pixel resolution. Instead of describing the sub-pixel modulation every pixel is assumed to have well defined parameters $a(x)$, $b(x)$ and $\Delta\phi(x)$.

Sufficient coherence appears to be a prerequisite to exploit beam modulation through grating interferometry based on the Talbot Lau effect [1], because the interference pattern of waves induced by a grating establishes the spatial beam modulation. Hence, the period of that grating implies the coherence requirement. For geometries with insufficient coherence of the beam at the grating the requirement is efficiently circumvented or solved by creating a partially coherent beam with a source grating, i.e. several beams with a coherence satisfying the requirement and superimposing through geometrical alignment constructively at the detector, respectively the analyzer grating. Coherence is here interpreted as the spatial coherence width of a beam at the grating generating the modulation achieved through collimation by a slit (slit grating, i.e. source grating) at a significant distance upstream. Different realizations and details of the set-ups shall not be discussed here but can be found in literature. It has to be noted that meanwhile a great number of variations exists, but the principles generally stay the same for modulated beam techniques. Here we shall focus on the principles of the detected signals and their origins.

Attenuation contrast

Attenuation contrast is the most common and conventional as well as apparently best known contrast mechanism. It can in most cases be simply described by the Beer-Lambert law [2]

$$a(x) = a_0(x)e^{-\int \mu(x,y,\lambda)dy} \tag{2}$$

Including the geometry of the beam and the sample as well as the energy, hence, also wavelength λ, dependent linear attenuation coefficient μ. Note, that the geometrical description is limited to the plane of the beam direction y and the transversal coordinate x, as for a beam assumed parallel this is sufficient, even for tomographic considerations. In addition, this will also suffice for the principle considerations on other grating based contrast signals, due to the grating symmetry. It is useful to present the well-known Beer Lambert equation as it will be a point of reference and analogy for the other, one might refer to as novel, contrast mechanisms. For this reason we also look a bit closer at the attenuation coefficient and what it is constituted of. For neutrons the linear attenuation coefficient μ, a reciprocal length, can be expressed as the product of the neutron cross section $\sigma(\lambda)$ and the particle density N

$$\mu = \sigma N \tag{3}$$

In general the linear attenuation coefficient is assumed orientation invariant, a density property. However, attenuation is constituted not only by absorption, as very often wrongly assumed or inconsiderately expressed, but also by scattering. Correspondingly $\sigma = \sigma_s + \sigma_a$ features a scattering and absorption term, respectively. Scattering which does not exempt directions close to the forward direction can lead to significant bias in measured cross sections, depending on the measurement geometry, in particular the sample to detector distance [3-12]. More interestingly, for neutron wavelengths utilized in grating interferometric imaging the scattering term, or even the total cross section, for crystalline materials is in many cases dominated by coherent elastic scattering from the crystal lattice. This implies that depending on crystallite sizes and microstructural isotropy, the linear attenuation coefficient can become orientation dependent as the coherent elastic part of σ_s can then be written as

$$\sigma_{coh,elas}(\lambda, \Omega) = \sum_n \left[\frac{(2\pi)^3 N}{2k\tau_{hkl}} \sum_{hkl} |F_{hkl}|^2 \delta_D (\tau_{hkl}^2 - 2k\tau_{hkl} \sin\theta_{n,hkl}) \right] \tag{4}$$

where $k=2\pi/\lambda$ is the modulus of the wavevector and correspondingly $\tau=2\pi/d$ is the length of the reciprocal lattice vector. Ω symbolizes the sample orientation in space, n is the number of individual crystal grains, while F_{hkl} is the structure factor and δ_D is representing the Dirac Delta function in expressing the Bragg condition with θ being the Bragg angle. A fact that leads to anisotropic behavior of samples as attenuation of certain volumes will differ for different projection angles e.g. due to texture effects (Fig. 1). In particular for grain mapping with neutrons from the transmission data [13] but also other investigations of local crystalline structure [14], these orientation dependent attenuation effects can be exploited while they are ignored in most standard applications. For 3D phase mapping for example, but given the example in Fig. 1 below, even in conventional white beam tomography, such orientation dependent effects can lead to significant bias, which needs to be considered.

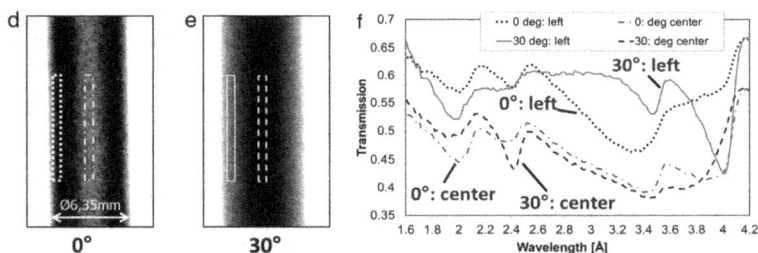

Fig. 1 Significantly textured but otherwise homogeneous cylindrical steel sample exposed at two different projection angles (left) and wavelength dependent transmission profiles according to the highlighted areas in the field of view on the left hand side radiographies (right).

Differential phase contrast

The second contrast modality is referred to as differential phase contrast, sometimes also just phase contrast. In contrast to the transmitted intensity, the phase of the radiation is not accessible straightforwardly. It constitutes a fundamentally different feature of the applied radiation than the detected intensity. The measurement of phase relations is the domain of interferometry and in particular neutron Mach-Zehnder interferometers [15] have been used for such purpose. In some cases such interferometers were also utilized for imaging based on phase shifts induced by samples with both x-rays [16] and neutrons [17]. However, due to the significant coherence and stability requirements of perfect crystal interferometers phase contrast became relevant for broad application, especially with x-rays, only later with the introduction of differential phase contrast [18,19] and propagation based techniques [20,21]. These techniques exploit the fact, that phase objects distort the wavefront and the respective locally induced angular deviations are measured directly with significant angular resolution or through the induced spatial intensity shift at different detector distances, respectively. The angular deviation induced, as described by refraction, corresponds to the induced local phase gradient $\nabla\varphi$, lending the name differential phase to this contrast modality. The rather small angles (arcsec) were first resolved by the use of analyser crystals in monochromator/analyser arrangements (i.e. perfect crystal double crystal arrangements), and methods called Schlieren imaging, diffraction enhanced or refraction contrast imaging [18,19,22-26]. However, grating interferometers introduced around the millennium [27-31] outperformed the other methods significantly due to relaxed coherence requirements, both in real and momentum space. The Talbot Lau grating interferometers resolve small angles due to

Neutron Radiography - WCNR-11   Materials Research Forum LLC
Materials Research Proceedings 15 (2020) 117-128   https://doi.org/10.21741/9781644900574-19

induced spatial beam modulation, for which meanwhile several methods are available. Interestingly in grating interferometers the differential phase translates into a phase shift of the real space modulation according to

$$\nabla j = \partial j / \partial x \propto Df(x) \tag{5}$$

The thus measured orientation dependent refraction, i.e. differential phase (Fig. 2 mid), can easily be transferred into the orientation independent phase contrast (Fig. 2 right) related to the real part of the refractive index δ by integration

$$\varphi(x) = \int \delta(x,y)dy \propto \int_0^x \Delta\phi(x)dx \tag{6}$$

Note that orientation dependence and independence here refers to the local interaction. The relation in eq. 6 provides a straightforward opportunity for tomographic reconstruction of the refractive index distribution $\delta(x,y)$, a scalar quantity. This constitutes an analogy to the linear attenuation coefficient μ or the imaginary part of the refractive index β retrieved in conventional attenuation or absorption contrast imaging, respectively (Fig. 2). The local orientation dependence of the measured refraction effects is neutralized through the integral which turns the differential phase into a phase contrast signal.

Because descriptions appear ambiguous and misleading sometimes it shall be clarified that refraction, i.e. the differential phase effect does not require specific coherence, due to the differential nature, and can be described by geometrical optics. Independent of the fact that differential phase effects do not necessarily preserve coherence characteristics of the affected beam, refraction in general also causes loss of modulation visibility and hence a dark-field contrast signal. In general differential phase and dark-field contrast are very closely related and interlinked in parts, but not equivalent. Clear distinction and understanding of the measured nature of interaction is key for correct interpretations of results in this regime as shall be discussed in the following section on dark-field contrast.

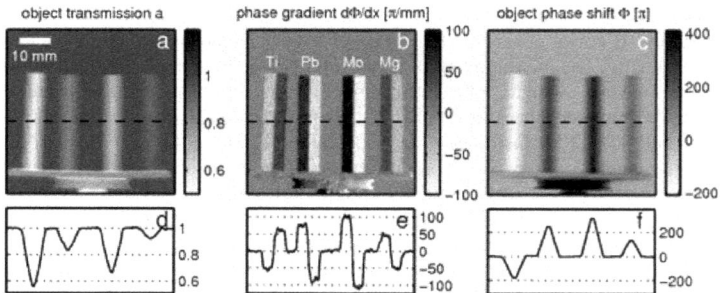

Fig. 2 This reprint with permission of AIP from Ref. [31] illustrates the analogy of attenuation and phase contrast (left and right), as well as the relation of differential phase contrast (mid) and phase contrast through transversal integration (right).

Dark-field contrast

The third contrast modality measured in a grating interferometer set-up for imaging has maybe caused the highest amount of ambiguity. It is based on the measure of the local modulation amplitude b(x), which needs to be normalized by the attenuation a(x) or simply expressed as the visibility $V=(I_{max}-I_{min})/(I_{max}+I_{min})=b(x)/a(x)$. Naturally, only the amplitude relative to the transmission constitutes a useful independent measure. Again, historically this kind of contrast has been utilized already long before grating interferometers were introduced in imaging with x-rays and neutrons, but was referred to with different terminology. In fact first neutron examples can be found already in the 1980ties [17,32] for magnetic structures, which for neutrons still constitute one of the main applications. In x-ray imaging one could consider the first works to be described in Ref [33,34] while first tomography was reported with x-rays in [35] shortly after first neutron dark-field tomography, back then referred to as USANS tomography, was reported [36].

While in summary the dark-field contrast measured always relates to an additional local angular spread of the incident beam, the contrast's origin has so far been attributed to (ultra) small angle scattering ((U)SAS)[34,36,37,38], Fresnel diffraction [17], refraction [37,39], incoherent scattering [40] and decoherence [41,42]. The term decoherence, which is generally in physics defined as the "process in which a system's behaviour changes from that which can be explained by quantum mechanics to that which can be explained by classical mechanics"[43], might intend to mean a certain loss of coherence, though e.g. small angle scattering and Fresnel diffraction are coherent scattering processes and refraction does e.g. in the differential phase contrast not necessarily alter the coherence significantly either.

However, a loss of modulation visibility can have a number of reasons, not necessarily related to a loss of coherence. It is e.g. obvious that a slight alteration of a three grating set-up where multiple beams and hence interference patterns are superimposed constructively, could lead to a destructive superposition. In contrast to a loss of interference of wavefunctions (e.g. through a loss of coherence) this is, however, a classical superposition adding intensities, not phase dependent amplitudes of a wavefunction.

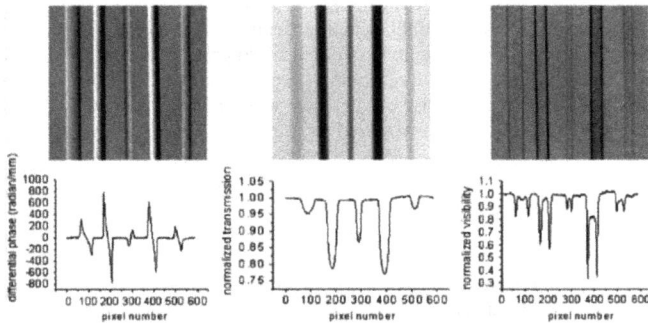

Fig. 3 Differential phase, attenuation and dark-field contrast profiles of a number of cylindric samples (reprinted with permission from [62])

Bearing such classical addition of interference patterns in mind, it is useful to reconsider the conditions of differential phase contrast, where within the spatial detector resolution a phase shift of modulation pattern is expected in one defined direction and magnitude. However, within a pixel there are or might be a number of modulation periods and a structure below this resolution limit might refract the beam to different directions, causing one part of the modulations within a pixel to shift significantly different to the other(s). The integrated measurement of the modulation within a pixel, however, will cause a loss of modulation amplitude, which in the actual pattern might not even exist. (It will indeed always exist in a certain region, where affected and non-affected or differently affected parts of the modulation overlap as well as due to divergence and intrinsic spatial resolution limitations). However, the actual coherence characteristics cannot depend on the pixel size, and hitherto a simple loss of coherence or even decoherence argument can obviously not apply to explain the measured result in general.

With this refraction argument it already becomes clear that, similar to attenuation contrast which can consist of scattering and/or absorption, the measured loss of amplitude referred to as dark-field contrast, does not yet allow a unique interpretation of the physical process measured. Correspondingly, it has been expressed already in the initial neutron grating dark-field publication [37] that "In the region of sharp edges refracted and non-refracted contributions are superimposed and hence decrease the fringe visibility. This accounts for enhanced contrast in the reconstruction at corners and curves, i.e., wherever an edge is not well defined due to limited resolution", and in Ref [44] it is stated, that "measurements show that due to the limited spatial resolution, refraction, i.e., differential phase contrast, also always affects the dark-field image"(compare Fig. 3). And therefore :"Different origins of the dark-field signal have to be taken into account, namely, ultra-small-angle scattering and refraction, which might complicate a straightforward interpretation of specific results with respect to the quantification of the size of inhomogeneities and defects." In X-ray dark-field imaging consciousness is raised only recently by a dedicated work [38] which similarly reads "Results of numerical calculations for monochromatic X-rays show that an unresolvable sharp edge generates not only differential-phase contrast but also visibility contrast."

To some extent, hence, the dark-field contrast measured due to resolution limits can be compared to attenuation values biased by resolution, either at interfaces or when structures are significantly below resolution and hence affect the measured density, where one cannot distinguish whether the overall density changes in this area or whether the very same material with the same density includes distinct structural features like e.g. pores, that are not resolved. (In fact a differentiation of the attenuation contrast could provide a contrast similar to differential phase contrast at edges, including the orientation dependence – a differential attenuation contrast so to speak (compare Fig. 2). However, when assuming refraction (i.e. differential phase contrast) and small angle scattering, referred to in most cases, to be the main sources of dark-field contrast, it is required to take a deeper look into these and how they are distinguished.

In general, small angle scattering is described by scattering theory [45] and in accordance to the used kinematic or first Born approximation [46] phase effects are neglected. This means in small angle scattering it is assumed that phase shifts between scattering centers are negligible, due to the small size and weak phase interaction. In ultra-small angle scattering, where structures can have sizes in the micrometer range, this assumption does not necessarily hold anymore, and interestingly refraction effects are measured in the very same angular range [47]. While in small angle scattering, described by a coherent superposition of waves scattered at different locations within an object, the subsequent angular distribution of the emerging interference, i.e. scattering pattern, is strongly size and shape dependent, the angular pattern of refraction is only shape dependent. Generally it is assumed, that when the approximation of negligible phase shifts does not hold anymore the SAS description breaks down and refraction dominates. Correspondingly it is to be expected, that a different quantification in terms of phase objects is required in this case. As this is not a problem only to dark-field imaging but also scattering experiments in this regime a number of works and investigations exist. Earlier works include N. F. Berk and K. A. Hardman-Rhyn [48], who come to the conclusion "that strong multiple scattering, as in very thick samples, tends to render beam broadening insensitive to the cross over from diffractive to refractive single-particle scattering." But "On the other hand, (...), say for scattering from large particles dilutely dispersed in thin samples, departures from the diffractive limit should be important." More recent considerations in particular for neutron refraction measured in USAS can be found e.g. in Ref [49,50].

Based on this discussion some caution is required characterizing dark-field contrast based on a "material constant" ε referred to as "linear diffusion coefficient" [51,52]. However, it is clear that the visibility due to convolution of the resulting angular profile with the modulation function [37] can be described by an exponential contrast behavior based on a path integral accounting for subsequent (multiple) interactions approximated by Gaussian angular distributions [36,53]. In analogy to attenuation and Beer Lamberts law this can hence be written as

$$V(x) = V_0(x)e^{\int \varepsilon(x,y,\lambda)dy} \tag{7}$$

and consequently enables tomographic reconstruction analogue to attenuation contrast tomography [35,36,37,51].

For small angle scattering the coefficient ε could be specified as [54]

$$\varepsilon(r) = \Sigma_{SAS}(G(r)-1) \tag{8}$$

where Σ_{SAS} is the total small angle scattering cross section, dependent on characteristic structure sizes, volume fractions and scattering contrast (for neutrons scattering length density contrast, for X-rays electron density contrast) and $G(r)$ is the projected real space correlation function of the scattering structure [54-56]. The microscopic real space parameter r probed in a measurement is equivalent to the autocorrelation length parameter

$$\xi = \lambda L_S/p \tag{9}$$

of the applied modulated beam measurement [54], where L_S is the sample to modulation detection distance.

It is worth noting, that the measured effects are described by simple superpositions of intensities and hence by geometric optics. While the interference pattern created by the respective grating and the small angle scattering have to be based on coherent wave optics, the grating and the scattering sample are not in any coherent correlation, and hence the results from both interference effects can be considered separately and combined through geometrical optics i.e. through intensity superposition [37,54]. Hence, coherence and decoherence considerations appear apart from the separated local coherent processes obsolete in this context.

However, there is a contribution to dark-field contrast from incoherent scattering indeed. Like in SAS incoherent scattering has to be assumed to generate a background signal, for neutrons in particular when hydrogen is involved, e.g. for particles in aqueous solution. The fraction of incoherently scattered neutrons will not contribute to the interference modulation on the detector, but such way forward scattered neutrons will contribute to the transmission a(x), while the rest will contribute to the attenuation $(1-a(x))/a_0(x)$. Corresponding corrections required for the quantification of the coherent SAS signal have been taken into account for various quantitative neutron works [40,57].

For refraction contributions to the dark-field contrast no general model is available and so far it is treated phenomenologically and qualitatively, i.e. as indicator for irregularities beyond spatial resolution that are not further analyzed. However, in analogous (U)SAS methods without

Neutron Radiography - WCNR-11

Materials Research Proceedings **15** (2020) 117-128

Materials Research Forum LLC

https://doi.org/10.21741/9781644900574-19

spatial resolution some considerations for regular structures exist [49,50], which can directly be transferred. For example for aligned cylindrical shapes with axis parallel to the modulation

$$V(r)/V_0(r) = 2\delta r\, K_1(\lambda r)\, /\pi\lambda \tag{10}$$

has been found [50], where K_1 is the first order modified Bessel function of second kind. Nevertheless, in most cases so far, where only the identification, but not the quantification and characterization, of defects beyond direct spatial resolution is in the center of attention, this will not play a major role. Whether equ.(7) applies in such cases, and how reliable tomographic reconstructions from such kind of artifacts is, requires to be clarified yet as well. However, a significant difference between refraction and SAS has to be considered carefully in all assumptions of detected effects and contributions. In particular it has to be regarded, that while SAS only concerns a small fraction of incident radiation and hence requires a certain volume (fraction) to be measureable in particular in transmission, phase shift and hence refraction at inclined interfaces affect all passing radiation. Hence, differential phase effects enable detection of single individual structures (of sizes competitive with, i.e. of significant fraction of the resolution limit), while SAS requires a large number and density of similar structures, hence significantly beyond the resolution limit to provide an effective signal. A straightforward distinction from the measurement does not appear obvious and hence attempts for quantitative structure determination through dark field imaging still require a priori knowledge and/or additional means of characterization. An orientation dependence is possible for both signals, depending on the geometry of the structures, but is not necessarily given, like e.g. in the cases of dark-field signal from refraction at macroscopic edges. A careful consideration of expected structure sizes and corresponding phase shifts induced appears absolutely necessary.

For neutrons in particular the refractive nature of dark-field contrast has another dimension and plays a central role for the maybe most exploited dark-field imaging application, namely of magnetic structures [17,41,58-60]. The magnetic field contributes to the real part of the refractive index for neutrons due to their magnetic moment μ_n coupled to the neutron spin. The neutron spin aligns parallel and antiparallel to an external magnetic field and the neutron correspondingly gains or looses momentum in the field. At interfaces with a strong magnetic field gradient, where the spin cannot adiabatically follow the magnetic field thus refraction takes place according to the refractive index contribution

$$\delta_B = \pm 2\mu_n mB\lambda^2/h^2$$

where m is the mass of the neutron and h the Planck constant. Due to the fact that for an unpolarised beam this implies symmetric refraction from the same local structure to opposite directions for the two spin eigenstates, this will in such case always lead to dark-field contrast through the superposition of opposite differential phase contrast signals from the same location. This has already been measured in interferometers and analyser based set-ups [17,58] visualizing domain walls. With a polarized beam however, the undisturbed local differential phase contrast can be measured and has been demonstrated to be usable for the reconstruction of strong magnetic fields through the magnetic refractive index in full analogy to differential phase contrast tomography [61].

References

[1] Henry Fox Talbot: LXXVI. Facts relating to optical science. No IV. In: The London Edinburgh Philosophical Magazine and Journal of Science. Band 9, Nr 56, 1836, Par.2 Experiments on Diffraction, pp 401. https://doi.org/10.1080/14786443608649032

[2] Pierre Bouguer: Essai d'optique, Sur la gradation de la lumiere. Claude Jombert, Paris 1792, pp 164

[3] P. M. Joseph, "The effects of scatter in x-ray computed tomography," Med. Phys. 9 (1982) 464–472. https://doi.org/10.1118/1.595111

[4] G. H. Glover, "Compton scatter effects in CT reconstructions," Med. Phys. 9 (1982) 860–867. https://doi.org/10.1118/1.595197

[5] L. A. Love and R. A. Kruger, "Scatter estimation for a digital radio- graphic system using convolution filtering," Med. Phys. 14 (1987) 178 – 185. https://doi.org/10.1118/1.596126

[6] J. A. Seibert and J. M. Boone, "X-ray scatter removal by deconvolution," Med. Phys. 15 (1988) 567–575. https://doi.org/10.1118/1.596208

[7] D. G. Kruger, F. Zink, W. W. Peppler, D. L. Ergun, and C. A. Mistretta, "A regional convolution kernel algorithm for scatter correction in dual- energy images: Comparison to single-kernel algorithms," Med. Phys. 21 (1994) 175–184. https://doi.org/10.1118/1.597297

[8] Kardjilov, N., de Beer, F., Hassanein, R., Lehmann, E. & Vontobel, P. Nucl. Instrum. Methods Phys. Res. A, 542 (2005) 336–341. https://doi.org/10.1016/j.nima.2005.01.159

[9] Pekula, N., Heller, K., Chuang, P. A., Turhan, A., Mench, M. M., Brenizer, J. S. &Uenlue , K., Nucl. Instrum. Methods Phys. Res. A, 542 (2005) 134–141. https://doi.org/10.1016/j.nima.2005.01.090

[10] Hassanein, R., Lehmann, E. & Vontobel, P., Nucl. Instrum. Methods Phys. Res. A, 542 (2005) 353–360. https://doi.org/10.1016/j.nima.2005.01.161

[11] Tremsin, A. S., Kardjilov, N., Dawson, M., Strobl, M., Manke, I., McPhate, J. B., Vallerga, J. V., Siegmund, O. H. W. & Feller, W. B., Nucl. Instrum. Methods Phys. Res. A, 651 (2011) 145–148. https://doi.org/10.1016/j.nima.2011.01.066

[12] M. Raventos, E. H. Lehmann, M. Boin, M. Morgano, J. Hovind, R. Harti, J. Valsecchi, A. Kaestner, C. Carminati, P. Boillat, P. Trtik, F. Schmid, M. Siegwart, D. Mannes, M. Strobl and C. Gruenzweig, A Monte Carlo approach for scattering correction towards quantitative neutron imaging of polycrystals, J. Appl. Cryst. 51 (2018). https://doi.org/10.1107/S1600576718001607

[13] A. Cereser, M. Strobl, S. A. Hall, A. Steuwer, R. Kiyanagi, A. S. Tremsin, E. B. Knudsen, T. Shinohara, P. K. Willendrup, A. Bastos da Silva Fanta, S. Iyengar, P. M. Larsen, T. Hanashima, T. Moyoshi, P. M. Kadletz, P. Krooß, T. Niendorf, M. Sales, W. W. Schmahl, and S. Schmidt Time-of-Flight Three Dimensional Neutron Diffraction in Transmission Mode for Mapping Crystal Grain Structures, Scientific Reports 7 (2017) 9561. https://doi.org/10.1038/s41598-017-09717-w

[14] Robin Woracek, Javier Santisteban, Anna Fedrigo, Markus Strobl, Diffraction in neutron imaging—A review, Nucl. Inst. Meth. A 878 (2018) 141–158. https://doi.org/10.1016/j.nima.2017.07.040

Materials Research Forum LLC
https://doi.org/10.21741/9781644900574-19

[15] Ludwig Zehnder: Ein neuer Interferenzrefraktor. In: Zeitschrift fuer Instrumentenkunde. Nr. 11, 1891, pp 275; Ludwig Mach: Ueber einen Interferenzrefraktor. In: Zeitschrift fuer Instrumentenkunde. Nr. 12, 1892, pp 89

[16] M. Ando and S. Hosoya, in Proc. 6th Intern. Conf. on X-ray Optics and Microanalyses, eds. G. Shinoda, K. Kohra and T. Ichinokawa (Univ. of Tokyo Press, Tokyo, 1072) 63

[17] M. Schlenker, W. Bauspiess, W. Graeff, U. Bonse, H. Rauch, J. Magn. Magn. Mat.15-18 (1980) 1507. https://doi.org/10.1016/0304-8853(80)90387-X

[18] K. Goetz, M.P. Kalashnikov, Yu. A. Mikhailov, G. V., Sklizkov, S. I. Fedotov, E. Foerster,P., Zaumseil: Preprint Nr. 159, FIAN UdSSR (1978)

[19] K.M. Podurets et al. Zh. Tekh. Fiz.59 (1989) 115-121

[20] Snigirev, A., Snigireva, I., Kohn, V., Kuznetsov, S. & Schelokov, I. On the possibilities of x-ray phase contrast microimaging by coherent high-energy synchrotron radiation. Rev. Sci. Instrum. 66 (1995) 5486–5492 . https://doi.org/10.1063/1.1146073

[21] B. E. Allman, P. J. McMahon, K. A. Nugent, D. Paganin, D. L. Jacobson, M. Arif & S. A. Werner, Imaging: Phase radiography with neutrons, Nature 408 (2000) 158–159. https://doi.org/10.1038/35041626

[22] Ingal, V. N. & Beliaevskaya, E. A. X-ray plane-wave topography observation of the phase contrast from a non-crystalline object. J. Phys. D 28 (1995) 2314–2317. https://doi.org/10.1088/0022-3727/28/11/012

[23] Davis, T. J., Gao, D., Gureyev, T. E., Stevenson, A. W. & Wilkins, S. W. Phase-contrast imaging of weakly absorbing materials using hard X-rays. Nature 373 (1995) 595–598. https://doi.org/10.1038/373595a0

[24] Chapman, L. D. et al. Diffraction enhanced x-ray imaging. Phys. Med. Biol. 42 (1997) 2015–2025. https://doi.org/10.1088/0031-9155/42/11/001

[25] W. Treimer, M. Strobl, A. Hilger, C. Seifert, U. Feye-Treimer, Refraction as imaging signal for computerized (neutron) tomography, Applied Physics Letters, 83, 2 (2003) 398-400. https://doi.org/10.1063/1.1591066

[26] M. Strobl, W. Treimer, A. Hilger, First realisation of a three-dimensional refraction contrast computerised neutron tomography, Nucl. Instr. Meth. B 222, 3-4 (2004) 653-658. https://doi.org/10.1016/j.nimb.2004.02.029

[27] J. F. Clauser, U.S. Patent No. 5,812,629 (1998);

[28] A. Momose, S. Kawamoto, I. Koyama, Y. Hamaishi, K. Takai and Y. Suzuki: Jpn. J. Appl. Phys. 42 (2003) L866. https://doi.org/10.1143/JJAP.42.L866

[29] T. Weitkamp, B. Noehammer, A. Diaz and C. David: Appl. Phys. Lett. 86 (2005) 054101. https://doi.org/10.1063/1.1857066

[30] F. Pfeiffer, T. Weitkamp, O. Bunk, and C. David, Nature Phys. 2, 258 (2006). https://doi.org/10.1038/nphys265

[31] F. Pfeiffer, C. Gruenzweig, O. Bunk, G. Frei, E. Lehmann, and C. David, Phys. Rev. Lett. 96, 215505 (2006). https://doi.org/10.1103/PhysRevLett.96.215505

[32] K.M. Podurets et al Physica B 156 & 157 (1989) 694-697. https://doi.org/10.1016/0921-4526(89)90766-7

[33] Pagot, E. et al. A method to extract quantitative information in analyzer-based X-ray phase contrast imaging. Appl. Phys. Lett. 82, 3421–3423 (2003). https://doi.org/10.1063/1.1575508

[34] L. Rigon, H. -J. Besch, F. Arfelli, R.-H. Menk, G. Heitner, and H. P.Besch, "A new DEI algorithm capable of investigating sub-pixel structures,"J. Phys. D 36, 107–112 (2003). https://doi.org/10.1088/0022-3727/36/10A/322

[35] J.G. Brankov, M.N. Wernick, Y. Yang, J. Li , C. Muehleman, Z. Zhong and M.A. Anastasio A computed tomography implementation of multiple-image radiography Med. Phys. 33 (2006) 278. https://doi.org/10.1118/1.2150788

[36] M. Strobl, W. Treimer, A. Hilger, Small angle scattering signals for (neutron) computerized tomography Appl. Phys. Lett., 85, 3 (2004) 488-490. https://doi.org/10.1063/1.1774253

[37] M. Strobl, C. Grünzweig, A. Hilger, I. Manke, N. Kardjilov, C. David, F. Pfeiffer, Neutron dark-field tomography, Phys. Rev. Lett. 101, 123902 (2008). https://doi.org/10.1103/PhysRevLett.101.123902

[38] F. Pfeiffer, M. Bech, O. Bunk, P. Kraft, E. F. Eikenberry, C. Broennimann, C. Gruenzweig and C. David, Hard-X-ray dark-field imaging using a grating interferometer, Nat. Mat. 7 (2008) 134. https://doi.org/10.1038/nmat2096

[39] W. Yashiro et al., Opt. Exp. 9233, 23 (2015) 7. https://doi.org/10.1364/OE.23.009233

[40] B. Betz, R. P. Harti, M. Strobl, J. Hovind, A. Kaestner, E. Lehmann, H. Van Swygenhoven and C. Grünzweig, Quantification of the sensitivity range in neutron dark-field imaging, Rev. Sci. Instrum. 86, 123704 (2015). https://doi.org/10.1063/1.4937616

[41] 5. C. Grünzweig, C. David, O. Bunk, M. Dierolf, G. Frei, G. Kühne, J. Kohlbrecher, R. Schäfer, P. Lejcek, H. Rønnow, and F. Pfeiffer, Neutron decoherence imaging for visualizing bulk magnetic domain structures, Phys. Rev. Lett. 101, 025504 (2008). https://doi.org/10.1103/PhysRevLett.101.025504

[42] T. Lauridsen, M. Willner, M. Bech, F. Pfeiffer, R. Feidenhans'l Detection of sub-pixel fractures in X-ray dark-field tomography Applied Physics A 121 (2015). https://doi.org/10.1007/s00339-015-9496-2

[43] http://www.dictionary.com/browse/decoherence

[44] A. Hilger, N. Kardjilov, T. Kandemir, I. Manke, and J. Banhart, D. Penumadu, A. Manescu, M. Strobl, Revealing micro-structural inhomogeneities with dark-field neutron imaging, J. Appl. Phys. 107, 036101 (2010). https://doi.org/10.1063/1.3298440

[45] Feigin, L. & Svergun, D. Structure Analysis by Small-Angle X-ray and Neutron Scattering (New York Plenum Press, 1987).

[46] Born, Max Quantenmechanik der Stossvorgänge, Zeitschrift für Physik. 38: 803 (1926). https://doi.org/10.1007/BF01397184

[47] W. Treimer, M. Strobl, A. Hilger Observation of edge refraction in ultra small angle neutron scattering Phys. Lett. A 305, 1-2 (2002) 87-92. https://doi.org/10.1016/S0375-9601(02)01391-9

[48] N.F. Berk and K. A. Hardman-Rhyn, Analysis of SAS Data Dominated by Multiple Scattering, J. Appl. Cryst. 21 (1988) 645-651. https://doi.org/10.1107/S0021889888004054

[49] Victor-O. de Haan et al. J. Appl. Cryst. (2007). 40, 151–157. https://doi.org/10.1107/S0021889806047558

[50] J. Plomp et al. / Nuclear Instruments and Methods in Physics Research A 574 (2007) 324–329. https://doi.org/10.1016/j.nima.2007.02.068

[51] M. Bech, O. Bunk, T. Donath, R. Feidenhans'l, C. David and F. Pfeiffer, Quantitative x-ray dark-field computed tomography, Phys. Med. & Bio. 55, 18 (2010). https://doi.org/10.1088/0031-9155/55/18/017

[52] 3. C. Grünzweig, J. Kopecek, B. Betz, A. Kaestner, K. Jefimovs, J. Kohlbrecher, U. Gasser, O. Bunk, C. David, E. Lehmann, T. Donath, and F. Pfeiffer, Quantification of the neutron dark-field imaging signal in grating interferometry, Phys. Rev. B 88, 125104, (2013). https://doi.org/10.1103/PhysRevB.88.125104

[53] PhD thesis, M. Strobl, TU Wien 2003

[54] M. Strobl General solution for quantitative dark-field contrast imaging with grating interferometers. Scientific Reports 4 (2014) 7243. https://doi.org/10.1038/srep07243

[55] Andersson, R., van Heijkamp, L. F., de Schepper, I. M. & Bouwman, W. G. Analysis of spin-echo small-angle neutron scattering. J. Appl. Cryst. 41, 868–885 (2008). https://doi.org/10.1107/S0021889808026770

[56] J. Kohlbrecher and A. Studer, Transformation cycle between the spherically symmetric correlation function, projected correlation function and differential cross section as implemented in SASfit, J. Appl. Cryst. (2017). 50, 1395-1403. https://doi.org/10.1107/S1600576717011979

[57] M. Strobl, B. Betz, R. P. Harti, A. Hilger, N. Kardjilov, I. Manke, C. Gruenzweig Wavelength dispersive dark-field contrast: micrometer structure resolution in neutron imaging with gratings J. Appl. Cryst. 49 (2016). https://doi.org/10.1107/S1600576716002922

[58] K.M. Podurets, V.A. Somekov, R.R. Chistyakov and S.Sh.Shilstein, Visualization of internal domain structure of Silicon Iron crystals by using neutron radiography with refraction contrast, Physica B 156 & 157 (1989) 694-697. https://doi.org/10.1016/0921-4526(89)90766-7

[59] I. Manke, N. Kardjilov, R. Schäfer, A. Hilger, M. Strobl, M. Dawson, C. Grünzweig, G. Behr, M. Hentschel, C. David, A. Kupsch, A. Lange, J. Banhart, Three-dimensionl imaging of magnetic domains, Nature Commun. 1, 125 (2010) . https://doi.org/10.1038/ncomms1125

[60] B. Betz, P. Rauscher, R. P. Harti, R. Schäfer, H. Van Swygenhoven, A. Kaestner, J. Hovind, E. Lehmann, and C. Grünzweig, Frequency-Induced Bulk Magnetic Domain-Wall Freezing Visualized by Neutron Dark-Field Imaging Phys. Rev. Applied 6, 024024 (2016). https://doi.org/10.1103/PhysRevApplied.6.024024

[61] M. Strobl et al. Appl. Phys. Lett. 91, 254104 (2007). https://doi.org/10.1063/1.2825276

[62] S. W. Lee, D. S. Hussey, D. L. Jacobson, C. M. Sim, M. Arif, Development of the grating phase neutron interferometer at a monochromatic beam line, Nuclear Instruments and Methods in Physics Research A 605 (2009) 16–20. https://doi.org/10.1016/j.nima.2009.01.225

Neutron Radiography - WCNR-11   Materials Research Forum LLC
Materials Research Proceedings **15** (2020) 129-135   https://doi.org/10.21741/9781644900574-20

Origin of Pseudo-Variation in High Resolution Neutron Grating Interferometry

Tobias Neuwirth[1,2, a *], Michael Schulz[1,b] and Peter Böni [2,c]

[1]Technical University of Munich, Heinz Maier-Leibnitz Zentrum (MLZ), Lichtenbergstr. 1, 85748 Garching, Germany

[2]Technical University of Munich, Department of Physics, Chair for Neutron Scattering (E21), James-Franck-Str. 1, 85748 Garching, Germany

[a]Tobias.Neuwirth@frm2.tum.de, [b]Michael.Schulz@frm2.tum.de, [c]Peter.Boeni@frm2.tum.de

Keywords: Neutron Imaging; Neutron Grating Interferometry; Quantitative nGI

Abstract. During neutron grating interferometry measurements with a highly collimated neutron beam a pseudo-variation becomes visible in the acquired data. This pseudo-variation prevents the quantitative analysis of the acquired data, as the measured dark field data depends now on both the ultra-small-angle scattering as well as the properties leading to the pseudo-variation. In the following, the origin of this variation and dependence on the collimation of the neutron beam is explained. It will be shown how by changing the stepped grating in an interferometry scan this variation can be eliminated.

Introduction

Neutron grating interferometry (nGI) is a technique used in neutron imaging allowing to indirectly resolve magnetic and material structures in the range of 100 nm to 20 μm by analyzing the ultra-small-angle scattering off such structures [1, 2, 3]. While resolving these structures is independent of the spatial resolution of the imaging instrument, the ability to locate these structures is still dependent on the spatial resolution. The spatial resolution of the instrument denotes the minimum resolvable structure size and is a combination of the geometric resolution of the instrument (L/D-ratio), the scintillator thickness and the effective pixel size of the detector [4]. In this paper we will investigate the influence of the pinhole size on the signal of the dark field image (DFI) measured with nGI. For this purpose we look at the interaction of a highly collimated neutron beam with the movement of the source grating during an nGI-scan. We will show that in this case two intensity oscillations, one generated by the nGI and one generated solely by the source grating, overlap with each other and generate a periodic pseudo-variation in the nGI-signal. Therefore, a quantitative analysis of the sample parameters becomes difficult. This effect depends strongly on the detailed geometry of the nGI setup. We will concentrate on the case of a highly asymmetric Talbot-Lau interferometer, which is the most common implementation for neutron imaging [1, 5, 6]. Lastly we will show how the problem of pseudo-variation can be resolved.

Principles of neutron grating interferometry

In the following we discuss a strongly asymmetric implementation of a Talbot-Lau-interferometer [1, 7] at a neutron imaging beamline. We denote this type of interferometer as a neutron grating interferometer (nGI). It allows to recover information about the transmission (TI), the differential phase shift (DPCI) and the scattering under ultra-small-angles (DFI) of a sample in the neutron beam.

An nGI consists of two absorption gratings (G_0) and (G_2) and a phase grating (G_1) [6]. The phase grating generates a complex interference pattern called the Talbot-carpet. At odd fractional

Talbot-distances this interference pattern shows a maximum in the contrast of the intensity modulation [8]. Due to the periodicity of the interference pattern in the micrometer range, the interference pattern cannot be directly resolved by a standard neutron imaging detector [9] Hence, the analyzer grating G_2 is required which has the same period as the interference pattern. It is positioned at an odd fractional Talbot-distance. The source grating G_0 is needed to provide the required spatial coherence, without decreasing the neutron flux too much [10].

By a translation x_{gi} of one of the gratings perpendicular to the grating lines the intensity at the detector oscillates, as either the minima or maxima of the interference pattern are blocked by grating G_2. This intensity oscillation can be approximated by:

$$I(x_{gi}, m, n) = a_0(m, n) + a_1(m, n) \cos\left(\frac{2\pi x_{gi}}{p_i} - \varphi(m, n)\right). \tag{1}$$

Here p_i is the period of the moving grating, $a_0(m,n)$ the mean value of the oscillation in pixel (m,n) $a_1(m,n)$ the amplitude and $\varphi(m,n)$ the phase. In many cases grating G_0 is stepped as it typically has the largest period of the three gratings (typically $\approx$ 1mm) in the most commonly used highly asymmetric nGI-setups. Furthermore, G_0 is typically located relatively close to the pinhole of the imaging instrument in order to fit the nGI setup into the beam line.

To calculate the TI, DPCI and DFI, the values of a_0, a_1 and φ have to be extracted for every pixel from both a scan without a sample (open beam) and a scan with a sample [11]. Afterwards by normalization of these values to their respective open beam values the TI, DPCI and DFI can be extracted [1].

Origin of pseudo-variation

In a neutron grating interferometer, the intensity of every pixel changes when performing an nGI-scan. In an ideal interferometer this change depends only on the position of the translated grating with respect to its periodicity. Hence, the intensity on the detector does not vary spatially for a particular position. In a real interferometer, however, small errors in the positioning of the gratings and grating manufacturing imperfections lead to a variation in the intensity on the detector. The nGI data used in this paper has been acquired with the nGI of the ANTARES imaging beamline [1, 9, 12] at the FRM-II. A detailed description of the used nGI-setup may be found in [1]. In Table 1 the experimental parameters for the different scans are given.

Table 1: Parameters of the scans performed for analysis of the pseudo-variation. P_D denotes the pinhole diameter, while the different L_i's denote the distance of the respective components to the pinhole. Above it is also shown which scan parameters have been used in the figures presented in this paper.

L/D-ratio	P_D [mm]	L_{G0} [m]	L_{G1} [m]	L_{G2} [m]	L_{Det} [m]	FoV_{Det} [mm]	Figure
305|1219	35.68|8.92	1.850	10.830	10.853	10.862	diam. 100	1, 2, 3
257|1031	35.68|8.92	1.650	8.850	8.868	8.878	71 x 76	4

In Fig. 1(a, b) raw open beam images at four different steps of two different stepping scans of G_0 are shown. The scan in (a) was performed at L/D = 305 while for the scan in (b) a higher collimation of L/D = 1219 was used. In (a) the expected intensity oscillation which is homogeneous over the entire image can be seen. This scan has been performed with a low collimation of L/D=305. In (b) the four raw open beam images show an additional spatial oscillation, which also moves during the stepping scan. In contrast to the scan in (a), a high collimation of L/D=1219 was chosen for this scan, while all other geometric parameters were kept constant. This spatial oscillation is aligned with the orientation of the source grating. In the following, the intensity variation seen in both scans is called the Talbot-oscillation, while we

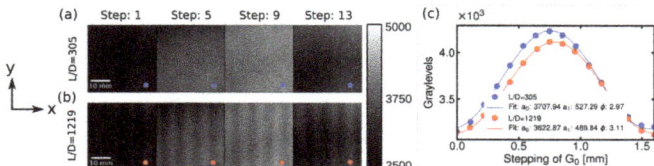

Figure 1: In (a) four steps of a L/D=305 and in (b) a L/D=1219 scan are shown. The influence of G_0 is visible in the lower row as a vertical spatial oscillation in the image. In c) the respective curves for the L/D=1219 scan (red marker) and the L/D=305 scan (blue marker) have been plotted. Decrease of amplitude and mean value of the L/D=1219 curve compared to L/D 305 curve. This is an effect of the influence of the two oscillations on each other.

denote the rapid oscillation only visible in the high collimation scan as G_0-oscillation.

Fig. 1(c) shows the oscillation and the fit parameters of the data of Fig. 1(a), (b) to Eq. 1 for the high collimation (red marker) and low collimation scan (blue marker). The curve of the high collimation scan shows a slightly lower amplitude a_1 and mean value a_0 than the low collimation scan. This difference is caused by the two overlapping oscillations in the nGI-scan. We have verified that the secondary oscillation is still visible after removing G_1 and G_2 from the beamline, proving that G_0 must be the cause for the oscillation.

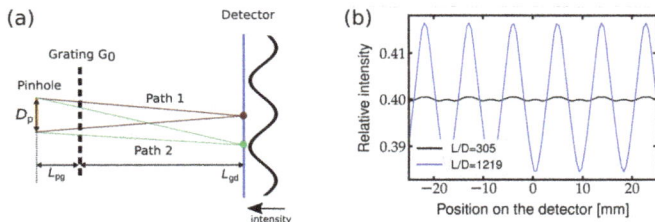

Figure 2: (a) Sketch of the intensity change caused by the introduction of G_0 into an highly collimated neutron beam. (b) Calculated horizontal intensity curve caused by G_0 for the L/D=305 (black curve) and the L/D=1219 (blue curve) at the detector position for an arbitrary vertical position on the detector.

In Fig. 2 an explanation for the oscillation caused by G_0 is presented. Panel (a) shows a 2D sketch of the different parts of the pinhole area that are covered by the absorbing lines (black), as seen from different positions on the detector. For this sketch it has been assumed that G_0 is a binary grating. Path 1 shows the minimum neutron flux possible, while path 2 shows the maximum neutron flux reaching the detector. The number of grating lines illuminated for a point on the detector depends on the chosen pinhole diameter. Consequently also the strength of the variation depends on the pinhole diameter.

In Fig. 2(b), the oscillations on the detector for L/D=305 and L/D=1219 have been calculated. For this calculation a binary grating G_0 with a duty cycle of 0.4, a homogeneous neutron beam, a

Neutron Radiography - WCNR-11   Materials Research Forum LLC
Materials Research Proceedings 15 (2020) 129-135   https://doi.org/10.21741/9781644900574-20

distance between pinhole and G_0 (L_{pg} = 1.95 m) and a distance between G_0 and detector (L_{gd} = 8.92 m) were assumed.

Due to the assumption of a homogeneous neutron flux, the visible (non-shadowed) area of the pinhole is directly proportional to the intensity on the detector. In agreement with the duty cycle of 0.4, the resulting variation has a mean value of 40 % of the maximum intensity. The variation has a period of approximately 9 mm which is independent of the pinhole diameter. For L/D=1219 the calculated amplitude relative to the mean value is 3.75%, which is slightly larger than the relative amplitude measured in the L/D=1219 scan which was around 3.4 %. The deviation may be due to the assumption of a perfect G_0-grating in the simulation, which is not true for the experiment. Moreover, the effects of the other gratings, being present in the measurement, are neglected. Furthermore, the simulation ignored scattering from the air, which may also contribute to the lower amplitude in the scan. For L/D=305 the simulation yields an amplitude relative to the mean value of 0.1 %. Comparably the noise in the measurement was around 1.5 % of the mean value. Hence, an extraction of the amplitude of the L/D=305 scan cannot be performed.

In the following, we develop an analytic description of the influence of the G_0-oscillation on the acquired nGI-signal. This will allow us to rapidly estimate the influence of the G_0-oscillation for different nGI configurations. Due to the round shape of the pinhole, the expected triangular function, which would appear for a square pinhole, is slightly rounded and can be approximated by a cosine function. Similarly to a position change on the detector, a movement of G_0 will cause an intensity variation for every pixel dependent on the position of the grating. Therefore the oscillation in one pixel caused by G_0 is described similar to Eq. 1:

$$I^{G0}(x_{g0}, m, n) = a_0^{G0}(m, n) + a_1^{G0}(m, n) \cos\left(\frac{2\pi x_{g0}}{p_0} - \varphi^{G0}(m, n)\right). \tag{2}$$

$a_0^{G0}(m,n)$ is the mean value, $a_1^{G0}(m,n)$ the amplitude and $\varphi^{G0}(m,n)$ the phase of the G_0-oscillation. The main difference between Eq. 1 and 2 is that $I^{G0}(x_{g0},m,n)$ has a periodic variation perpendicular to the G_0-grating lines, which is described in the simulation as an oscillation of $\varphi^{G0}(m,n)$. Grating G_0 causes, as shown in Fig. 2 a spatial intensity modulation in the neutron beam which changes the incoming intensity on the other components of the nGI. This incoming intensity is dependent on the stepping position of G_0. Hence, to describe the oscillation seen in a pixel, Eq. 1 is modified accordingly:

$$I^{mod}(x_{g0}, m, n) = I^{G0}(x_{g0}, m, n) + V^{real}(m, n) * I^{G0}(x_{g0}, m, n) \cos\left(\frac{2\pi x_{g0}}{p_0} - \varphi(m, n)\right). \tag{3}$$

Here the previously constant offset value a_0 has been replaced by the incoming intensity $I^{G0}(x_{g0},m,n)$ and a_1 has been replaced by $V^{real}(m,n)* I^{G0}(x_{g0},m,n)$. $V^{real}(m,n)$ is the intrinsic visibility which describes the quality of the gratings used, the used setup parameters and the amount of scattering inside the sample.

Comparison of simulated and measured data

Using Eq. 3 and the parameters extracted from the L/D=1219 scan, a data set has been simulated and compared with the L/D= 1219 scan. The extracted parameters were a_0^{G0} = 2600, a_1^{G0} = 78, V^{real} = 0.19 and $\varphi(m,n)$ = 0. φ^{G0} has a periodic variation over the image as caused by grating G_0. In Fig. 3, the a_0-map, the a_1-map and the visibility-map of the simulation (a-c) and the L/D=1219 measurement (d-f) are shown.

Neutron Radiography - WCNR-11   Materials Research Forum LLC
Materials Research Proceedings **15** (2020) 129-135   https://doi.org/10.21741/9781644900574-20

The a_0-map (a_1-map) is the image representation of the mean value a_0 (amplitude a_1) evaluated for every pixel. Accordingly, the visibility-map is the pixel-wise division of the a_1-map by the a_0-map. Table 2 lists the standard deviations of the simulated and measured maps for the areas marked by the red contour in Fig. 3.

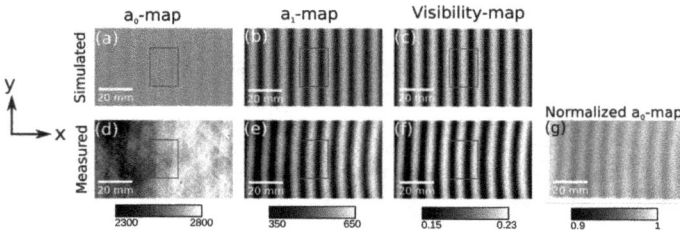

Figure 3: (a-f) Comparison of the a_0-map, the a_1-map and the visibility-map simulated using the parameters extracted from the L/D=1219 scan and the measured data from the L/D=1219 scan. The a_1-map and the visibility-map show good agreement of simulation and measurement. In Table 2, the standard deviations of the areas marked by the red contours are presented. The bending of the stripes in the measurement are caused by non-constant phase of the Talbot-oscillation over the image. For the simulation the phase of the Talbot-oscillation has been chosen to be constant. In the measured a_0-map the periodic variation is strongly suppressed compared to the simulation map, due to inhomogeneities in the neutron beam and the scintillator. In (g) the discernibility of the structure is enhanced by normalizing the L/D=1219 a_0-map to the a_0-map of the L/D=305 scan.

Table 2: Standard deviation inside the red contour marked in Fig. 3.

Standard deviation [arb. unit.]	a_0-map	a_1-map	visibility-map
Simulated	14	52	0.020
Measured	77	64	0.023

While the standard deviations for the a_1-map and the visibility-map are in a similar range, the standard deviation for the a_0-map differs strongly. The standard deviations for the measured maps are constantly higher than the simulated maps. This is caused by neglecting the effects of the scintillator structure and the beam inhomogeneity in the simulation. The a_0-map of the simulation shows the periodic pseudo-structure while in the a_0-map of the measurement this periodic variation is strongly suppressed. This suppression is caused by the inhomogeneity of the measured a_0-map, which is considerably stronger than the structure seen in the simulated map. The inhomogeneity is again an effect of the scintillator structure and the beam inhomogeneity. To effectively visualize the structure in the measured a_0-map, it is normalized to the a_0-map of the L/D=305 measurement. This normalization eliminates the inhomogeneities caused by the scintillator structure and the general beam inhomogeneity. In Fig. 3 (g), the resulting image is shown. In this image, it can be clearly seen that the pseudo-structure is also visible in the mean value of the oscillation. Therefore, the TI is also affected by the pseudo-structure, but the effect is suppressed by the general inhomogeneity of the neutron beam.

Neutron Radiography - WCNR-11
Materials Research Proceedings **15** (2020) 129-135

Materials Research Forum LLC
https://doi.org/10.21741/9781644900574-20

Lastly, the simulation shows a perfectly vertical pseudo-structure while the structure in the measurement is slightly bent. From Eq. 3 it follows that the resulting variation depends on the relative phase shift of the G_0-oscillation and the Talbot-oscillation. Due to a Moiré-stripe, caused by imperfections in the setup, the phase of the Talbot-oscillation is not static but varies over the image, causing the bending of the resulting pseudo-structure. The direction of this bending depends on the specific form of the Moiré-stripe, which in turn depends on the alignment of the gratings.

Elimination of the pseudo-variation

As shown above the main problem of the nGI-technique using a tight collimation is the simultaneous occurrence of two intensity oscillations, which both depend on the position of the source grating G_0. Hence, by stepping either grating G_1 or G_2 instead of G_0 the G_0-oscillation changes to a static intensity variation that can be removed via normalization. We expect a similar problem of two simultaneous oscillations to occur if G_2 is stepped, as it is also an absorption

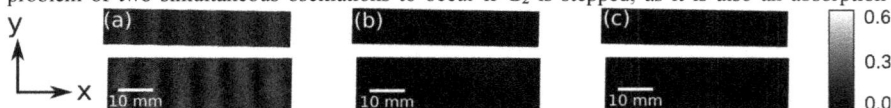

Figure 4: Comparison of the DFI of (a) the G_0-L/D=1031 scan, with (b) the G_1-L/D=1031 scan and (c) the reference G_0-L/D=257 scan.

grating. Hence, G_1 has been chosen to be stepped, as it does not produce an intrinsic intensity modulation, but rather a phase modulation. To verify the validity of this approach three nGI-scans with different collimations and stepping methods were performed. Four electrical steel sheets were used as test samples.

In Fig. 4, the DFI (a) of a high collimation nGI-scan with G_0 stepping (G_0-L/D=1031) is presented and compared to the DFI (b) when stepping G_1 (G_1-L/D=1031). Additionally, in (c) a reference scan performed by stepping G_0 at low collimation is presented (G_0-L/D=257). This scan has been chosen as the reference scan as this is the standard type of nGI-scan performed. For this type of scan the oscillation introduced by G_0 is below the noise level and does not affect the scan. It can immediately be seen that the pseudo variation visible in the G_0-L/D=1031 scan is eliminated in the G_1-L/D=1031 scan. Additionally, the G_0-L/D=1031 shows a drift in the setup between the sample and open beam scan, as the pseudo-variation is even visible in the open beam area. Hence, to allow for quantitative measurements with high spatial resolution the stepping of the G_1 grating is the most promising approach.

Conclusion

We have shown that nGI measurements with a highly collimated neutron beam can result in a periodic pattern appearing in the visibility and phase map, caused by a projection of the G_0 grating. Due to the movement of G_0 during an nGI-scan this periodic pattern is not static, but also varies during the scan. This variation combined with the Talbot-oscillation causes an emergence of a pseudo-variation of the visibility-map. In contrast to the Talbot-oscillation, the G_0-oscillation is unaffected by ultra-small-angle scattering and refraction of neutrons passing through the sample, which leads to a corruption of the evaluated signal. We have shown that the combination of the G_0-oscillation and the Talbot-oscillation experiences a phase shift inside the sample, which in turn leads to the pseudo-structure inside the sample in the DFI.

In order to effectively perform quantitative nGI measurements with high collimation, the stepping scan has been performed by stepping G_1 instead of G_0. As a result the periodic pattern

Neutron Radiography - WCNR-11   Materials Research Forum LLC
Materials Research Proceedings **15** (2020) 129-135   https://doi.org/10.21741/9781644900574-20

generated by G_0 becomes static and can be easily dealt with by the standard normalization procedure used during data evaluation.

References

[1] T. Reimann et al. "The new neutron grating interferometer at the ANTARES beamline: design, principles and applications". In: Journal of Applied Crystallography 49.5 (2016), pp. 1488–1500. https://doi.org/10.1107/S1600576716011080

[2] T. Reimann et al. "Visualizing the morphology of vortex lattice domains in a bulk type-II superconductor". In: Nature communications 6 (2015), p. 8813. https://doi.org/10.1038/ncomms9813

[3] C. Grünzweig et al. "Bulk magnetic domain structures visualized by neutron dark-field imaging". In: Applied Physics Letters 93.11 (2008), p. 112504. https://doi.org/10.1063/1.2975848

[4] I. S. Anderson et al. Neutron Imaging and Applications: A Reference for the Imaging Community 2009. Tech. rep. ISBN 978-0-387-78692-6.

[5] Y. Seki et al. "Development of Multi-colored Neutron Talbot–Lau Interferometer with Absorption Grating Fabricated by Imprinting Method of Metallic Glass". In: Journal of the Physical Society of Japan 86.4 (2017), p. 044001. https://doi.org/10.7566/JPSJ.86.044001

[6] C. Grünzweig et al. "Design, fabrication, and characterization of diffraction gratings for neutron phase contrast imaging". In: Review of Scientific Instruments 79.5 (2008), p. 053703. https://doi.org/10.1063/1.2930866

[7] E. Lau. "Interference phenomenon on double gratings". In: Ann. Phys. 6 (1948), p. 417. https://doi.org/10.1002/andp.19484370709

[8] A. Hipp et al. "Energy-resolved visibility analysis of grating interferometers operated at polychromatic X-ray sources". In: Opt. Express 22.25 (Dec. 2014), pp. 30394–30409. https://doi.org/10.1364/OE.22.030394

[9] M. Schulz et al. "ANTARES: Cold neutron radiography and tomography facility". In: Journal of large-scale research facilities JLSRF 1 (2015), p. 17. https://doi.org/10.17815/jlsrf-1-42

[10] F. Pfeiffer et al. "Neutron phase imaging and tomography". In: Physical Review Letters 96.21 (2006), p. 215505. https://doi.org/10.1103/PhysRevLett.96.215505

[11] S. Marathe et al. "Improved algorithm for processing grating-based phase contrast interferometry image sets". In: Review of Scientific Instruments 85.1 (2014), p. 013704. https://doi.org/10.1063/1.4861199

[12] E. Calzada et al. "New design for the ANTARES-II facility for neutron imaging at FRM II". In: Nuclear Instruments and Methods in Physics Research Section A: Accelerators, Spectrometers, Detectors and Associated Equipment 605.1 (2009), pp. 50–53. https://doi.org/10.1016/j.nima.2009.01.192

Neutron Radiography - WCNR-11   Materials Research Forum LLC
Materials Research Proceedings 15 (2020) 136-141   https://doi.org/10.21741/9781644900574-21

Conversion from Film Based Transfer Method Neutron Radiography to Computed Radiography for Post Irradiation Examination of Nuclear Fuels

Glen C Papaioannou[1, a *], Dr. Aaron E Craft[2, b], Michael A Ruddell[3, c]

[1]Idaho National Laboratory, P.O. Box 1625, Idaho Falls, ID 83415, USA

[2]Idaho National Laboratory, P.O. Box 1625, Idaho Falls, ID 83415, USA

[3]Idaho National Laboratory, P.O. Box 1625, Idaho Falls, ID 83415, USA

[a]glen.papaioannou@inl.gov, [b]aaron.craft@inl.gov, [c]michael.ruddell@inl.gov

Keywords: Transfer Method Neutron Radiography, Computed Radiography, Conversion Foils, Dysprosium, Indium, Image Plates, Image Resolution, Post Irradiation Examination, Nuclear Fuels, NRAD, Idaho National Laboratory

Abstract. The Idaho National Laboratory supports multiple programs that are actively developing, testing, and evaluating new nuclear fuels including advanced commercial nuclear fuels, accident tolerant fuels, reduced-enrichment research reactor fuels, transmutation fuels, and advanced reactor fuels. Post irradiation examinations (PIE) of nondestructive and destructive techniques are performed at INL to research nuclear fuel behavior and performance under various conditions. The Neutron Radiography Reactor (NRAD) at INL is utilized to perform nondestructive examinations of irradiated nuclear fuels and the results of these examinations are used to aid in the decision making process for subsequent destructive exams. NRAD has historically performed transfer method neutron radiography with film as the imaging medium. This paper documents the effort to convert from film to computed radiography for PIE. Equipment selection, system characterization, and testing were performed as part of phase I. Phase II, as described in this paper, focuses on determining image resolution of the computed radiography system.

Introduction

The current neutron radiography method used for evaluating irradiated fuel at NRAD is the foil-film transfer technique [1]. Conversion foils (dysprosium and indium) absorb neutrons from a neutron beam that pass through the fuel and become temporarily radioactive in inverse proportion to their absorption in the fuel. The foils are then removed from the beam and film is placed in contact with the activated foil and exposed to the decay radiation from the foil overnight. After this exposure, the film is chemically-processed and scanned to produce the final radiographic image. This transfer technique is time consuming, but is one of very few techniques that can provide high quality radiographic images of irradiated fuel. INL is pursuing multiple efforts to advance its neutron imaging capabilities for evaluating irradiated fuel and other applications, including converting from film to photo-stimulated phosphor (PSP) image plates (IP) for neutron computed radiography (nCR). Neutron nCR is the current state-of-the-art for neutron imaging of highly-radioactive objects [2].

The nCR process uses the transfer method but an IP is coupled to the activated foil instead of film (Figure 1).

Neutron Radiography - WCNR-11   Materials Research Forum LLC
Materials Research Proceedings **15** (2020) 136-141   https://doi.org/10.21741/9781644900574-21

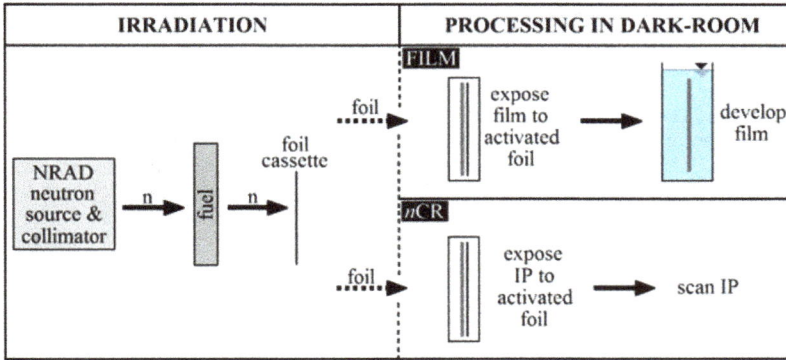

Fig. 1. Transfer Method Neutron Radiography Process

Computed Radiography Equipment and Characterization

A complete nCR system includes PSPs, IPs, an nCR scanner, a computer control workstation, and a high resolution computer monitor. nCR is based on PSP powder deposited on a substrate to form an IP. The substrate provides structural support for the IP. Information on the composition and structure of the phosphor layer is closely guarded by manufacturers and often proprietary, and thus largely unavailable. The protective layer protects the PSP from damage, and is an abrasion-resistant material that is transparent to the light-spectra of interest. The IPs chosen for this application are the flexible Carestream (formerly Kodak) plates, which were chosen based on their image quality and relatively robust protective layer. Commercial manufacturers typically provide three grades of IP: 1) standard or general purpose (GP), 2) high-resolution (HR), and 3) ultra-sharp or "blue" plate. The GP plates have a thicker PSP layer than the HR and blue plates, and thus they require less exposure dose at the expense of lower spatial resolution. HR and blue plates have a thinner PSP layer than GP plates, and thus require more exposure dose but provide higher resolution. The PSP layer in blue IPs contain a blue dye, which absorbs the red scanner laser, reducing the size of the laser interaction volume from which the blue light is emitted, producing higher sharpness in the resulting image. Different commercial manufacturers of IPs may use different substrate materials, have differing PSP chemistries and thicknesses, and use different concentrations of blue dye in the blue IPs, all generally provide the same three grades of IPs. The GP plates provide adequate spatial resolution for general purpose applications, in the range of 350 – 400 microns. If higher spatial resolution is required than the GP plates can provide, the HR plates may be desirable, ~250 microns. While the HR plates offer improved spatial resolution, they require almost twice the exposure dose compared to GP plates. Blue plates offer higher resolution capabilities than even the HR IPs, ~175 microns, but require around 60% more exposure dose compared to the HR plates to produce the same output signal. When comparing these plates to decide which type of IP is most appropriate to use, the desired image quality and exposure times, including both the neutron beam exposure and decay times should be the determining factors.

There are a variety of CR scanners available from commercial manufacturers, and their specific features may vary widely. Two styles of scanners include flat-bed scanners where the IP is flat while scanned and drum-type scanners where the IP is curved, but the fundamental mechanism of acquiring the image signal from an exposed IP is the same regardless of scanner style. A mechanical transport mechanism moved an exposed IP through the scanner. A red laser rasters over the surface of the IP and stimulates the activated PSP, which then releases the stored energy as blue light through photo-stimulated luminescence. The blue light signal is amplified by a photo-multiplier tube into a measureable signal, then converted to a digital signal that is read by a computer. The position of the laser and the output signal are both known, allowing for the formation of an image. Many scanners provide an erasure function that erases the IP using an array of red LEDs positioned after the scanner laser. The laser spot size is often the driving factor for the effective spatial resolution that a scanner can provide. Scanners with 50 µm diameter laser spot size are commonly available, and high-resolution scanners with laser diameters of around 10 µm are also available. The scan resolution is determined by the mechanical transport speed and the laser rastering rate. The PMT voltage may be manipulated to amplify the signal from an IP to utilize the bit-depth provided by the scanner. Manipulation of these and other scanner parameters may be available to the operator depending on the particular software and manufacturer.

Conversion to nCR consists of characterization and qualification of an nCR system using the current foil transfer technique with the goal of maintaining similar image quality compared to the current technique with film. Preliminary studies measured the image quality of radiographs acquired using the existing film-based method and determined CR system equipment capable of providing the desired image quality. The hardware selected for this application included a ScanX Discover HR scanner made by ALLPRO Imaging and image plates from Carestream.

Neutron radiography takes significantly more resources (time and cost) than x-ray radiography, and acquisition of multiple radiographs is required to test and characterize the response of the CR system to various input parameters. As part of phase I, x-ray characterization of the CR system was performed to provide a better understanding of the CR system and inform subsequent image acquisitions using neutrons. Additionally, x-ray CR (xCR) was performed to provide a baseline for the performance of the scanner that could be used in the future to test whether the scanner is still performing to its capacity or if maintenance is required.

Neutron Computed Radiography of Irradiated Fuel

A set of nCR radiographs were taken of an irradiated fuel pin, and the IPs were scanned and the grayscale intensities of the resulting images evaluated. From these intensities, the change in PMT voltage required to achieve the desired intensity was calculated. The second set of exposed IPs was scanned using these calculated values provided the expected grayscale values. Thus, x-ray characterization provided the understanding needed to significantly reduce the number of neutron radiographs required to determine the scanner settings for nCR.

Figure 2 and 3 shows neutron radiographs of an irradiated fuel pin acquired using film and nCR with a XL-Blue plate and an HR plate. The exposure parameters for these radiographs are listed in Table 1 and were performed at an L/D of 125. The histograms of these images have been adjusted to display a similar grayscale range to allow for a more meaningful comparison between film and nCR. From a qualitative perspective, the image quality of nCR closely matches that of film, and features such as cracks and gaps can be visualized with similar contrast. Film radiographs exhibit higher sharpness than nCR but lower latitude for the thermal-neutron radiographs.

Materials Research Forum LLC
https://doi.org/10.21741/9781644900574-21

Table 1. Image acquisition parameters: L/D 125, 5 half-life decay (overnight)

Conversion Foil	Imaging Medium	Neutron Beam Exposure Time (min.)	PMT Voltage	Scan Resolution
Dysprosium 125 μm thick metal foil (unfiltered neutron beam)	AGFA D3-SC film	22:00	-	47 lp/mm
	*n*CR, XL-Blue plate	20:00	625 V	25 μm
Indium 125 μm thick metal foil (cadmium-filtered beam)	Kodak T200	22:00	-	47 lp/mm
	*n*CR, HR plate	20:00	725 V	25 μm

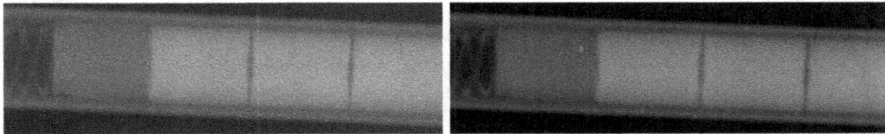

Fig. 2. Thermal neutron radiographs of an irradiated fuel pin acquired using film (left) and nCR with XL-Blue plate (right)

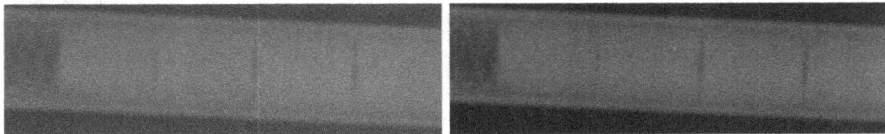

Fig. 3. Epithermal neutron radiographs of an irradiated fuel pin acquired using film (left) and nCR HR plate (right)

Determining Image Resolution
To determine the resolution of the NRAD system utilizing CR, a new Resolution Test Piece, RTP, was designed for use in the NRAD elevator. The new design consists of several standards to provide qualitative and quantitative image resolution. A Siemens Star furnished by the Paul Scherrer Institute for resolution determination in multiple directions, with the outer ring starting at 500 microns and the inner ring measuring 20 microns. See Figure 4. Edge specimens positioned at 2-5 degrees, one of gadolinium and one of hafnium for MTF determination from thermal and epithermal neutron energies, respectfully. Gadolinium doped, dimensionally inspected rulers mounted in the X and Y directions. Lastly, a beam purity indicator and a sensitivity indicator per ASTM E545 neutron radiography image quality indicators.

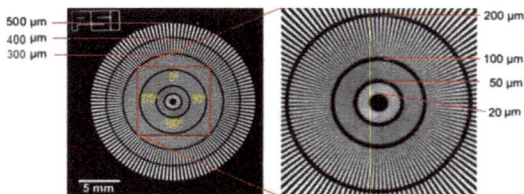

Figure 4 Siemens Star

The RTP was positioned remotely in the beamline to acquire baseline images, measure MTF, visible resolution, and contrast. Several exposures were taken with film and nCR for evaluation, see Figure 5. Images of the sensitivity indicator exhibit a Category-I facility according to ASTM E-545 [3, 4]

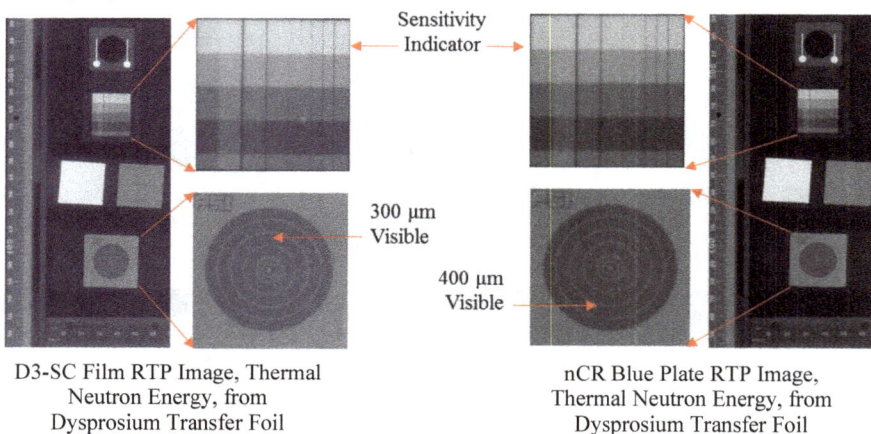

| D3-SC Film RTP Image, Thermal Neutron Energy, from Dysprosium Transfer Foil | nCR Blue Plate RTP Image, Thermal Neutron Energy, from Dysprosium Transfer Foil |

Figure 5 Resolution Test Piece

It is important to note that the transfer method of irradiated nuclear fuels requires shielding and confinement. These factors determine the maximum resolution possible for the given system due to geometric unsharpness. That is the distance from the specimen to the conversion foil. Irradiated specimens are contained inside a hot cell, and lowered down and elevator shaft into the neutron beam. The shaft acts as confinement for the irradiated specimen, and is located inside a shielded subcell room, similar to an X-ray cave. The transfer foil is remotely positioned outside the elevator shaft, as close as possible to the shaft wall. Based on the irradiated specimen size, shape, and fixturing, the distance between the transfer foil and the specimen can range upwards of 10 cm. The further away the specimen is from the foil, the more out of focus it will appear in the image negatively impacting the image resolution.

Summary

The final qualification of the nCR system requires written justification containing system characterization results and baseline image interpretation between film and nCR of image quality indicators and irradiated fuel experiments. Data from phase I characterization efforts and phase II image analysis will be documented in a report stating the range of operating parameters and resolution of the system. Currently, the NRAD nCR system is capable of 400 micron visible resolution. Continuing efforts directed at improving resolution involve; improving geometric unsharpness with new specimen handling fixtures that can be positioned closer to the transfer foil, increasing L/D ratio to 300, and improving foil surface geometry to enhance contact with the IP during decay.

References

[1] Craft, A.E., Chichester, D.L., Williams, W.J., Papaioannou, G.C., and Wachs, D.M., 2015. Conversion from radiographic film to photo-stimulable phosphor plates for neutron computed radiography of irradiated nuclear fuel. ASNT Annual Conference 2015, pp. 23-27.

[2] Vontobel, P., Tamaki, M., Mori, N., Ashida, T., Zanini, L., Lehmann, E. H., & Jaggi, M., 2006. Post-irradiation analysis of SINQ target rods by thermal neutron radiography. Nuclear Materials, 356(1), 162-167. https://doi.org/10.1016/j.jnucmat.2006.05.033

[3] Craft, A.E., Hilton, B.A., Papaioannou, G.C., 2015. Characterization of a neutron beam following reconfiguration of the Neutron Radiography Reactor (NRAD) core and addition of new fuel elements. Nuclear Engineering and Technology, Submitted for publication. https://doi.org/10.1016/j.net.2015.10.006

[4] ASTM International, 2005. Standard method for determining image quality in direct thermal neutron radiographic examination. American Society for Testing and Materials, ASTM E545-05.

Neutron Radiography - WCNR-11
Materials Research Proceedings 15 (2020) 142-148

Materials Research Forum LLC
https://doi.org/10.21741/9781644900574-22

Epithermal Neutron Radiography and Tomography on Large and Strongly Scattering Samples

Burkhard Schillinger[1,a*], Aaron Craft [2,b]

[1] Heinz Maier-Leibnitz Zentrum and Physics E21, Technische Universität München, Lichtenbergstr.1, 85748 Garching, Germany

[2] Advanced Characterization and Post-Irradiation Examination Department Idaho National Laboratory, USA

[a]Burkhard.Schillinger@frm2.tum.de , [b]Aaron.Craft@inl.gov

Keywords: Neutron Imaging, Neutron Computed Tomography, Cadmium Filter, Epithermal Neutrons, Penetration

Abstract. Standard neutron imaging is usually performed with a thermal or cold spectrum, but sometimes, penetration of thermal or cold neutrons is not sufficient for thick samples, so higher energies should be tried e.g. on technical machine parts or on rock samples containing crystal water. A 1-2 mm cadmium filter effectively absorbs all neutrons with energies lower than 0.4 eV, but is relatively transparent for higher energies, including epithermal neutrons. Radiography and computed tomography measurements were recorded at Idaho Nation Laboratory in the USA and Heinz Maier-Leibnitz Institut (MLZ) at the FRM II reactor in Germany. Image quality was improved by better penetration and by filtering out scattered and thermalized neutrons.

Introduction
Standard neutron imaging is usually performed with a thermal or cold spectrum, but large samples may either attenuate these energies too much, or generate too much scattering that blurs the obtained images. This is illustrated exemplarily on an electric motor and pump worth thick steel parts, and with a ferrous rock sample containing crystal water. To alleviate this problem, higher energies should be tried. Instead of fission or fusion neutrons of several MeV, the epithermal energy range above the so-called cadmium edge can be tried.

A 1-2 mm cadmium filter cuts off all neutrons with energies lower than 0.4 eV (Fig.1a) [1], but is relatively transparent for higher energies. For energies just above 0.4 eV, a standard 6LiF+ZnS neutron screen can be used, as opposed to ZnS screens that use an organic binder to generate recoil protons triggered by fast (~1 MeV) neutrons. But since the 6Li cross section is inversely proportional to the neutron velocity and as such follows the 1/v law (Fig.1b) [1], the visible neutron energies may be in the range 0.4 - 100eV, rather lower than higher, as sensitivity decreases with increasing energy. Radiography and computed tomography measurements were recorded at Idaho Nation Laboratory (INL) in the USA and Heinz Maier-Leibnitz Institut (FRM II) in Germany. Image quality was improved by better penetration through the sample and by filtering out down-scattered and thermalized neutrons behind the sample.

Materials Research Forum LLC
https://doi.org/10.21741/9781644900574-22

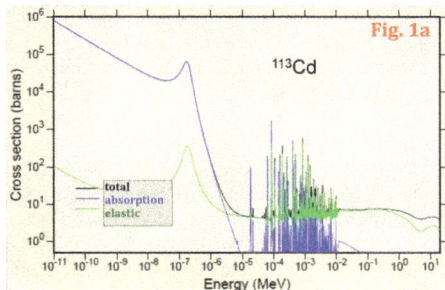

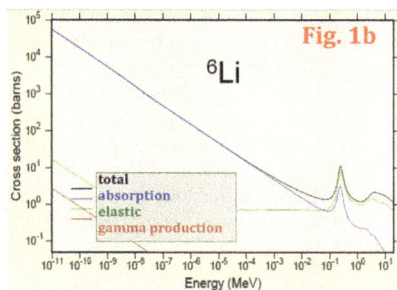

Fig. 1a,b: Neutron cross sections for ^{113}Cd and ^{6}Li from the Los Alamos database [1]

Measurements at the NRAD reactor of Idaho National Lab

The 250 kW NRAD reactor at INL is optimized for radiography on highly radioactive samples (spent fuel) using the transfer method with dysprosium and cadmium-filtered indium foils, which are sensitive in the thermal and epithermal ranges, respectively [2]. For this, the beam tubes are mounted close to the cladding of the core (Fig.2) with only a few millimeters of water between the fuel and beginning of the beam tube, so the epithermal flux is maximized because only minimal neutron moderation happens on the way into the beam tubes. The thermal flux, if unfiltered, is still high enough for standard thermal neutron imaging. The neutron energy spectrum is shown in Fig. 3 [3]. Because of the high radiation level in the measuring chamber, electronic detectors were introduced only very recently [4], [5]. For the measurements, a 1 mm Cd filter was placed before the sample (Fig.4), and for the double-filtered measurements described below, a second filter was placed between sample and detector.

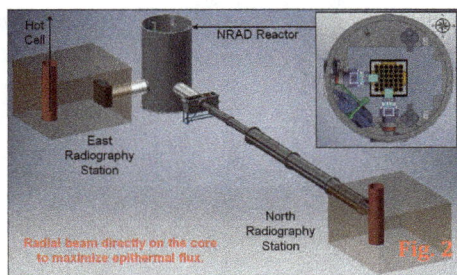

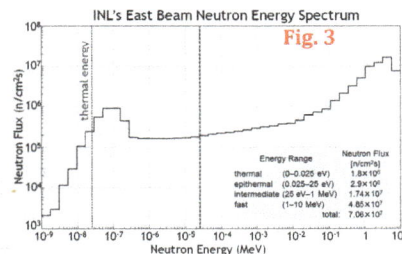

Fig. 2,3: The NRAD radiography facilities and the neutron spectrum at NRAD

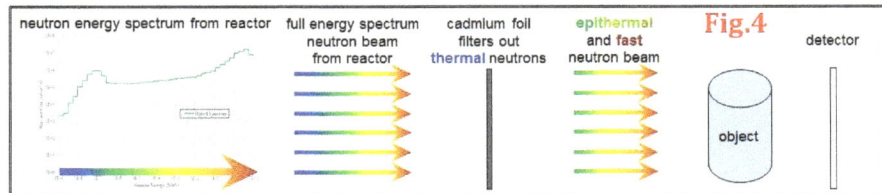

Fig. 4: Filtering the beam with Cd

Fig. 5a and 5b show epithermal measurements at INL. Due to the low sensitivity of the ^{6}LiF+ZnS-screen for epithermal neutrons, exposure time was 10 minutes. A fist-sized rock containing iron and a lot of crystal water was impenetrable for thermal neutrons, and also for epithermal neutrons, it showed an unclear and smeared radiography image. This is of special interest because many fossils are found in this type of rock, which is impenetrable for X-rays. A brass screw placed behind the rock became visible only after a second Cd filter was placed between sample and detector – the crystal water had apparently downscattered and thermalized the epithermal neutrons, which could then be absorbed by the second Cd filter. Fig 5b proves that the image truly shows epithermal neutrons – the Siemens star in its holder, made from Gadolinium, is completely transparent. A thermal radiography is shown further down with the MLZ measurements. The boron nitride discs in the Beam Purity Indicator on top attenuate much more than the lead discs, proving it is mainly a neutron, not a gamma image. A 1 cm stripe of lithiated polyethylene attenuates more than 1 cm of lead.

The employed LiF+ZnS scintillation screen is not sensitive to fast fission neutrons, so they remain unused in this setup, but first tests with a fast neutron screen are promising.

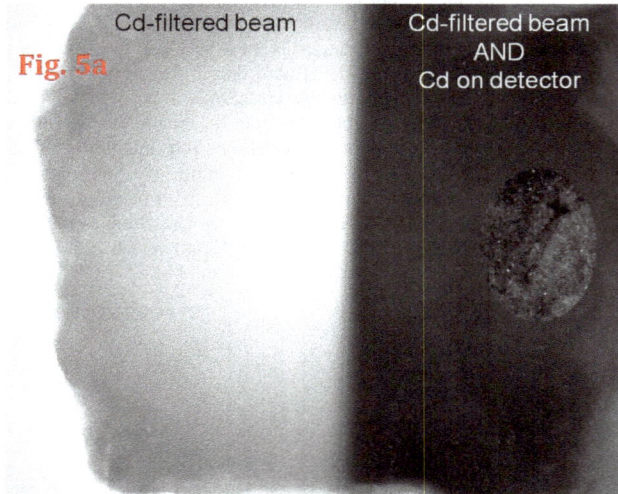

Fig. 5a: Fist-sized ferrous rock with crystal water single and double Cd-filtered. The image is enhanced in the oval showing a brass screw behind the rock.

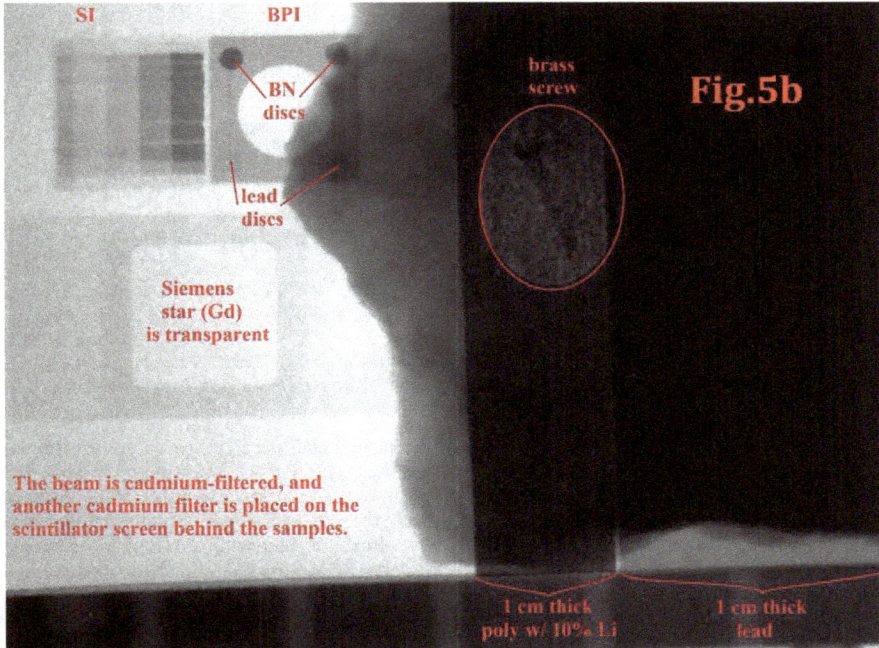

Fig. 5b: Fist-sized ferrous rock with crystal water single and double Cd-filtered with additional samples of Gd, lead and lithiated polyethylene.

Measurements at the FRM II reactor of Heinz Maier-Leibnitz Zentrum at Technische Universität München

The 20 MW FRM II reactor uses a compact core and provides the highest neutron flux per MW worldwide. For tumor treatment and fast neutron imaging, FRM II has a converter plate made of uranium fuel that can be moved in front of a beam tube to produce extra unmoderated fission neutrons (Fig.6). With 80 kW power of the converter plate, a third of the epithermal output of NRAD at INL was expected at the NECTAR neutron imaging facility, but unfortunately, the gamma flux at this facility is so high that in a Cd-filtered beam, polyethylene appears transparent, and lead appears dark, so the resulting images are gamma images, even in spite of the low gamma sensitivity of the LiF+ZnS screen.

The only alternative is the cold neutron imaging facility ANTARES, which looks directly at the cold source, so no epithermal flux was expected. However, the cold source is optimized for maximum output of cold neutrons, so it is undermoderated, because a fully thermalized cold source would have too much self-absorption. Surprisingly, a significant amount of epithermal neutrons penetrates through the cold source, so epithermal radiographs need only one minute of exposure time. Fig. 7 shows a thermal and an epithermal radiography of a Gd Siemens star. Gd is transparent for epithermal neutrons, so it remains invisible.

Neutron Radiography - WCNR-11

Materials Research Forum LLC

Materials Research Proceedings **15** (2020) 142-148

https://doi.org/10.21741/9781644900574-22

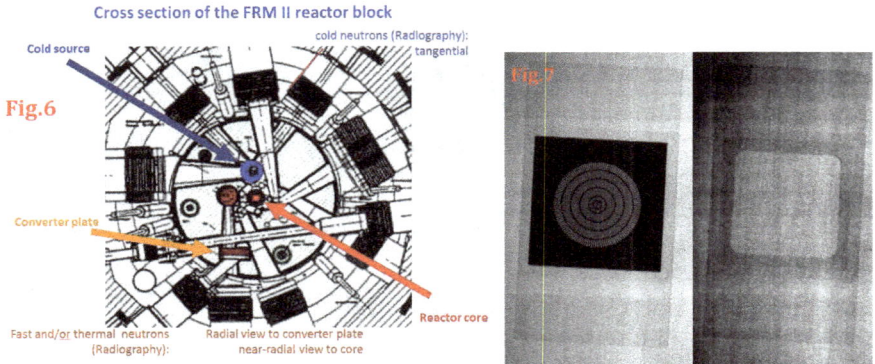

Fig. 6 (left): Cross section of the FRM II reactor block, Fig.7 (right): A thermal and an epithermal radiograph of a Gd Siemens star. Gd is transparent for epithermal neutrons.

The experiment with the ferrous rock was also repeated at FRM II. Fig.8 shows the single-filtered image of a rock with a plastic sleeve and a brass screw and bolt on the left, while the double-filtered image on the right shows increased contrast for the small samples. If the distance between the detector and the samples is increased, the contrast for the small samples also increases in the single-filtered image, which proves that the blurring is caused by scattered neutrons which go in all directions, and have less chance of hitting the detector in increased distance.

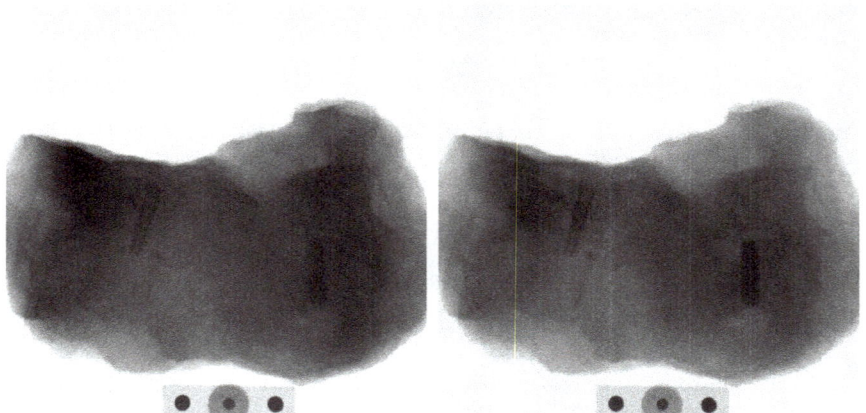

Fig. 8: The ferrous rock experiment repeated with one (left) and two (right) Cd filters: The double-filtered image shows increased contrast for a plastic sleeve end a brass screw and bolt.

With only 60 seconds for one epithermal image, the intensity was sufficient to perform epithermal computed tomography of a technical sample, in this case an electric motor with an

Neutron Radiography - WCNR-11
Materials Research Proceedings **15** (2020) 142-148

Materials Research Forum LLC
https://doi.org/10.21741/9781644900574-22

attached water pump. The material thickness of the pump can in some places not even be penetrated by epithermal neutrons, but on the whole, penetration is much better than for cold neutrons, i.e. without the Cd filter. Fig. 9 shows a cold and an epithermal radiography of the pump, with a small photograph of the setup top center. The steel of the electro magnet of the motor and parts of the impeller of the pump remain impenetrable even for epithermal neutrons, but the rest shows better penetration for epithermal neutrons.

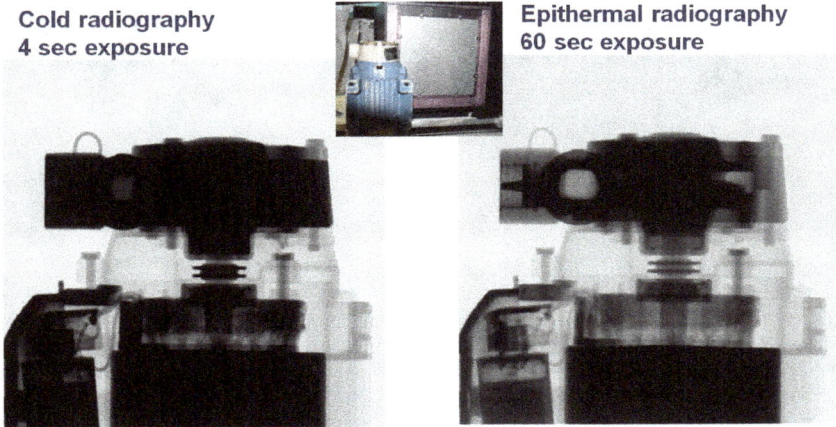

Fig. 9: Cold and epithermal neutron radiography of an electric motor and pump.

In the reconstruction in Fig. 10 and 11, two tomographic slices are shown each for cold and epithermal neutrons. For cold neutrons, the pump impeller is not reconstructed at all, for epithermal neutrons, it works at least in part. For regions of good penetrability, there is no loss in resolution visible for epithermal neutrons compared to cold neutrons, as can be seen in the 3D views in Fig. 12.

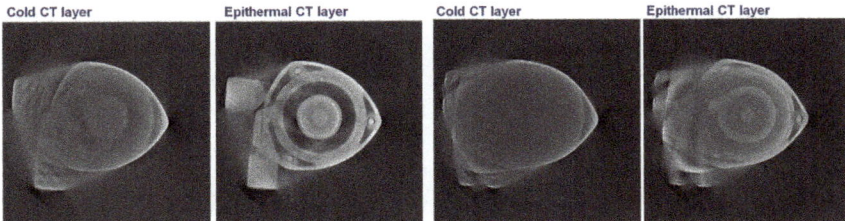

Fig. 10 (left two) and 11 (right two): Two tomographic slices each for cold and epithermal neutrons. For cold neutrons, the pump impeller is not reconstructed at all, but it works to some extent for epithermal neutrons.

Cold 3D CT   Epithermal 3D CT

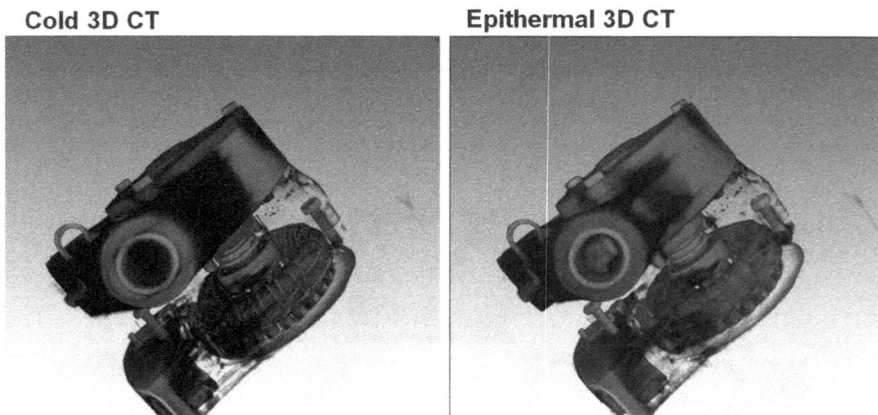

Fig. 12: 3D views of the cold end epithermal neutron CT reconstruction

Conclusions and Outlook

Epithermal neutron imaging will perform no wonders, but can significantly increase penetrability on thick samples. Work at INL and FRM II will continue by inserting bismuth gamma filters into the beams at NECTAR and NRAD. Further improvement is expected by new scintillation screens containing neutron converters with higher sensitivity for epithermal neutrons, and even up to higher energies. Many experiments need to be large enough to be meaningful, which is often too thick for thermal neutrons. Potential applications may include large fuel bundles, large fossils that are simply too thick for thermal neutrons to penetrate, smaller fossils in rock containing highly-attenuating minerals for thermal neutrons such as crystal water content and high iron concentration, or the study of water migration through large soil or rock specimens.

References

[1] Neutron cross sections on https://t2.lanl.gov/nis/data/endf/endfvii.1-n.html

[2] A.E. Craft, D.M. Wachs, M.A.Okuniewski, et al., Neutron radiography of irradiated nuclear fuel at Idaho National Laboratory, Physics Procedia 69 (2015) 483-490. https://doi.org/10.1016/j.phpro.2015.07.068

[3] S.H. Giegel, A.E. Craft, G.C. Papaioannou, and C.L. Pope, Neutron beam flux and energy spectrum characterization at the Neutron Radiography Reactor at Idaho National Laboratory, in this issue

[3] A. Craft, B. Schillinger, First neutron computed tomography with digital neutron imaging systems in a high-radiation environment at the 250 kW Neutron Radiography Reactor at Idaho, in this issue

[4] B. Schillinger, J. Krüger, A quadruple multi-camera neutron computed tomography system at MLZ, in this issue

Neutron Radiography - WCNR-11   Materials Research Forum LLC
Materials Research Proceedings **15** (2020) 149-153   https://doi.org/10.21741/9781644900574-23

Feasibility Study of Two-Dimensional Neutron-Resonance Thermometry using Molybdenum in 316 Stainless-Steel

Tetsuya Kai[1, a *], Kosuke Hiroi[1], Yuhua Su[1], Mariko Segawa[1],
Takenao Shinohara[1], Yoshihiro Matsumoto[2], Joseph D. Parker[2],
Hirotoshi Hayashida[2] and Kenichi Oikawa[1]

[1]Japan Atomic Energy Agency, Tokai-mura Naka-gun Ibaraki 319-1195, Japan

[2] Comprehensive Research Organization for Science and Society,
Tokai-mura Naka-gun, Ibaraki 319-1106, Japan

[a] tetsuya.kai@j-parc.jp

Keywords: Temperature Measurement, RADEN, GEM Detector, Neutron Transmission, Reliability, Energy Dependent Imaging, Width, Time of Flight

Abstract. The energy-dependence of neutrons were measured through a 3-mm thick 316 stainless-steel with homogeneous temperatures from room temperature to about 500 degrees Celsius to investigate whether molybdenum contained in 316 stainless-steel was available as a sensor material for neutron resonance thermometry. Dips in the energy spectra around the 44.8 eV resonance of molybdenum were broadened with the increasing temperature, and a calibration line from width to temperature was obtained. A neutron measurement was also carried out for a 316 stainless-steel plate having a temperature distribution. By analyzing the width of the resonance at each position, a reasonable temperature distribution was obtained. Molybdenum contained in 316 stainless-steel was found to be useful for neutron resonance thermometry.

Introduction

A two-dimensional thermometry technique based on neutron resonance reactions derives the temperature of specified elements in an object by analyzing the Doppler broadening of a neutron resonance absorption width measured by a time-analyzing neutron imaging-detector. This technique is expected to be one of the important applications of the energy-resolved neutron imaging techniques. As a part of ongoing feasibility studies, the authors discussed the reliability of this technique using tantalum and tungsten from 26 to 285°C in a previous study [1]. Tantalum is the most commonly used sensor material for neutron resonance thermometry due to its suitable resonance properties (width, energy and magnitude), however, a more widely-available material has an advantage to promote applications of this technique.

The authors focused on molybdenum contained in 316 stainless-steel (with a weight fraction of 2-3 wt%), which is a well-known and widely-used material in various fields. One molybdenum isotope, molybdenum-95 (natural abundance is 15.9%), exhibits resonances of 15 kilo-barns at a neutron energy of 44.8 eV. The magnitude of this resonance cross section is large enough for resonance absorption imaging, and the thinner relative width of the resonance than that of the tantalum largest resonance at 4.3 eV may bring about higher temperature sensitivity, while the resonance energy is about 10 times higher according to the evaluated nuclear data library, JENDL-4 [2]. This consideration of the resonance properties, and a calculation using the REFIT code [3,4], indicated that molybdenum could be used for neutron resonance thermometry. Figure 1 shows neutron transmission rates of 50 μm thick molybdenum around the resonance at temperatures between 25 and 500°C calculated by the REFIT code. The thickness of molybdenum is equivalent to the 3-mm thick 316 stainless-steel containing 2.1% molybdenum.

The broadening of the resonance shape with increasing temperature was simulated. It should be noted that the temperature effect might not be the same between pure molybdenum (used in the REFIT calculation) and molybdenum contained within an alloy since the Doppler broadening caused by the increase of thermal vibrations of the molybdenum atoms can be affected by the crystallographic structure of the composite material. Thus, the authors were motivated to measure the broadening of the molybdenum resonance in 316 stainless-steel to confirm the availability of such temperature measurement method. A demonstration measurement of the temperature distribution of a 316 stainless-steel plate was carried out.

Experimental

Calibration measurement. The measurements in this study were performed at the energy-resolved neutron imaging system, RADEN [5], installed at J-PARC. For the measurement, the proton beam power was 400 kW (25 Hz) with a single-bunch operation mode. Thermal and cold neutrons were eliminated from the incident beam by a 1-mm thick cadmium filter, and the neutron beam size was collimated to 50×50 mm^2 using stainless-steel and boron-impregnated polyethylene blocks. The broadening of the resonance of molybdenum in 316 stainless-steel was measured by using a 3 mm thick plate at temperatures of 23, 99, 146, 194, 244, 293, 343, 393, 444 and 494°C. The stainless-steel was set inside a vacuum quartz tube (inner diameter: ca. 36 mm) of a furnace [1] having an Inconel heater surrounded by a gold-coated reflector, and the temperatures of the plate were measured by a thermocouple attached on the plate. A copper plate (1 mm in thickness) was attached to the stainless-steel plate to ensure a homogeneous temperature distribution. The energy-dependent transmitted neutron intensity was imaged using a gas-electron multiplier (GEM) detector [6] located at a neutron flight path of 23.8 m. The sample area within the neutron beam was 30×50 mm^2.

Neutron intensities through the sample at 23 and 494°C are shown in Fig. 2 as a function of neutron energy. The statistical uncertainties at the resonance energy (bottoms of dips) were 0.3 and 0.4% for the 23 and 494°C, respectively. It was clearly recognized that dips corresponding to the 44.8 eV

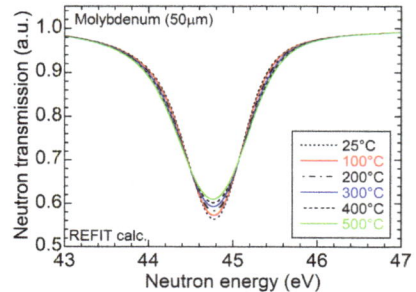

Fig. 1 Neutron transmission rates of 50 μm thick molybdenum calculated with the REFIT code at various temperatures.

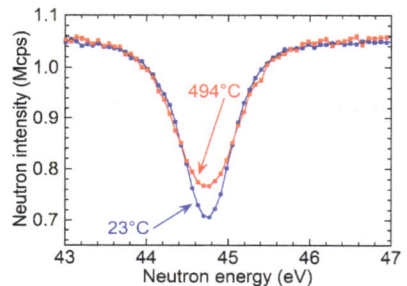

Fig. 2 Neutron intensities around the 44.8 eV molybdenum resonance through a 3 mm-thick 316 stainless steel at 23 and 494°C.

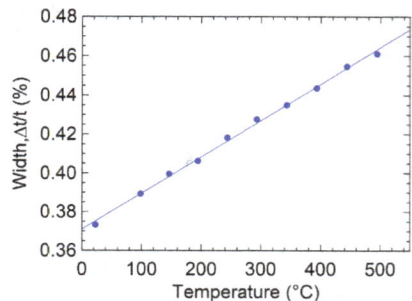

Fig. 3 Measured widths of molybdenum resonance in a 3 mm thick 316 stainless-steel at various temperatures. The solid line is a linear fit result.

molybdenum resonance were broadened at the higher temperature. To estimate the broadening effect, the widths of the resonance relative to the temperature were obtained by fitting the measured spectra assuming a Gaussian-shaped resonance cross section as shown in Fig. 3. The fitted Gaussian width was adopted as the measured width. A calibration line to derive temperature from the width of the resonance was obtained by fitting these results. A reasonable linearity was recognized between temperature and width although fitting errors (not shown in Fig. 3) were rather large due to using the simple Gaussian function without taking into account pulse shapes of incident neutron beam.

Demonstration measurement. A stainless-steel plate of 3-mm thickness was prepared to demonstrate a measurement of the two-dimensional distribution of temperature. The same material as the calibration measurement was used. A schematic view of the plate is shown in Fig. 4 (a). A slit of 1 mm in width was made from the top to the center of the viewed area of 50×50 mm^2 expecting a temperature gap. An aluminum nitride heater was attached at the right upper part of the plate, and was kept at 300°C. The lower part of the plate was used to secure it on a thermal insulation board. The temperatures of the lower part were monitored by two thermocouples. Infrared thermography was also used to monitor the temperature distribution. The plate was coated with black heat-resistant paint to ensure high temperature emissivity of infrared rays. Figure 4 (b) shows an example of the temperature distribution obtained by the infrared thermography. A temperature distribution caused by the heat transfer from the heater and an isolation by the slit was observed. The temperature increased in 20 minutes after turning on the heater, and then became stable. Under these stable temperature conditions, the transmitted neutrons were measured by the GEM detector located at the same position as the calibration measurement. The distance from the sample plate to the detector was 275 mm. The GEM detector produced neutron spectra in the field-of-view of 100×100 mm^2 divided by 128×128 pixels. As shown in a neutron radiograph at 44.8 eV in Fig. 4 (c), neutrons were collimated to 67 $\times$ 67 pixels (52×52 mm^2), and the slit in the plate was confirmed to be in the field-of-view. The cadmium filter was used as in the calibration measurement. The L/D was 400. The neutron measurements were carried out for both heater ON and OFF conditions.

The proton beam power was 500 kW with a double-bunch operation mode, in which two proton bunches were contained in a pulse injected to the neutron source at 25 Hz. The second bunch was delayed by 0.6 µs relative to the first one, and the width of each bunch was 0.1 µs.

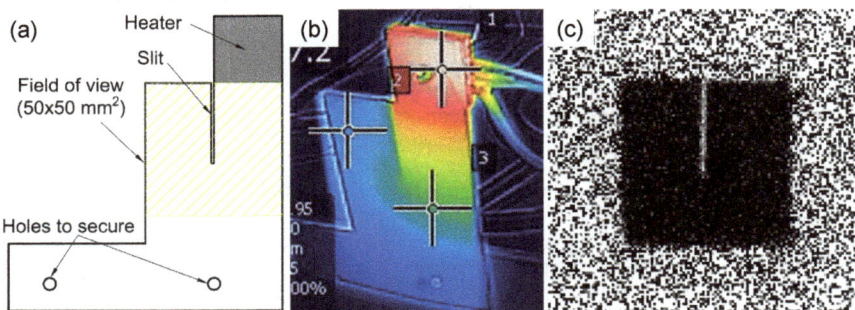

Fig. 4 (a) Schematic view of sample plate made by 316 stainless-steel of 3-mm thickness, (b) temperature distribution during heating measured by infrared thermography. The sample plate was coated with black heat-resistant paint to ensure high emissivity, (c) neutron radiograph at 44.8 eV.

The neutron spectrum measured in the heater OFF condition was used to correct the difference of the operation mode. The normalized neutron intensities around the 44.8 eV resonance at room temperature measured in the single-bunch (calibration measurement) and double-bunch (demonstration) modes are compared in Fig. 5. The width at 22°C for the double-bunch mode was 0.021% (0.05 μs) larger than that expected by the calibration line obtained in single-bunch mode. Although the difference of operation mode effected somewhat the width of the measured resonance dip, it was not as large as the temporal difference of the bunches. The calibration line for the single-bunch mode was modified to increase width by 0.021% in order to obtain the calibration line for the double-bunch mode as shown in Fig. 6. This discrepancy was accidentally caused in upgrading accelerator power from 400 kW to 500 kW.

Results and discussions

The measured neutron intensity spectra in 20 × 20 pixels (15.6 × 15.6 mm^2) were averaged by scanning the position in one-pixel steps to improve the statistic accuracy. The averaged neutron spectra in a 48 × 48 matrix were obtained from the 67 × 67 neutron spectra. The resonance dips in the spectra were fitted in the same manner as in the calibration measurement, and the two-dimensional distribution of widths was produced. Figure 7 shows the distribution of temperature obtained by converting the width distribution at a heater temperature of 300°C by applying the calibration line in Fig. 6. The temperature was highest near the heater, and decreased with increasing distance from the heater to the bottom. The effect of the slit was also shown at the left side of the figure, where the temperature rapidly decreased at the slit. The derived temperature at the upper right corner near the heater was about 210-220°C, while that measured by the infrared thermography was about 230°C. The derived temperature at the upper left corner, the lowest temperature area, was 20-40°C, while about 50°C was indicated by the infrared thermography. The temperature difference seemed to be larger for the lower temperature, although a more detailed, quantitative comparison was difficult because the averaging area was rather large compared to the gradient of the temperature. The temperature distribution derived by analyzing the neutron

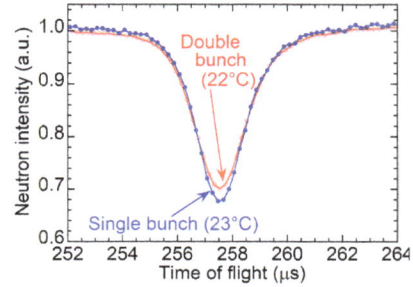

Fig. 5 Normalized neutron intensities around molybdenum resonance measured in proton beam operation modes of single and double bunch through room temperature samples.

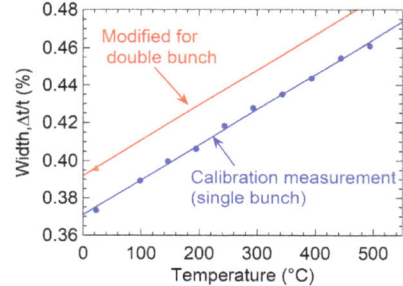

Fig. 6 Calibration lines obtained in the single-bunch mode and modified for the double-bunch mode.

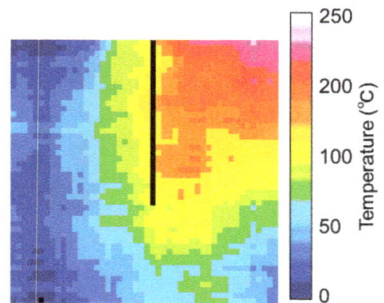

Fig. 7 Two-dimensional distribution of temperature of 316 stainless-steel obtained by analyzing molybdenum 44.8 eV resonance. Black line indicates the position of slit.

resonance, however, qualitatively reproduced the temperature distribution of the stainless-steel well.

Summary

The energy-dependent neutron spectra were measured through a 316 stainless-steel plate of 3-mm thickness at homogeneous temperatures between 23 and 494°C. The Doppler broadening of the 44.8 eV resonance of molybdenum in 316 stainless-steel was observed, and the widths of the resonance were obtained as a function of temperature by fitting a Gaussian-shaped resonance cross section. It was found that the width increased linearly with the temperature in this range, and the calibration line was obtained. This calibration measurement was carried out in the single-bunch mode of the proton accelerator, while the demonstration measurement was performed in the double-bunch mode. The effect of the different operation modes was corrected by the result of the demonstration measurement at room temperature to produce a modified calibration line.

The demonstration measurement was carried out using a 3-mm thick stainless-steel plate with a heater (set at 300°C) attached at the upper corner, just outside the field-of-view. Although some difference was observed between the temperatures derived by analyzing the neutron resonance spectra and those measured by the infrared thermography, the temperature distribution was qualitatively well reproduced considering the rather large averaging area of 15.6×15.6 mm^2 for the neutron resonance measurement. Based on this result, it was concluded that molybdenum in 316 stainless-steel could be used as a sensor material for the neutron resonance thermometry technique. A Gaussian-shaped resonance cross section was assumed for simplicity in this study, however, a function taking into account pulse shapes of neutron beam, intrinsic shape of resonance cross section, background, etc. would have great advantages in reproducing neutron transmission spectra. Such sophisticated function will be utilized to conduct detailed reliability estimation and to promote practical applications.

Acknowledgement

These measurements were carried out under the "Instrument Group Use" proposal (Proposal No.: 2018I0022) of the MLF J-PARC.

References

[1] T. Kai, K. Hiroi, Y. Su, T. Shinohara, J. D. Parker, et al., Reliability estimation of neutron resonance thermometry using tantalum and tungsten, Phys. Procedia 88 (2017) 306-313. https://doi.org/10.1016/j.phpro.2017.06.042

[2] K. Shibata, O. Iwamoto, T. Nakagawa, N. Iwamoto, A. Ichihara, S.et al., JENDL-4.0: A New Library for Nuclear Science and Engineering, J. Nucl. Sci. Technol. 48 (2011) 1-30.

[3] M.C. Moxon, T. C. Ware, C. J. Dean, Users' guide for REFIT-2009-10, UKNSF(2010)P243, UK Nuclear Science Forum, 2010.

[4] H. Hasemi, M. Harada, et al., Evaluation of nuclide density by neutron resonance transmission at the NOBORU instrument in J-PARC/MLF, Nucl. Instrum. Method. A 773 (2014) 137-149. https://doi.org/10.1016/j.nima.2014.11.036

[5] T. Shinohara, T. Kai, K. Oikawa, M. Segawa, et al., Final design of the energy-resolved neutron imaging system "RADEN" at J-PARC, J. Phys.: Conf. Series 746 (1) (2016) 012007. https://doi.org/10.1088/1742-6596/746/1/012007

[6] S. Uno, T. Uchida, M. Sekimoto, T. Murakami, K. Miyama, M. Shoji, et al., Two-dimensional neutron detector with GEM and its applications. Physics Procedia 26 (2012) 142. https://doi.org/10.1016/j.phpro.2012.03.019

Neutron Radiography - WCNR-11　　　　　　　　　　　Materials Research Forum LLC
Materials Research Proceedings 15 (2020) 154-159　　https://doi.org/10.21741/9781644900574-24

Development of kfps Bright Flash Neutron Imaging for Rapid, Transient Processes

R. Zboray [1,a*] Ch. Lani[1,b], A. Portanova[2,c]

[1]Department of Mechanical and Nuclear Engineering, The Pennsylvania State University, 233 Reber Building, University Park, PA 16802, USA

[2]Radiation Science and Engineering Center, College of Engineering, 101 Breazeale Nuclear Reactor, The Pennsylvania State University, University Park, PA 16802, USA

[a]rzz65@psu.edu, [b]chadlani17@gmail.com, [c]arh6@psu.edu

Keywords: Bright Flash Neutron Radiography, Time-Resolved Neutron Imaging, kfps, TRIGA Pulse

Abstract. High-speed neutron radiography is limited by the available flux even on the strongest spallation sources. Therefore, capturing rapid, transient processes by neutron imaging remains difficult. TRIGA reactors have the capability due to their special fuel composition to produce extremely bright neutron pulses for a short duration. This opens the possibility to image short, very rapid transient processes at very high rates. We have developed bright flash thermal neutron radiography at the beam line of the 1 MW Penn State Breazeale research reactor and demonstrated imaging rates up to 4 kfps. Here we discuss and analyze some aspects of the technique.

Introduction

High temporal resolution neutron radiography is limited by the available flux at beam lines even on the strongest neutron sources like ILL, HFIR, J-PARC (typically up to 1×10^8 n/cm^2/s). At several hundred frames per second (fps) or even kfps rates and the corresponding exposure times high neutron flux is needed to obtain images with acceptable signal-to-noise ratio and statistics. The flux limitation can be alleviated using phase-lock, ensemble averaging techniques for periodic, repeating processes as it has been demonstrated on several samples like engines, pumps etc. However, capturing rapid, transient processes by neutron imaging remains difficult as truly high frame rates are required to avoid significant motion blur. At high enough frame rate, the motion artifacts decrease, and the corresponding image blurring can be minimized.

High frame rate imaging has been proven feasible by Zboray and Trtik up to 800 fps [1] on the ICON beam line of the SINQ spallation source. Such frame rate was achieved at the cost of reducing the field of view (FOV) due to partial chip read out. Nakamura et al. [2] imaged at 125 fps and 500 fps the process of molten-fuel behavior by injecting liquid metal into water relevant for severe accident scenarios in nuclear reactors. Sibamoto et al. [3] were looking into water injected into molten metal as potential heat exchange and removal modality for next generation molten metal cooled reactors at 1125 fps. Pulses created in TRIGA nuclear reactors might overcome the flux limitations and offer the possibility of high-speed radiography. Such pulses can produce huge power and neutron flux surges that exceed the levels available at the strongest available spallation sources or steady reactor-based beam lines. This flux can potentially enable imaging rates at a few kilo frames per second (kfps). As the pulse duration is typically around 20-40 milliseconds, the technique is suitable to applications for examining very rapid, transient processes at such times scales. Pulsed imaging has been looked at before back in the 1980's by Bossi et al. utilizing an analog film-based system at Oregon State University [4]. They claim that it was possible to take images with frame rates up to 10 kfps, although the paper contains only

Neutron Radiography - WCNR-11   Materials Research Forum LLC
Materials Research Proceedings **15** (2020) 154-159   https://doi.org/10.21741/9781644900574-24

images taken at 1 and 2 kfps. Related to Bossi's work, Wang used the same TRIGA reactor to obtain the necessary flux and used analog film to capture the images at 1000 fps [5]. More recent work with bright flash imaging was performed by Tremsin et al. [6] and Lerche et al. [7]. However, they only imaged static objects using the high neutron flux of the pulsed McClellan Nuclear Research Center's TRIGA reactor instead of looking into dynamic processes.

We have recently presented bright flash neutron radiography utilizing pulses from the Breazeale TRIGA reactor at the Pennsylvania State University with up to 4 kfps imaging rate [8]. First test results on a simple adiabatic, air-water two-phase flow in a bubbler demonstrated the capabilities of the technique. It can be potentially useful for any fast-transient process where one can leverage the penetrating power of neutrons through high-Z materials and their sensitivity for low-Z materials. Note that imaging somewhat slower transient processes needing only a few hundred fps can also benefit from the high flux in the pulses by having better image quality. Here we discuss and analyze further some aspects of the technique including signal-to-noise ratio, applicability of different scintillators and obtaining quantitative results for the process examined.

Imaging setup, pulses

The experimental setup is described in detail in [8], we repeat here the most important figures. The imaging beam port of the Penn State Breazeale, 1 MW TRIGA type, reactor was used. We have imaged air-water two-phase flow in an aluminum, flat bubbler with 5 mm internal thickness (in beam direction). The neutron imaging beam line features a tangential collimator and a steady neutron flux of 1.7×10^7 n/cm^{2}s at full power. It has an L/D collimation ratio of around 150 at the sample position. Two types of pulses, one with a 2 \$ and another with a 2.5 \$ reactivity worth, were used for the experiments (for the definition of reactivity in \$ see [9]). The former has a peak value of 284.0 ± 6.006 MW and the full width at half maximum (FWHM) of 24.47 ± 2.573 msec, while the latter has a peak value of 761.0 ± 12.44 MW and a FWHM of 15.38 ± 0.2727 msec. The 2 \$ pulse results in a peak neutron flux of about 4.75×10^9 n/cm2-s, whereas the peak flux from the 2.5 \$ pulse was estimated to be about 1.3×10^{10} n/cm2-s. During the imaging experiments using the pulses, the neutron flux was also measured independent of the imaging system using a boron coated, electrically compensated ion chamber (CIC) detector from Westinghouse [10] placed adjacent the reactor core. The typical 2 \$ and 2.5 \$ pulses measured by the CIC and by the imaging detector are shown on the right of Fig. 1. For the imaging detector, the time evolution of the gray intensity of the image averaged over its whole FOV is shown. It illustrates that the CIC measurement can be nicely fitted with the solution of a theoretical model of Fuchs-Nordheim [9] and the normalized pulse shape from the imaging detector fits nicely to the one of CIC except at low power ($\leq$10% pulse height). We attribute this to degraded camera performance at low signal-to-noise ratio (SNR) and high internal gain setting. The deviation from CIC decreases for 2kfps, i.e. for higher SNR, as is shown on the same figure. We discard showing and using data from such a low SNR parts of the pulses later in the paper (see Fig. 3) the data being not trustworthy.

We have used a 400 µm thick LiF/ZnS:Cu screen from Scintacor [11] on the camera-based imaging detector setup, which featured a legacy, CMOS camera Photron FASTCAM-Ultima 512. The camera's chip has 512 x 512 pixels and can be readout up to 30 kfps, with full frame read out up to 2 kfps. The camera has a variable internal gain, which has been set to second highest level (x4) during the experiments. The camera was affixed to a Fujinon f/0 .7 lens with a focal length of 50 mm resulting in a field of view of ca. 70 x 70 mm. We elaborate here on the applicability of ZnS scintillator for very high-speed imaging. Note that [11] specifies 85 µs as decay time to 10% of the light output (λ=ln2/85us), therefore the afterglow might have a non-negligible effect at a few kfps imaging rate. We have setup a simple afterglow model assuming a single decay constant for the scintillation light, giving the afterglow fraction, AF, in the measured intensity as:

$$AF(i) = \frac{\frac{[I(i-1)+AF(i-1)I(i-1)](1-\exp(-\lambda dt))}{\lambda dt}}{I(i)}, \tag{1}$$

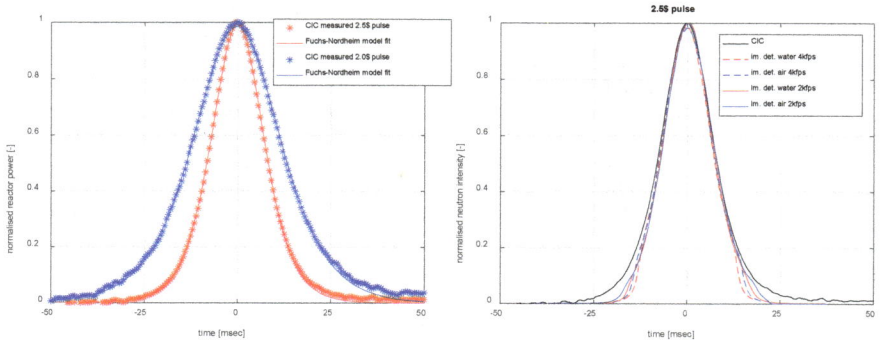

Fig. 1: (a) Fitting the pulse shape measured by the CIC detector for 2$ and 2.5$ pulses using the theoretical model of Fuchs-Nordheim [9]. (b) The normalized pulse shape measured by the imaging detector (with the bubbler filled with water or air, respectively) agrees well with the one obtained by CIC except at very low power.

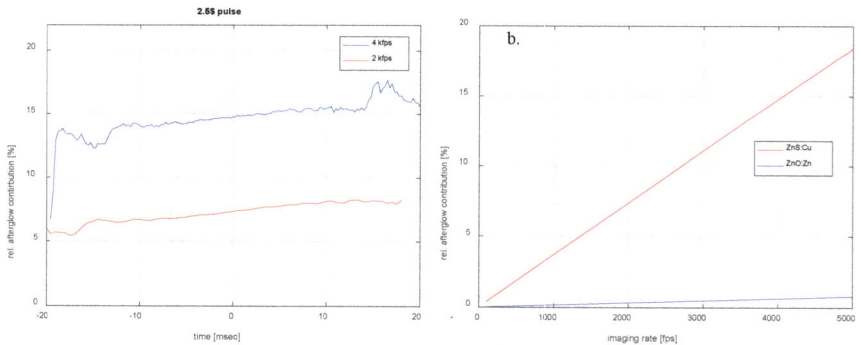

Fig. 2: (a) Fraction of the afterglow in the total measured intensity for a 2.5$ pulse (see [8]) and two different imaging rates for our ZnS:Cu screen given by Eq. (1). (b) The equilibrium fraction of afterglow at constant imaging intensity for ZnS:Cu and ZnO:Zn scintillators as given by Eq. (3).

where dt is the exposure time and $I(i)$ is the true recorded image intensity (i.e. without afterglow) at time step i. For bright flash imaging, the latter changes during the pulse. The afterglow contribution in a ZnS screen is indeed not negligible around 14% on average for 4kfps and decreases to around 6-7% for 2 kfps as is shown in Fig. 2a. Note that the leading half of the pulse has a slightly lower afterglow fraction than the trailing half as is expected. For the simpler case of a constant true image intensity, Eq. (1) simplifies to

$$AF(i) = \frac{(1+AF(i-1))(1-\exp(-\lambda dt))}{\lambda dt} \tag{2}$$

and it can be easily shown that this recursive equation after a few iterations (images) settles to an equilibrium value given by

$$AF_{eq} = \frac{1-\exp(-\lambda dt)}{\lambda dt+\exp(-\lambda dt)-1} \tag{3}$$

For future work, it might be worth considering ultra-low afterglow scintillator screens like e.g. LiF/ZnO:Zn offered by Scintacor with a decay time of 3.5 us [11]. Comparison of the equilibrium afterglow fraction for ZnS:Cu and ZnO:Zn are shown in Fig. 2b confirming values below 1 % for ZnO:Zn even at 4-5kfps. However, as is shown in [12, 13], ZnO:Zn has significantly lower scintillation efficiency and higher gamma sensitivity than ZnS. Therefore, the LiF/ZnS scintillator screen still represents a reasonable compromise for a few kfps image rate.

Quantitative results
Besides the visualization of the process, one might want to extract quantitative information from the images such as the instantaneous, pixel-wise gas volume fraction distribution for the case of the two-phase flow in the bubbler. For this the images have to be normalized using corresponding flat field images taken with the bubbler filled with water and with air as:

$$\varepsilon(x,y) = \frac{\ln(\frac{I_{2ph}(x,y)}{I_w(x,y)})}{\ln(\frac{I_g(x,y)}{I_w(x,y)})} \tag{4}$$

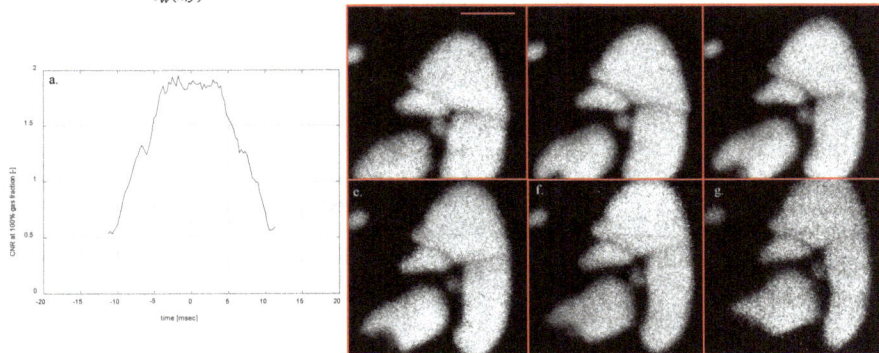

Fig. 3: (a) The variation in the CNR for 100% gas fraction for a 2.5 $ pulse acquired at 4 kfps. (b) to (g) Corresponding image sequence of the instantaneous gas volume fraction distribution, ε, shown for every 10th image taken, i.e. with a time gap of 2.5 ms. The gray scale of the images is [-0.5,1.5] and the frame are colored in red for a better visualization.

where $I(x,y)$ is the intensity of the image at pixel (x,y), ε is gas volume fraction and the subscripts $2ph$ stands for two-phase flow, w and g are for flat field images with the bubbler filled with water and with air, respectively. An image sequence of the instantaneous gas volume fraction distribution is shown in Fig 3b to 3g for a 2.5 $ acquired at 4 kfps. Although it is not very visible on the sequence shown, the image quality changes significantly during the pulse with changing flux. Fig. 3a shows the contrast-to-noise ratio (CNR) between gas and liquid at

100% gas fraction for the same pulse and imaging rate. Obviously, the CNR increases for decreasing frame rate or for a higher pulse at the same acquisition rate.

The quantitative images of the instantaneous pixel-wise gas fraction distribution show non-negligible bias, i.e. values above 1 and below 0 ranging from -30% to +180%. This bias may be caused by physical processes such as beam hardening, sample and room scatter contributions, etc. Regarding scatter and beam hardening effect, it must be noted that in our earlier work using the same bubbler [1], although on a different beam line, we did not find such a strong bias in the gas fraction. That other beam line had a cold spectrum, this however is not expected to have a significant effect on beam hardening and sample scatter compared to our thermal beam. A more significant difference might stem from different levels of room scatter. Our beam line does not feature any evacuated neutron flight tubes, although our beam has about 2 m of free flight in air before reaching the sample and has a much compacter beam cave compared to the other beam line. Both these features can contribute to higher room scatter levels. A simple room scatter model assuming a constant room scatter contribution all over the image was used to try improve the results. According to the model the measured intensity of flat field images with the bubbler filled with air and water, respectively, can be written as:

$$I_{g,m} = I_{g,o} + S, \text{ and } I_{w,m} = I_{g,o} \exp(-\Sigma t) + S \tag{5}$$

where S is the constant room scatter contribution, subscript o denotes the true beam intensity without scatter and t is the thickness of the water layer. Σ is the effective neutron attenuation coefficient of the 5 mm thick water layer in the bubbler. Its value, including beam hardening and detector sensitivity effects, has been estimated to be around 3.5/cm based on Monte Carlo simulations in a fashion similar to [14]. Taking the ratio of the two above equations

$$\frac{I_{w,m}}{I_{g,m}} = \frac{\exp(-\Sigma t)+S/I_{g,o}}{1+S/I_{g,o}}, \tag{6}$$

and using the measured ratio of flat field image intensities, the scattered fraction, $S/I_{g,o}$, can be determined. Typically, quite high values of 50-60% are found based on the data. This can be used in Eq. 4 to correct the measured intensities. This indeed improves the quantitative results quite a bit but does not fully eliminate the bias. Note that apart from the above correction method, room scatter can and will be diminished by relatively simple physical countermeasures on the beamline and on the imaging detector. Other potential bias effects like the low-count bias and the statistical uncertainty of the images have been analyzed in [8].

Summary

We have developed and tested a bright flash neutron radiography setup at the Penn State TRIGA reactor. We have proven using a simple two-phase flow device for illustration purposes that imaging with frame rates up to 4 kfps are feasible. The imaging detector is shown to follow the intensity variation of the pulses well and agree with the predictions of the theory. Although ZnS-based scintillator has a relatively long decay time, it is shown to be still a reasonable compromise for kfps imaging due mainly to the lack of high-light-yield, fast scintillators. The quantitative results revealed some bias effects very likely due to room scatter contribution, which can be diminished by countermeasures in future work.

The technique opens up applications for visualizing and quantifying any rapid, transient processes, where one can leverage the speed of imaging combined with the penetrating power of neutrons through high-Z material and their sensitivity to low-Z materials. Such processes will be explored in detail in the future.

Acknowledgement
The authors are thankful to Dr. Jim Turso for providing the CIC measurement results.

References

[1] R. Zboray, P. Trtik, "800 fps neutron radiography of air-water two-phase flow", MethodsX https://doi.org/10.1016/j.mex.2018.01.008

[2] H. Nakamura, Y. Sibamoto, Y. Anoda, Y. Kukita, K. Mishima & T. Hibiki, "Visualization of Simulated Molten-Fuel Behavior in a Pressure Vessel Lower Head Using High-Frame-Rate Neutron Radiography", Nuclear Technology, 125:2, (1999), 213-224. https://doi.org/10.13182/NT99-A2943

[3] Y. Sibamoto, Y. Kukita, H. Nakamura. "Visualization and Measurement of Subcooled Water Jet Injection into High-Temperature Melt by Using High-Frame-Rate Neutron Radiography", Nuclear Technology, 139:3, (2002),205-220. https://doi.org/10.13182/NT02-A3314

[4] R.H. Bossi, A.H. Robinson, J.P. Barton, "High-Speed Motion Neutron Radiography," Nucl. Technol., 59, 363, (1982). https://doi.org/10.13182/NT82-A33039

[5] S-H. Wang, "High Speed Motion Neutron Radiography of Two-Phase Flow", Oregon State University (1981)

[6] A.S. Tremsin, M. Lerche, B. Schillinger, W. B. Feller, Bright flash neutron radiography capability of the research reactor at McClellan Nuclear Research Center, Nuclear Instruments and Methods in Physics Research A, 748 (2014), 46-53. https://doi.org/10.1016/j.nima.2014.02.034

[7] M. Lerche, A.S. Tremsin, B. Schillinger, Bright Flash Neutron Radiography at the McClellan Nuclear Research Reactor, Physics Procedia, 69, (2015), 299-303. https://doi.org/10.1016/j.phpro.2015.07.042

[8] Ch. Lani & R. Zboray, "Development of a high frame rate neutron imaging method for two-phase flows", Nucl. Inst. Meth. A. (2019), In Press, Corrected Proof. https://doi.org/10.1016/j.nima.2018.12.022

[9] D. L. Hetrick, Dynamics of Nuclear Reactors, University of Chicago Press, 1971

[10] Westinghouse, WL-8074 Boron coated, electrically compensated ion chamber.

[11] Scintacor, Neutron Screens, available at https://scintacor.com/wp-content/uploads/2015/09/Datasheet-Neutron-Screens-High-Res.pdf, (accessed on July 20th, 2018)

[12] N. Kubota, M. Katagiri, K. Kamijo, H. Nanto, Evaluation of ZnS-family phosphors for neutron detectors using photon counting method, Nuclear Instruments and Methods in Physics Research A, 529, (2004), 321–324. https://doi.org/10.1016/j.nima.2004.05.004

[13] G. Jeff Sykora, Erik M. Schooneveld, Nigel J. Rhodes, "ZnO:Zn/6LiF scintillator - A low afterglow alternative to ZnS:Ag/6LiF for thermal neutron detection", Nuclear Inst. and Methods in Physics Research A, 883, (2018), 75–82. https://doi.org/10.1016/j.nima.2017.11.052

[14] R. Zboray, H-M. Prasser, "Optimizing the performance of cold-neutron tomography for investigating annular flows and functional spacers in fuel rod bundles", Nucl. Eng. Des., 260, (2013), 188-203. https://doi.org/10.1016/j.nucengdes.2013.03.026

Neutron Radiography - WCNR-11
Materials Research Proceedings 15 (2020) 160-164

Materials Research Forum LLC
https://doi.org/10.21741/9781644900574-25

Neutron Transmission Spectrum of Liquid Lead Bismuth Eutectic

Yojiro Oba[1, a *], Daisuke Ito[2,b], Yasushi Saito[2], Yohei Onodera[2],
Joseph Don Parker[3], Takenao Shinohara[4], and Kenichi Oikawa[4]

[1]Materials Sciences Research Center, Japan Atomic Energy Agency, 2-4, Shirakata, Tokai, Ibaraki, Japan

[2]Institute for Integrated Radiation and Nuclear Science, Kyoto University, 2, Asashiro-nishi, Kumatori, Osaka, Japan

[3]Comprehensive Research Organization for Science and Society, 162-1, Shirakata, Tokai, Ibaraki, Japan

[4]J-PARC Center, Japan Atomic Energy Agency, 2-4, Shirakata, Tokai, Ibaraki, Japan

[a]ohba.yojiro@jaea.go.jp, [b]itod@rri.kyoto-u.ac.jp

Keywords: Neutron Transmission Imaging, Pulsed Neutron, Lead Bismuth Eutectic

Abstract. Energy-resolved neutron transmission imaging experiment was carried out to characterize the liquid phase of lead bismuth eutectic (LBE). The neutron transmission image confirmed that the LBE is homogeneous in a sample vessel. The obtained neutron transmission spectrum of the liquid LBE shows a wavy behavior in the wavelength dependence. This behavior is attributed to neutron attenuation by the neutron diffraction of the liquid LBE, which can give information about the atomic structure of the liquid LBE similar to Bragg edge transmission in the crystalline solid phases.

Introduction

Lead bismuth eutectic (LBE, 44.5 mass% Pb 55.5 mass% Bi) is a candidate material of coolant for LBE-cooled accelerator driven systems (ADS) and fast breeder reactor (FBR) [1,2]. Understanding of thermal properties and flow behavior of the LBE are crucial for design and safety analysis of such ADS and FBR facilities. To characterize those properties of the LBE, structural information plays a key role because it is highly related to those properties. Recently, Ito *et al.* have clearly demonstrated that an energy-resolved neutron transmission imaging experiment is useful to study the structure of the LBE [3]. The analysis of Bragg edge transmission, which results from the neutron attenuation by neutron diffraction, provides the structural information about crystalline solid phases of the LBE [4].

However, the structures of liquid phases in the LBE were not considered in the previous study even though they are closely linked to the flow behavior in operation as the coolant. If the neutron diffraction contributions caused by the liquid phases of the LBE can be detected in the neutron transmission spectra, they probably give beneficial information about the atomic structures in the liquid phases in the same manner as the Bragg edge transmission. Therefore, we performed the energy-resolved neutron imaging measurements of the liquid phase of the LBE to observe the neutron diffraction contribution of the liquid LBE in the neutron transmission spectra.

Experiment

The experiments were carried out at the energy-resolved neutron imaging instrument BL22 RADEN and at the neutron beamline for observation and research use BL10 NOBORU in J-

Neutron Radiography - WCNR-11   Materials Research Forum LLC
Materials Research Proceedings **15** (2020) 160-164   https://doi.org/10.21741/9781644900574-25

PARC [5,6]. The LBE was sealed in a rectangular sample vessel made of austenitic stainless-steel plates with the thickness of 1 mm. The thickness of the LBE was 10 mm. The sample vessel was then covered with glass wool and aluminum tape for thermal insulation. The temperature of the LBE was measured using five thermocouples and controlled using cartridge heaters and cooling fans. Details are described elsewhere [3]. A neutron gas electron multiplier (nGEM) detector was used to obtain the neutron transmission images. The dimensions of each pixel and the effective area of the detector were 0.8×0.8 mm and 100×100 mm, respectively. The sample was placed just before the detector.

Results and discussion

Fig. 1 (a) shows the neutron transmission image of the LBE in the sample vessel at 220 °C integrated between the wavelength of 0.1 and 0.6 nm. Surface of the LBE is observed at 105 pixels = 84 mm from the bottom of the vessel (shown by a white broken line). In the neutron transmission image, five bar-shaped shadows extended from the side vessel walls are neutron attenuation by the thermocouples. Other than these features, the neutron transmission image shows horizontal stripe-like inhomogeneity. This is not originated by the LBE, but by the sample vessel. Fig. 1 (b) shows the neutron transmission image of the identical sample vessel after removing the LBE. The same horizontal stripe-like inhomogeneity is observed in the image. Since the neutron transmission image of only the sample vessel without the thermal insulator, *i.e.*, only the stainless steel, exhibits flat neutron attenuation [Fig. 1 (c)], this inhomogeneity results from the glass wool and aluminum tape horizontally wound around the vessel.

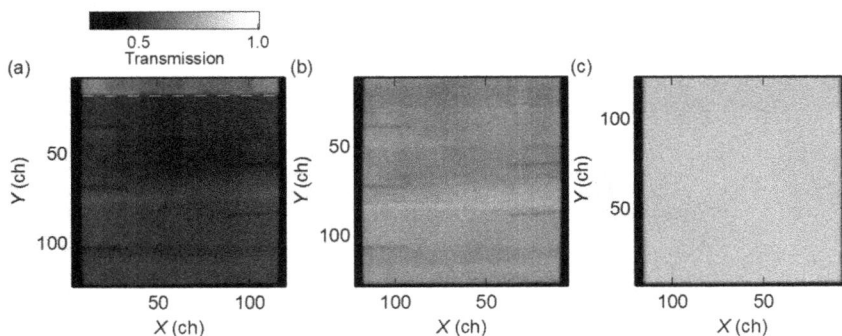

Fig. 1 Neutron transmission images integrated between wavelength of 0.1 and 0.6 nm of (a) LBE in sample vessel, (b) sample vessel after removing LBE, and (c) stainless steel plates of sample vessel. Horizontal white broken line indicates the surface of LBE. Black edge regions correspond to vessel wall made of stainless steel plates.

Based on the Beer-Lambert law [7], the neutron attenuation contribution of the sample vessel can be removed by dividing the neutron transmission of the LBE in the sample vessel by that of the empty vessel. Fig. 2 shows the divided neutron transmission image, which should be composed of only the LBE contribution. The horizontal stripe-like inhomogeneity disappears in the divided image. However, the white and black shadows of the thermocouples remain. This is probably caused by the moving of the thermocouples during the removal of the LBE from the sample vessel. Hereafter, only center parts of the neutron transmission images (shown by a white broken square in Fig. 2) are discussed to avoid unexpected effects of the thermocouples. The divided image of the LBE shows uniform neutron attenuation in the center parts. This reflects the uniformity of the LBE in the sample vessel.

Fig. 3 indicates the neutron transmission spectra averaged in the center parts of the neutron transmission images of the LBE in the sample vessel, the sample vessel after removing the LBE, and the sample vessel without the thermal insulator (= the austenitic stainless steel). The vertical dotted line indicates the Bragg cutoff of the austenitic stainless steel, where the wavelength is equal to twice the largest lattice spacing $2d_{max}$ in corresponding crystal structure. Below the Bragg cutoff, all the neutron transmission

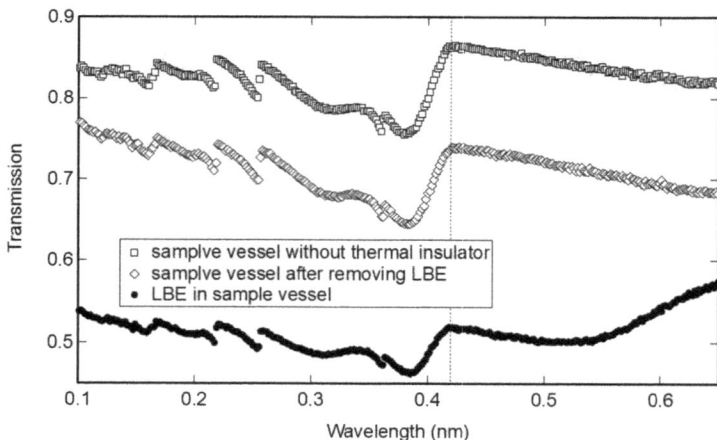

Fig. 2 Divided neutron transmission image of LBE. The white broken square is the region to calculate average neutron transmission.

spectra show the Bragg edge transmission from the austenitic stainless steel, while no Bragg edge transmission from the crystalline LBE is observed in the spectrum of the LBE. This indicates that the LBE is completely in the liquid phase. At the wavelength longer than the Bragg cutoff, although the Bragg edge transmission of the austenitic stainless steel no longer occurs, only the spectrum of the LBE curves gently. This is probably due to the liquid LBE.

To characterize the liquid LBE, the neutron transmission spectra was separated into the contributions of the liquid LBE, the austenitic stainless steel (= the sample vessel without the thermal insulator), and the thermal insulator. Dividing the neutron transmission of the LBE in the sample vessel by that of the empty sample vessel gives the contribution of the liquid LBE.

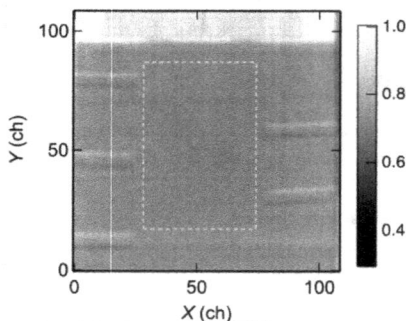

Fig. 3 Neutron transmission spectra. Black circles, open diamonds, and open squares denote LBE in sample vessel, sample vessel after removing LBE, and sample vessel withou thermal insulator. Vertical dotted line is Bragg cut-off of austenitic steel.

Similarly, the contribution of the thermal insulator can be obtained by dividing the neutron transmission of the empty sample vessel by that of the austenitic stainless steel. Fig. 4 shows these contributions of the liquid LBE, the austenitic stainless steel, and the thermal insulator in the neutron transmission spectra. The Bragg edge transmission occurs merely in the contribution of the austenitic stainless steel. The contribution of the thermal insulator is featureless and almost monotonically decreases with increasing the wavelength. In contrast, the contribution of the liquid LBE has obviously different features. In addition to the gentle curve observed in Fig. 3, whole the spectrum shows a wavy behavior. Fig. 4 also shows an absorption contribution calculated from the chemical composition and the density of the LBE using the database of total cross-sections [1,7]. Since the absorption contribution is negligibly small, the wavy behavior of the liquid LBE is attributed to the attenuation by the neutron diffraction, which reflects the atomic structure of the liquid LBE. Therefore, the homogeneous neutron transmission image in Fig. 2 means that the atomic structure of the liquid LBE is uniform.

A recent study has demonstrated that the neutron attenuation coefficient is proportional to the scattering intensity integrated in all solid angles [8]. Although this study discussed only a contribution of small-angle neutron scattering (SANS), this idea can be simply expanded to neutron diffraction. Hence, broadened diffraction pattern of the liquid LBE [9] probably brings about the wavy behavior instead of the sharp Bragg edge transmission, which is connected with sharp Bragg peaks in the crystalline solid phases.

Based on the ref. [8], conventional analytical techniques for the neutron diffraction of the liquid phases can be also applied to the neutron transmission spectra. Such application enables to obtain the structural information about the liquid phases from the neutron transmission images. This will be useful to study the liquid LBE in flow channels. However, contributions of background, such as the sample vessel, have to be precisely removed to obtain meaningful information from the neutron transmission spectra.

Summary
In this study, the energy-resolved neutron transmission imaging experiment of the liquid LBE

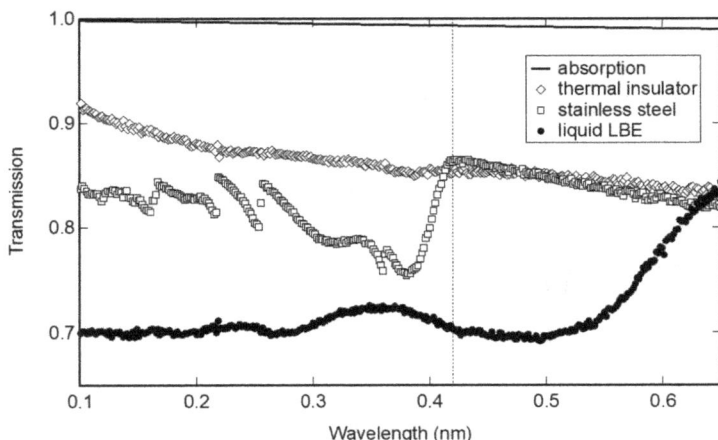

Fig. 4 Contributions of liquid LBE, stainless steel, and thermal insulator in the neutron transmission spectra. Black circles, open squares, and open diamonds denote contributions of liquid LBE, stainless steel, and thermal insulator, respectively. Solid line is calculated absorption contribution.

was performed. The neutron transmission image shows the uniformity of the liquid LBE. The neutron transmission spectrum of the liquid LBE was successfully extracted from the neutron transmission spectra of the LBE in the sample vessel and the empty sample vessel. The wavy behavior of the neutron transmission spectrum of the liquid LBE is probably resulted from the broadened diffraction of the liquid LBE.

Acknowledgement

Neutron experiments at the MLF J-PARC were performed under Proposal Nos. 2016A0272, 2017A0144, and 2017B0120.

References

[1] C. Fazio, et al., Handbook on Lead-bismuth Eutectic Alloy and Lead Properties, Materials Compatibility, Thermal-hydraulics and Technologies 2015 Edition, OECD, 2015.

[2] V. Sobolev, Properties of Liquid Metal Coolants, in: R. J. M. Konings (Ed.), Comprehensive Nuclear Materials, Elsevier, Amsterdam, 2012, pp. 373-392. https://doi.org/10.1016/B978-0-08-056033-5.00130-0

[3] D. Ito, Y. Saito, H. Sato, T. Shinohara, Visualiz ation of solidification process in lead-bismuth eutectic, Phys. Proc. 88 (2017) 58-63. https://doi.org/10.1016/j.phpro.2017.06.007

[4] O. Takada, T. Kamiyama, Y. Kiyanagi, Study on phase transition of Pb-Bi eutectic alloy by neutron transmission spectroscopy, J. Nucl. Mater. 398 (2010) 129-131. https://doi.org/10.1016/j.jnucmat.2009.10.022

[5] T. Shinohara, T. Kai, K. Oikawa, M. Segawa, M. Harada, T. Nakatani, M. Ooi, K. Aizawa, H. Sato, T. Kamiyama, H. Yokota, T. Sera, K. Mochiki, Y. Kiyanagi, Final design of the Energy-Resolved Neutron Imaging System "RADEN" at J-PARC, J. Phys.: Conf. Ser. 746 (2016) 012007. https://doi.org/10.1088/1742-6596/746/1/012007

[6] K. Oikawa, F. Maekawa, M. Harada, T. Kai, S. Meigo, Y. Kasugai, M. Ooi, K. Sakai, M. Teshigawara, S. Hasegawa, M. Futakawa, N. Watanabe, Design and application of NOBORU–NeutrOn Beam line for Observation and Research Use at J-PARC, Nucl. Instrum. Met. Phys. Res. A 589 (2008) 310-317. https://doi.org/10.1016/j.nima.2008.02.019

[7] A.-J. Dianoux, G. Lander, Neutron Data Booklet, Old City Publishing, Philadelphia, 2003.

[8] Y. Oba, S. Morooka, K. Ohishi, J. Suzuki, S. Takata, N. Sato, R. Inoue, T. Tsuchiyama, E. P. Gilbert, M. Sugiyama, Energy-resolved small-angle neutron scattering from steel, J. Appl. Cryst. 50 (2017) 334-339. https://doi.org/10.1107/S1600576717000279

[9] I. Kaban, W. Hoyer, Y. Plevachuk, V. Sklyarchuk, Atomic structure and physical properties of liquid Pb-Bi alloys, J. Phys.: Condens. Matter 16 (2004) 6335-6341. https://doi.org/10.1088/0953-8984/16/36/001

Neutron Radiography - WCNR-11
Materials Research Proceedings **15** (2020) 165-173

Materials Research Forum LLC
https://doi.org/10.21741/9781644900574-26

Quantitative Crack Analysis using Indirect Neutron Radiography and Neutron Activation Analysis with Contrast Enhancement Agents

Russell Jarmer[1,a], Dr. Jeffrey King[1,b,*], Dr. Aaron Craft[2,c], Dr. Robert O'Brien[2,d]

[1]Colorado School of Mines 1500 Illinois St. Golden, CO 80401 United States of America

[2]Idaho National Laboratory 1955 N. Fremont Ave. Idaho Falls, ID 83415 United States of America

[a]jarmer@mymail.mines.edu [b]kingjc@mines.edu [c]aaron.craft@inl.gov [d]robert.OBrien@inl.gov

Keywords: Indirect Neutron Radiography, Contrast Agents, Quantitative Analysis, Neutron Activation Analysis

Abstract. This paper presents an analysis of infiltration and washing methods to develop a quantitative metric for crack features using indirect neutron radiography and neutron activation analysis. A gadolinium contrast agent enhances the visibility of crack features in digitally processed neutron radiographs. Neutron activation analysis of crack infiltrants using dysprosium as a surrogate for gadolinium directly measures the amount of infiltrant in crack features.

Introduction

Indirect neutron radiography is a valuable tool for the non-destructive examination of irradiated nuclear fuel; however, the ability to detect fine cracks in cladding materials is generally constrained by the spatial and contrast resolution limits of the imaging technique. Infiltrating fine cracks with a contrast agent can enhance the ability to image these cracks through neutron radiography. Establishing a documented standard for quantitative analysis of crack structures using neutron radiography with contrast-enhancing infiltrants will provide a valuable tool for examining irradiated materials, particularly advanced cladding materials. Phase One of this project tested three gadolinium infiltrant solutions containing varying amounts of methanol and ammonium lauryl sulfate to quantify the effect of lowering the surface tension of the infiltrant to enhance the ability of the infiltrant to penetrate small cracks. Scanning and digitally processing the resulting radiographic images provided a quantitative measurement of the crack density, extent of infiltrant penetration, and the effectiveness of washing to remove the residual gadolinium. Neutron Activation Analysis (NAA) of a dysprosium based infiltrant solution in Phase Two of the project directly measured the amount of infiltrant in the crack features to provide a comparison to the image-based crack densities in Phase One.

Image contrast is a primary limitation to crack detection through radiography. When filled with air, the cracks are difficult to distinguish from the surrounding material, as the difference between the neutron attenuation by the air and the neutron attenuation by the surrounding material is small. Increasing the imaging time can increase the contrast between the crack and the parent material. Alternatively, filling the crack with a highly attenuating contrast agent increases the contrast between the crack and the surrounding material without increasing imaging time [1].

A contrast agent for neutron radiography must provide a large neutron cross section, and must be easy to prepare into an infiltrant solution. Gadolinium (III) nitrate meets both requirements, as gadolinium provides the largest thermal neutron cross-section of any naturally occurring element [2], and gadolinium nitrate is readily soluble in water.

Phase One of this project investigated the ability of a gadolinium-based infiltrant to enhance the contrast of indirect neutron radiography images at the Neutron RADiography (NRAD) reactor at the Idaho National Laboratory. This reactor is a 250 kW TRIGA (Training, Research, Isotopes, General Atomics) reactor, and provides neutrons for two beamlines [3]. Two radiography stations at the NRAD allow for indirect neutron imaging using a cassette containing activation foils. Two activation foils, dysprosium for thermal neutrons and indium for epithermal neutrons, capture the neutrons that pass through a sample placed in the neutron beam in front of the cassette. The neutron attenuation by the materials in the sample creates an activated map of neutron intensity in the activation foils. After a sufficient irradiation time, placing the activation foil in contact with photographic film allows radiation produced by the decay of the activated atoms in the foil to expose the film. Developing the exposed film produces a radiographic image of the sample.

Phase Two of this project test the ability to infiltrate and remove contrast agent using Neutron Activation Analysis (NAA). NAA works by exposing a sample to a neutron field; materials within the sample absorb neutrons, creating new isotopes, some of which undergo radioactive decay. A High Purity Germanium (HPGe) detector counts gamma radiation emitted by these decays and bins the results by energy. These results determine the composition of the sample due to charactersitic decays of isotopes. Additionally, the number of decays of a particular isotope can be used to determine the mass of the parent isotope present in the sample [4].

Methodology

Neutron Radiography Methods. A set of ten numbered aluminum test blocks made following ASTM standards provided a documented standard object for investigating the ability of potential contrast agents to infiltrate cracks and appear in subsequent radiographs. Conventional dye penetrant testing provided a useful comparison to the imaging of cracks via neutron radiography. A 15 minute dwell period after application of the dye penetrant allowed full penetration of crack features. A white developing agent sprayed on the surface of the block pulled the dye out of the cracks to reveal the crack patterns. Optical photographs of the cracks, taken immediately after developer application and after a two minute development period, provided the reference images for the neutron radiography experiments (Fig. 1). Dye penetrant testing revealed no cracking in Blocks 4 and 8, excluding those blocks from further testing.

Table 1 lists the combinations of contrast agents and solvents used in infiltrant and washing solutions used in this experiment. The different solutions test how lowering the viscosity and surface tension of the infiltrant solutions improved the ability of the infiltrant to penetrate cracks under vacuum.

Figure 1. Reference image of cracks in Block 6 revealed by dye penetrant testing.

Table 2 details the infiltration and wash steps for each block. The steps in each test sequence of Phase One proceeded in the order presented in the first column of Table 2 (i.e. Wash 2 and Wash 3 (Steps 4 and 5) were followed by Infiltration 3 (Step 6)). The second and third columns in Table 2 give the step name and step ID (used in Figures 3-6). The next four columns list the solution from Table 1 used to perform the process listed in the final column for that step, as a function of the block being treated. The project obtained new radiographs of the blocks after each step in Table 2.

Table 1. Phase One (neutron radiography) infiltration and washing solutions.

	Contrast Agent	Solvent Composition
Solution 1	0.3M $Gd(NO_3)_3$	Water
Solution 2	0.3M $Gd(NO_3)_3$	50% water/50% methanol (vol%)
Solution 3	0.3M $Gd(NO_3)_3$	50% water/50% methanol (vol%) with 5 wt% ammonium lauryl sulfate
Solution 4	None	Water
Solution 5	None	50% water/50% methanol (vol%)

Table 2. Phase One (neutron radiography) infiltration and washing methods.

			Solution (From Table 1)			
Step	Step Name	ID	Blocks 1&2	Blocks 6&9	Blocks 5&7	Blocks* 3&10	Process
1	Infiltration 1	I1	2	1	1	2	Submerged in solution at atmospheric pressure for 5 minutes
2	Wash 1	W1	4	4	4	4	Submerged in solution in ultrasonic bath for 5 minutes followed by 5 minutes under -0.03 MPa vacuum
3	Infiltration 2	I2	3	2	2	3	Submerged in solution at atmospheric pressure for 15 minutes
4	Wash 2	W2	5	5	5	4	Submerged in solution in ultrasonic bath for 10 minutes
5	Wash 3	W3	5	4	4	4	Submerged in solution at atmospheric pressure for 15 minutes followed by 15 minutes under -0.03 MPa vacuum
6	Infiltration 3	I3	3	2	2	3	Submerged in solution under -0.03 MPa vacuum for 7.5 minutes, repeated 4x
7	Wash 4	W4	5	5	5	4	Submerged in solution under -0.03 MPa vacuum for 22.5 minutes then placed in an ultrasonic bath for 5 minutes, repeated 3 times
8	Wash 5	W5	4	4	4	4	Submerged in solution and placed under -0.03 MPa vacuum for 25 minutes. Then 3x 5 minute cycles under -0.03 MPa vacuum and 5 minutes in an ultrasonic bath

*Dye Penetrant testing did not reveal cracks in Blocks 4 and 8

As an example, the infiltration and washing sequence for Block 6 started with an infiltration in a water-based solution of the contrast agent (Solution 1 in Table 1) at atmospheric pressure for five minutes. The next step was Wash 1, which put the block in a solution of clean water (Solution 4 in Table 1), and cleaned it in an ultrasonic bath for five minutes followed by five minutes at −0.03 MPa vacuum. Infiltration 2 then infiltrated Block 6 a second time; this step submerged the block at atmospheric pressure for 15 minutes in a solution of the contrast agent in a 50/50 vol% mixture of water and methanol (Solution 2 in Table 1). Wash Step 2 used only a 10 minute ultrasonic bath using a cleaning solution of 50/50 vol% water and methanol (Solution 4 in Table 1). This was followed by another wash step (Wash 3) that used clean water and submerged the sample for 15 minutes at atmospheric pressure then for 15 more minutes under Table 2 details the infiltration and wash steps for each block. The steps in each −0.03 MPa vacuum. Infiltration 3 again used Solution 2 in Table 1 (50/50 vol% water and methanol mixture with a contrast agent), submerging the sample then subjecting it to −0.03 MPa vacuum for 7.5 minutes, after which the vacuum was released, then redrawn back to −0.03 MPa. This was repeated for a total of four vacuum cycles. Wash 4 for Block 6 used a washing solution of 50/50 vol% water and methanol to submerge the solution, followed by a vacuum of −0.03 MPa for 22.5 minutes and an ultrasonic bath for five minutes. The vacuum and ultrasonic cycles were then repeated 2 additional times. The final step for Block 6 (Wash 5 in Table 2) used clean water (Solution 4 in Table 1) and an initial vacuum of −0.03 MPa for 25 minutes followed by three cycles of five minutes under −0.03 MPa vacuum and five minutes in an ultrasonic bath.

The following process produced the neutron radiographs taken at the East Radiography station after each infiltration or wash step: After a 22 minute exposure of the samples in the beamline, the activation foils were placed in contact with the radiographic film overnight; then, the film was developed the next day by an automatic film processor. The crack metric is based on the images produced by the dysprosium foil.

Next, image processing of the digital scans of the film radiographs created a metric based on the percentage of black pixels in the selected crack boundary box. Fig. 2 shows a crack as it appears across each of the steps in the digital metric process. Step 1 took the cropped radiograph (Fig. 2a) and used Image J [5] to invert and shift the histogram to a constant value (Fig. 2b). Step 2 used a rolling ball algorithm to remove non-crack noise from the image (Fig. 2c), Step 3 applied a threshold to the crack features to make crack features black and the rest of the block white (Fig. 2d), and Step 4 isolated individual crack features in each block (Fig. 2e).

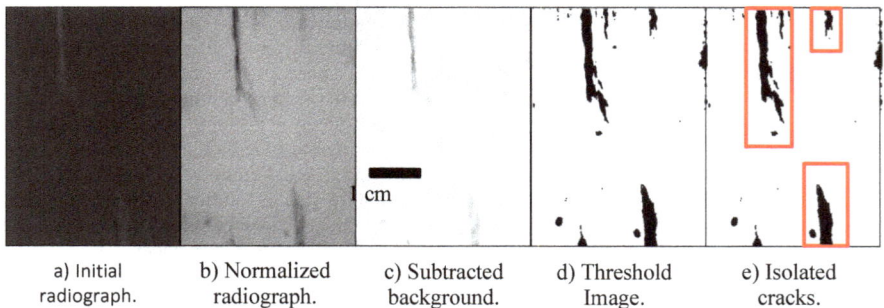

| a) Initial radiograph. | b) Normalized radiograph. | c) Subtracted background. | d) Threshold Image. | e) Isolated cracks. |

Figure 2. Images produced by each step of digital metric process (Block 6).

Neutron Activation Analysis Methods. Phase Two replaced the gadolinium contrast agent with dysprsoium as the primary neutron capture isotope of gadolinium creates a stable isotope, making NAA based on this isotope impractical. Dysprosium has similar chemistry to gadolinium, and dysprosium-164 has a substantial (2840 b) thermal neutron cross section, and the activation product dysprosium-165 has a 2.13 hour half-life [2,6]. The Geological Survey TRIGA Reactor (GSTR), a 1 MW_{th} reactor, owned and operated by the United States Geological Survey, irradiated the samples for the NAA testing. Eq. 1 relates the count rate of an activation product to the number of parent isotope atoms.

$$\frac{C}{\varepsilon y} = \frac{N_a \sigma \phi}{\lambda} [1 - e^{-\lambda t_i}][e^{-\lambda t_d}][1 - e^{-\lambda t_c}]. \tag{1}$$

where C is the number of gamma rays from the decay of the product isotope of interest recorded by the HPGe detector, ε is the detector efficiency for the corresponding gamma ray energies, and y is the branching ratio of the decay of interest. N_a is the number of atoms of the parent isotope present in the sample, σ is the thermal neutron cross section for parent isotope, ϕ is the thermal neutron flux experienced by the sample during irradiation, and λ is the decay constant of the product isotope. The times in the exponential terms are irradiation time (t_i), decay time (t_d) and count time (t_c).

Phase Two included four new crack samples, made using ASTM procedures as a guideline, and a control sample. Table 3 describes the mass, composition, and size of the samples, as well as a description of the cracks in the samples. Axial cracks propagate along the direction of extrusion from the manufacturing process. Through cracks are observable on both the front and the back of the sample, while surface cracks appear on a single sample surface.

Table 3. Phase Two (NAA) sample details.

Sample	Mass (g)	Composition	Sample Type	Crack Description
1	4.10	Aluminum 2024	2" wide, 0.375" thick bar	Axial through crack
2	3.23	Aluminum 2024	2" wide, 0.375" thick bar	Axial through crack
3	6.61	Aluminum 2024	1.5" outer diameter pipe with 0.25" wall thickness	Axial surface crack
4	1.52	Aluminum 6061	1 cm outer diameter tube with 1 mm wall thickness	Axial surface crack
control	3.17	Aluminum 2024	2" wide, 0.375" thick bar	No crack

Table 4. Phase Two (NAA) infiltration and washing methods.

Process Step	Method	Solution Composition
Infiltration	Submerged in solution at atmospheric pressure for 5 minutes	0.3M $Dy(NO_3)_3$ in de-ionized water
Wash 1	Surface scrub	De-ionized water
Wash 2	Submerged in solution in ultrasonic bath for 30 minutes, repeated once	De-ionized water
Wash 3	Submerged in solution in ultrasonic bath for 10 minutes, repeated four times	De-ionized water

Table 4 shows the methods of infiltration and washing as well as the composition of the solutions used in Phase Two. The infiltration step submerged the five samples in the infiltrant solution at atmospheric pressure for five minutes. Wash 1 attempted to remove residual infiltrant from the surfaces, while Wash 2 and 3 submerged the samples in de-ionized water with multiple cycles in an ultrasonic bath to remove deposited infiltrant from the cracks. Neutron Activation Analysis determined the mass of infiltrant remaining in the crack sample after each process step in Table 5, based on Eq. 1. Solving Eq. 1 also determined the neutron flux in each capsule during each irradiation based on the mass and resulting activity of a gold wire flux monitor included in each sample capsule.

Table 5. Phase Two (NAA) irradiation details.

Process Step	Capsule	Samples	Flux Monitor Mass (g)	Irradiation Time (s)	Calculated Flux (neutron/cm^{2}s)
Infiltration	1	1, 4	0.1210	5.1	1.42×10^{12}
	2	2, 3	0.1206	5.1	1.39×10^{12}
	3	control	0.1236	6.3	1.33×10^{12}
Wash 1	1	2, 3	0.1232	8.1	1.30×10^{12}
	2	1, 4, control	0.1237	8.1	1.34×10^{12}
Wash 2	1	1, 4, control	0.1246	8.5	1.36×10^{12}
	2	2, 3	0.1235	8.5	1.39×10^{12}
Wash 3	1	1, 4, control	0.0071	8.1	1.16×10^{12}
	2	2, 3	0.0080	8.1	1.17×10^{12}

Results and Analysis

Phase One demonstrated that infiltrating crack features with gadolinium enhanced the contrast of the features in resulting radiographs, allowing the development of a digital crack measurement. Phase Two directly measured the mass of a dysprosium infiltrant crack features using NAA. Changes in the size of the measured crack features between successive infiltration and washing steps in Phase One determined the effectiveness of each processing step. Changes in the mass of infiltrant measured between each step in Phase Two provided a check on the validity of the Phase One image-based metric.

Neutron Radiography Results. The Phase One image-based metric tests included eight blocks, yielding 21 distinct cracks. Each infiltration or washing step produced an image-based crack measurement as shown in Fig. 2. Figs. 3-6 present the percent change in the measured crack area as a function of the infiltration and washing steps described in Table 2.

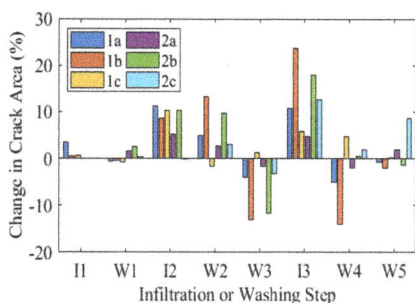

Figure 3. Changes in measured crack area as a function of processing step for blocks 1&2 in Table 2.

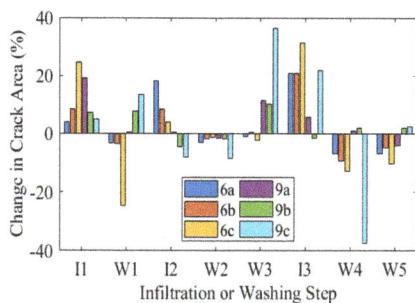

Figure 4. Changes in measured crack area as a function of processing step for blocks 6&9 in Table 2.

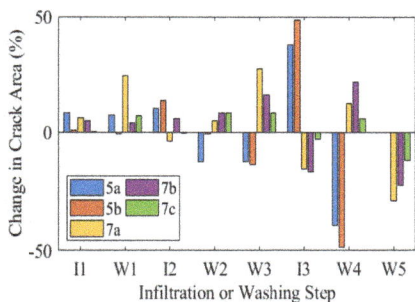

Figure 5. Changes in measured crack area as a function of processing step for blocks 5&7 in Table 2.

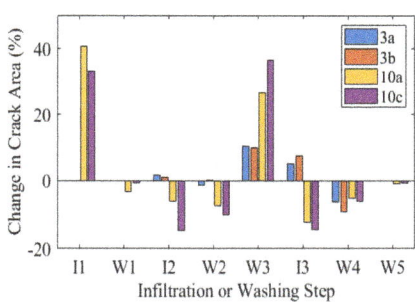

Figure 6. Changes in measured crack area as a function of processing step for blocks 3&10 in Table 2.

In general, infiltration steps that included multiple vacuum phases (Step I3 in Figs. 3-5) resulted in the largest increases in measured crack area, indicating that multiple vacuum phases were particularly effective at depositing gadolinium in cracks. This trend was not observed in Blocks 3 and 10 in Fig. 6, because Step W3 may have fully infiltrated the crack, preventing the subsequent infiltration step (I3) from further increasing the amount of gadolinium in the cracks. Infiltration Step 3, applied to Blocks 5, 7, 6, and 9 (Figs. 4 and 5), resulted in slightly more consistent infiltration than the same process applied to Blocks 1, 2, 3, and 10 (Figs. 3 and 6), indicating that adding a surfactant (ammonium lauryl sulfate) to the solution did not increase the amount of gadolinium deposited in the cracks. Wash Step 4 in Figs. 3-6, generally removed the largest amount of gadolinium, indicating that multiple cycles of vacuum and ultrasonic cleaning are the most effective way to remove the contrast agent from the cracks.

Neutron Activation Analysis Results. Figs. 7a and 7b show the mass of dysprosium in the crack samples measured by NAA after the initial infiltration (Fig. 7a) and the change in the mass indicated by NAA after each washing step in Table 4 (Fig. 7b). The cracks in Samples 1, 2, and 3 took up more dysprosium during the Infiltration Step than the crack in Sample 4 (which was also not indicated by dye penetrant testing). The mass of dysprosium indicated in the Control Sample in Figs. 7a and 7b was most likely trace infiltrant on the surface of the sample. Thus, the first wash step in Phase Two focused only on removing residual infiltrant from sample surfaces.

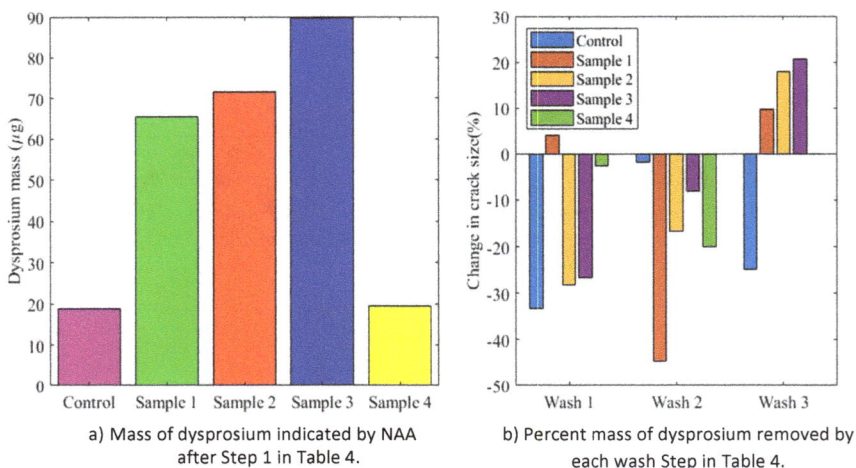

a) Mass of dysprosium indicated by NAA
after Step 1 in Table 4.

b) Percent mass of dysprosium removed by
each wash Step in Table 4.

Figure 7. Neutron Activation Analysis metric results.

Wash Step 1 in Table 4 removed ~30% of the dysprosium from the surface of the uncracked sample (see Fig. 7b). Therefore, a thorough surface washing was necessary to remove the residual infiltrant from the non-cracked portions of the sample after each infiltration step. Ideally, a thorough surface cleaning of residual infiltrant from an uncracked sample would remove 100% of the residual infiltrant, indicating that the control sample may contain extremely fine cracks or pores that the infiltrant was able to penetrate.

Wash Step 2 in Table 4 removed ~20% more dysprosium than Wash Step 3, indicating that longer cycles in the ultrasonic are more important than the number of cycles (see Fig. 7b). Since the washing solution was not changed between ultrasonic cycles, the additional cycles in Wash 3 may have deposited additional infiltrant in the cracks.

Summary and Conclusions

Phase One of this project tested the viability of a quantitative crack metric based on the analysis of digital scans of neutron radiographs of cracks infiltrated with gadolinium (III) nitrate. This phase also tested methods of infiltrating and washing the gadolinium contrast agent to/from cracks in standard aluminum test blocks. Gadolinium nitrate is an effective contrast agent, and image processing of the resulting radiographs produced a useable metric; however, the results from the washing methods were inconsistent and sometimes indicated increased crack extent after washing. Phase Two considered Neutron Activation Analysis as a secondary metric to directly measure the mass of an infiltrant present in crack samples. This metric also indicated inconsistencies in the effectiveness of the washing steps. Developing an effective and consistent

Materials Research Forum LLC
https://doi.org/10.21741/9781644900574-26

washing process will be necessary for the use of neutron imaging with a contrast enhancement agent as a pre-irradiation examination technique.

Future work will refine the infiltration and washing methods and may involve solutions containing both radiography contrast and NAA agents which will allow a direct comparison of the digital image processing data with the mass calculations to validate the neutron image based crack metric. The characteristic gamma ray emitted by dysprosium-165 is very low energy (94.7 KeV) and has a low branch frequency (3% of decays), leading to longer irradiation and count times to yield good statistics. An isotope with multiple, higher energy, and more frequent decays, may reduce the uncertainty in the mass calculations as several data points would be available to independently calculate the mass of infiltrant in each crack, allowing shorter irradiation and count times. The shorter irradiation times would produce lower amounts of activated materials, meaning less dose to personnel handling the samples after irradiation. Shorter count times also reduce potential dose and increase the number of countable samples for a given period.

References

[1] Brenizer J.S., Hosticka B., Berger H., Gillies G.T., The Use of Contrast Agents to Enhance Crack Detection Via Neutron Radiography, NDT & E International, 32 (1999) 37-42. https://doi.org/10.1016/S0963-8695(98)00024-3

[2] Information on https://www.ncnr.nist.gov/resources/n-lengths/

[3] Craft A.E, Wachs D.M., Okuniewski M.A., Chichester D.L., Williams W.J., Papaioannou G.C., Smolinski A.T., Neutron Radiography of Irradiated Nuclear Fuel at Idaho National Laboratory, Physics Procedia, 69 (2015) 483-490. https://doi.org/10.1016/j.phpro.2015.07.068

[4] Shulyakova O., Avtonomov P., Kornienko V., New Developments of Neutron Activation Analysis Applications, Procedia-Social and Behanvioral Sciences, 195 (2015) 2717-2725. https://doi.org/10.1016/j.sbspro.2015.06.380

[5] Rasband, W.S., ImageJ, U. S. National Institutes of Health, Bethesda, Maryland, USA, https://imagej.nih.gov/ij/, 1997-2016.

[6] Information on https://www.nndc.bnl.gov/chart/

Neutron Radiography - WCNR-11
Materials Research Proceedings 15 (2020) 174-179

Materials Research Forum LLC
https://doi.org/10.21741/9781644900574-27

Effect of Scattering Correction in Neutron Imaging of Hydrogenous Samples using the Black Body Approach

Chiara Carminati[1,a] *, Pierre Boillat[1,2], Sarah Laemmlein[3], Petra Heckova[4],
Michal Snehota[4], David Mannes[1], Jan Hovind[1], Markus Strobl[1]
and Anders Kaestner[1]

[1]Laboratory for Neutron Scattering and Imaging, Paul Scherrer Institut, CH-5232 Villigen-PSI

[2]Electrochemistry Laboratory, Paul Scherrer Institut, CH-5232 Villigen-PSI

[3]Cellulose & Wood Materials, Swiss Federal Laboratories for Materials Science and Technology (EMPA), CH-8600 Duebendorf, Switzerland

[4]University Centre for Energy Efficient Buildings, Czech Technical University in Prague, Bustehrad, Czech Republic

[a]chiara.carminati@psi.ch

Keywords: Scattering Artefacts, Systematic Biases, Quantitative Neutron Imaging

Abstract. The "black body" (BB) method is an experimental approach aiming at correcting scattering artifacts and systematic biases from neutron imaging experiments. It is based on the acquisition of reference images, obtained with an interposed grid of neutron absorbers (BB), from which the background including contaminations of scattering from the sample can be extrapolated. We evaluate in this paper the effect of the BB correction on two experimental datasets acquired with different setups at the NEUTRA and ICON beamlines at the Paul Scherrer Institut. With the two experiments we demonstrate the efficient utilization of the method for 2D as well as 3D data and in particular for kinetic studies. In the first dataset, differently varnished wood samples are studied through time resolved kinetic neutron radiography to evaluate the change in wood moisture content due to changes in relative humidity. In the second case study, engineered soil sample simulating a small experimental bioretention cell with rainfall, also known as rain garden, is imaged through on-the-fly neutron tomography.

Introduction

Together with spectral influence (beam hardening) and effects at pronounced edges, scattering from the sample and the detection system poses the biggest challenge for quantitative measurements of the linear attenuation coefficients in neutron imaging experiments with high spatial resolution.

We have recently introduced an efficient method and the corresponding data treatment for scattering correction that, compared to earlier attempts, has the advantage in neither requiring prior knowledge of the neutron spectrum nor of the sample composition [1, 2]. The method is based on the acquisition of additional reference images with an interposed grid of neutron absorbers, called "black bodies" (BB) which lend the method its name: BB correction. The main idea of the approach is that the signal measured behind a black body can be interpreted as the additive background of scattering components.

Here we present exemplary applications of the method to different imaging experiments and we discuss the effect on relevant quantifications. We consider two experiments that were performed at the Paul Scherrer Institut (PSI). In the first one, a collection of wood samples with different varnishing treatments are studied through kinetic neutron radiography in order to evaluate the changes in wood moisture content due to changes in relative humidity. In the second case study, an engineered soil sample simulating a small experimental bioretention cell with rainfall, also known as rain garden, is imaged. First, we briefly describe the type of images which are required for the BB correction, then we present the experimental setup for the two study cases and finally we show the effect of the applied correction.

BB correction

The BB correction is an experimental approach with the aim of mitigating artifacts due to systematic biases through scattering components [1,2]. Such artifacts consist of an increased transmission signal, often especially in the center of a bulk sample, resulting in a bias towards lower computed attenuation coefficients and "cupping type" effects in tomographic reconstructions, i.e. radially decreasing attenuation coefficients towards the center.

Additional images are required for the BB correction: the first one is the open beam with the BB grid (BB-OB), the second is with the sample and the BB grid in the beam (BB-S). From the BB-OB images, the systematic biases that are due to scattering by the experimental apparatus can be estimated through extrapolation between the black bodies in the grid. From the BB-S, the scattering contributions of the sample can be evaluated. Dedicated image processing was developed to interpolate for each pixel position of the field of view the neutron flux measured at the BB positions, thus estimating the background and scattering images [2,3].

Depending on the experiment type, BB-S images can and have to be acquired at different time steps during a full experimental run. In case of kinetic radiography studies, BB-S images can be acquired prior to and/or after the non-BB images, depending on whether the sample changes sufficiently during the measurement process that it affects the scattering background. Depending on the study, correspondingly, BB-S images may need to be interleaved with the non-BB images. For tomography, a sparse tomography with regular angular steps with interposed BB is generally recommended, with a number of projections on the order of the square root of the number of projections of the conventional tomography. In case of a highly symmetrical sample, for example cylindrical, this scheme can be furthermore relaxed, and a set of BB-S images can be acquired at the same tomographic angle before and/or after the tomographic scan.

Case 1: Wood

The aim of the first experiment is to study the development of moisture content in wooden musical instruments. Eight spruce wood samples with a varnished surface were produced, in order to reproduce violin characteristics. Sample dimensions of each block were 10x10x50 mm^3. The lateral surfaces were sealed to ensure that the origin of the sorption process was limited to the top and bottom surfaces of each sample.

The aim of the specific experiment was the analysis of the time dependent moisture content (MC) distribution over the wood cross section. In a climate chamber [4], the samples were put at a controlled temperature of 20°C, with relative humidity initially set at 35%. While keeping the temperature constant, relative humidity was raised up to 95%. After 20 min, the RH reached the 95% level, then the RH was kept constant at 95% for 5 h. The RH was reduced back to 35% again to be kept constant for another 5 h. During this process and within this sample environment, time resolved radiographs were obtained at the thermal neutron imaging beamline NEUTRA [5] at the neutron source SINQ at PSI using a scintillator/camera detector system with a field-of-view (FOV) of 150×150mm^2. The scintillator used was a 50μm thick LiF/ZnS based screen. The camera was an Andor Neo sCMOS, where the optics was set to a FOV of 161x136 mm and 2560x2160 pixel chip resulting in an effective pixel size of 63.1μm. Each image in the time series featured 15 s exposure time, with a time increment chosen to be 5 min. The total experiment duration for a single set of samples was 10.5 h during which 130 radiographs were acquired.

Neutron Radiography - WCNR-11   Materials Research Forum LLC
Materials Research Proceedings 15 (2020) 174-179   https://doi.org/10.21741/9781644900574-27

Open beam and sample images with a BB grid were obtained at the beginning of the experiment, with RH of 35%. The sample scattering and background were extrapolated and subtracted from the radiographs through an ImageJ plugin implementing the BB correction. From the normalized images, the change in wood MC at each time step was computed as: $\Delta MC_t = \rho_h \Delta d_h \big/ \rho_w d_w$,

where ρ_h and ρ_w are the densities of water and oven dry wood, d_w is the oven dry wood thickness and finally the difference in water thickness is expressed as $\Delta d_h = -\ln(T_t/T_0)/\Sigma_h$, with Σ_h being the attenuation coefficient for water with respect to the NEUTRA spectrum, and T_t and T_0 the transmission images in the wood regions for each time steps t and for the reference time 0 (when RH=35%), respectively.

Case 2: Soil
A soil sample was taken from a filter layer of a test bed simulating a bioretention cell, also called rain garden, which is a low-impact development construction that accumulates, infiltrates and treats storm water. The sample was composed by 50% of sand, 20% of topsoil and 30% of compost. The resulting soil mixture contained 12% mass fraction of particles smaller than 2 μm, 14% mass fraction of particles sized between 2 and 50 μm and 74% mass fraction of particles sized between 50 and 2000 μm. The particle density of the soil was 2563 kg/m^{-3}.

The sample was imaged at the cold neutron imaging beamline ICON [6] at PSI using a 100μm thick LiF/ZnS based scintillator coupled with an Andor Neo sCMOS camera with a FOV of 40×40 mm^2 and effective pixel size of 68 μm. During imaging, rainfall episodes were simulated by constant flux using heavy water as flowing fluid. Fluid drainage happened by gravity, while the inflow and outflow of fluid were balance monitored. As the drying and wetting are fast processes, the imaging technique has to be fast enough to capture the water flow. On-the-fly tomographies [7] were acquired with a continuous turning of 360°/min. Four 360° turns with 300 projections for each complete turn were performed with individual projection exposure times of 0.2 s (rate: 5 fps) and a corresponding neutron dose per projection of about 100 neutrons/pixel. BB images were also acquired on-the-fly at the beginning and at the end of the experiment, thus corresponding to the dry and wet conditions. CT reconstruction featuring BB correction was done in our in-house open source software MuhRec [2,3].

Results

Case 1: Wood
Fig. 1 shows the results of image data processing for the wood samples with and without BB correction in the two cases of minimal and maximal RH, i.e. 35% and 95%, respectively, obtained at time 0 and after 5h. In both cases, the effect of correction is clearly visible, with lower sample transmission values for the images normalized with BB correction. The extracted change in water mass within the wood samples at each time step is shown in Figure 2. Without BB correction, the mass content is underestimated up to 30% at the point of higher MC (95%, after 5 h). These results are confirmed by the measured variation of water mass before and after the experiment with a precision balance (circle in the picture), representing the reference measure.

Neutron Radiography - WCNR-11   Materials Research Forum LLC
Materials Research Proceedings 15 (2020) 174-179   https://doi.org/10.21741/9781644900574-27

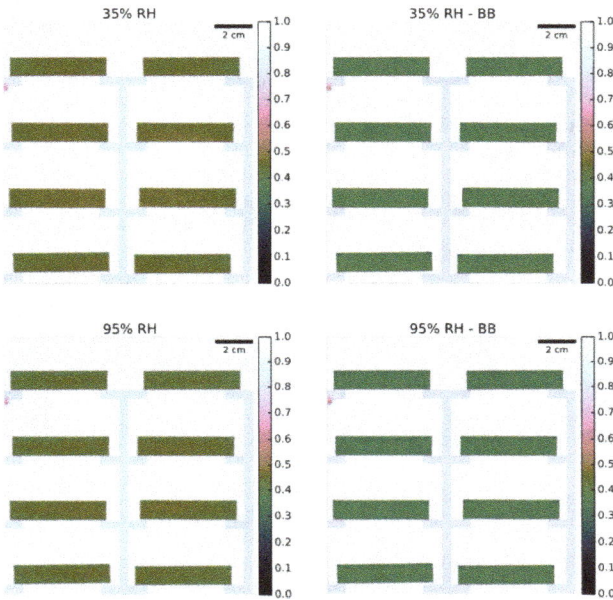

Figure 1: Transmission images computed at 35% RH (top) and 95% RH (bottom) without (left) and with (right) BB correction

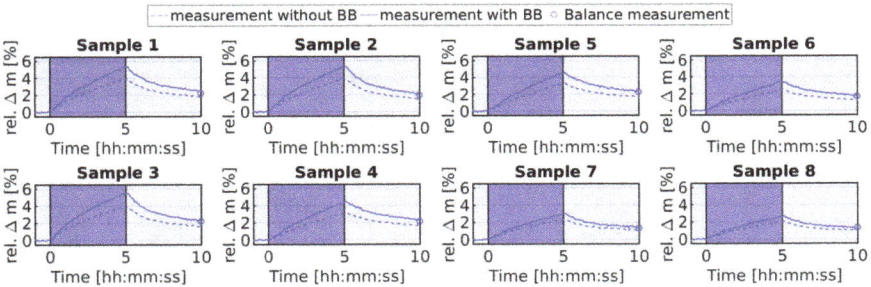

Figure 2: Relative change of water content computed from transmission images without (dashed line) and with (continuous line) BB correction, the reference balance measurement is also shown (circle). Samples 1 to 4 correspond to the top four samples, samples 5 to 8 are the four bottom samples in Fig. 1

Case 2: Soil

As a representative example of the results a cross-sectional slice at mid height of a sample reconstruction of the rain garden sample is shown in Fig.3. Even though these datasets are

Neutron Radiography - WCNR-11
Materials Research Proceedings **15** (2020) 174-179

Materials Research Forum LLC
https://doi.org/10.21741/9781644900574-27

affected by a high noise level due to very low neutron statistics, attenuation coefficients are noticeably different when adopting the BB correction, for both dry and wet conditions. In particular contrast appears improved notably.

Radial mean values of attenuation coefficients plotted in Fig. 4 show that correction with the BB approach results in higher attenuation coefficient (between 5 and 14%) for all radial position. A cupping effect does not appear prominent in the results, due to the small sample size and the in-homogeneity of the sample. However, when plotting the percent difference between the radial mean values obtained without and with BB correction (fig 4, right panels) with respect to the radial distance from the sample center, a pronounced decrease in difference when moving away from the sample center, as typical of a cupping effect, can be observed and as expected more pronounced for the wet condition.

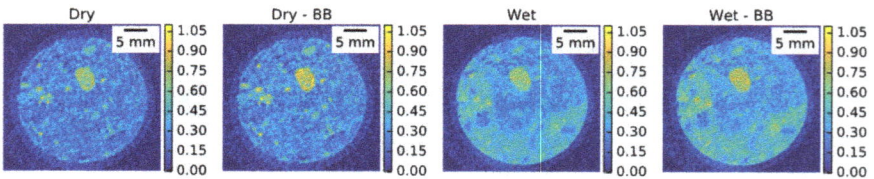

Figure 3: Reconstructed CT of the rain garden sample at dry and wet conditions without and with BB correction

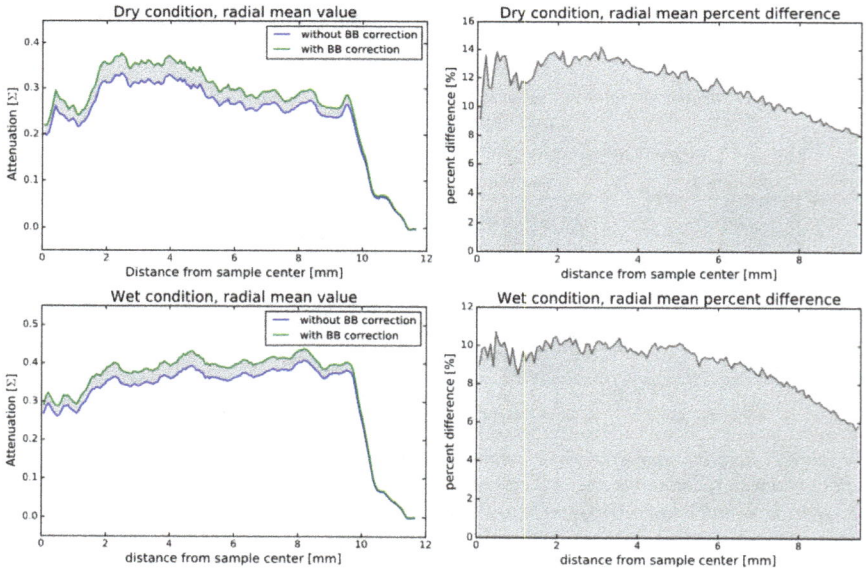

Figure 4: Radial mean values of attenuation coefficient, for dry and wet sample status.

Conclusions

We have presented the beneficial effect of the BB approach for scattering and systematic bias correction in two experimental datasets. In both cases, the additive background that results in underestimated attenuation coefficient appears to be mitigated through the BB correction. In the time resolved radiographic study of wood samples, uncorrected images resulted in clear underestimation of water mass, while the BB corrected measurements resulted in good agreement with an independent reference measurement. For the on-the-fly tomographies, the BB correction proved effective in successfully compensating cupping, resulting from scattering bias, even though the effect was weak and not easy to identify without the BB approach. The effect was found to contribute about 5% in mean attenuation coefficient error and with higher significance for the wet condition. These results demonstrate that the BB correction is well applicable to time resolved kinetic neutron imaging studies in 2D as well as in 3D. As hydrogenous materials are strong incoherent neutron scatterers, this type of correction is indispensable for sensitive quantitative studies in neutron imaging, for example when the aim is to quantify the amount of water.

Funding statements

This project receives funding from the European Union's Horizon 2020 research and innovation programme under grant agreement No 654000.

The experiment on engineered soil sample was supported by Czech Science Foundation under grant No 17-21011S.

References

[1] P. Boillat, C. Carminati, F. Schmid, C. Grünzweig, J. Hovind, A. Kaestner, et al. Chasing quantitative biases in neutron imaging with scintillator-camera detectors: a practical method with black body grids. Optics Express. 26 (2018) 15769. https://doi.org/10.1364/OE.26.015769

[2] C. Carminati, P.Boillat, F.Schmid, P. Vontobel, J. Hovind, M. Morgano, M. Raventos, M. Siegwart, et al. Implementation and assessment of the black body bias correction in quantitative neutron imaging. PLoS ONE. 14(1), e0210300. https://doi.org/10.1371/journal.pone.0210300

[3] MuhRec release 4.0.1. doi:10.5281/zenodo.1438402

[4] D. Mannes, F. Schmid, T. Wehmann, E. Lehmann. Design and Applications of a Climatic Chamber for in-situ Neutron Imaging Experiments. Physics Procedia, 88 (2017), 200-207. https://doi.org/10.1016/j.phpro.2017.06.028

[5] E.H. Lehmann, P. Vontobel, L. Wiezel. Properties of the radiography facility NEUTRA at SINQ and its potential for use as European reference facility. Nondestructive Testing and Evaluation. 16 (2001) 191-202. https://doi.org/10.1080/10589750108953075

[6] Kaestner AP, Hartmann S, Kühne G, Frei G, Grünzweig C, Josic L, et al. The ICON beamline – A facility for cold neutron imaging at SINQ. Nuclear Instruments and Methods in Physics Research Section A: Accelerators, Spectrometers, Detectors and Associated Equipment. 659 (2011) p. 387-393. https://doi.org/10.1016/j.nima.2011.08.022

[7] M. Zarebanadkouki, A. Carminati, A. Kaestner, D. Mannes, M. Morgano, S. Peetermans, E.H. Lehmann, P. Trtik. On-the-fly neutron tomography of water transport into lupine roots Physics Procedia, 69 (2015), 292-298. https://doi.org/10.1016/j.phpro.2015.07.041

Neutron Radiography - WCNR-11   Materials Research Forum LLC
Materials Research Proceedings 15 (2020) 180-184   https://doi.org/10.21741/9781644900574-28

Fast Neutron Imaging at a Reactor Beam Line

R. Zboray[1,a*], Ch. Greer[1,b], A. Rattner[1,c], R. Adams[2,d], Z. Kis[3,e]

[1]Department of Mechanical and Nuclear Engineering, The Pennsylvania State University, 233 Reber Building, University Park, PA 16802, USA

[2]Swiss Federal Institute of Technology Zurich, Department of Mechanical and Process Engineering, Sonnegstrasse 3, CH-8092, Zurich, Switzerland

[3]Hungarian Academy of Science, Centre for Energy Research, 29-33 Konkoly Thege Miklos street, 1121 Budapest, Hungary

[a]rzz65@psu.edu, [b]czg5155@gmail.com, [c]asr20@psu.edu, [d]adams@lke.mavt.ethz.ch, [d]zkis@iki.kfki.hu

Keywords: Fast Neutron Imaging, Reactor Beam Line, Plastic Scintillator, ZnS

Abstract. Though fast neutron contribution in a thermal imaging beam line is typically considered as a burden, we have investigated its application for imaging using the high energy tail of the fission spectrum at a reactor beam line. Fast neutron imaging is a promising non-destructive technique for testing dense and voluminous objects of practically any material composition. Fast neutron radiography and tomography have been performed using the RAD beam line of the 10 MW research reactor of the Budapest Neutron Centre (BNC), Hungary, using a camera-based imaging detector system on different bulky objects (up to 150 mm in diameter) and the results are presented here.

Introduction

Fast neutron imaging is a promising non-destructive technique for testing dense and voluminous objects of practically any material composition. Only fast neutrons can provide images with reasonable contrast and quality for dense and voluminous samples containing mixed low-Z/high-Z materials, where photon-based techniques or even thermal neutron imaging would fail. Several different applications have been attempted ranging from homeland/civil security problems like detecting heavily shielded explosives, or contraband, to looking at robust cultural heritage objects or at industrial processes in heavily attenuating containers.

While fast neutron contribution in a thermal imaging beam line is typically considered as a burden, here we investigate its application for imaging. In many thermal neutron imaging beam lines, the epithermal and fast neutrons are intentionally suppressed by using, typically, sapphire filters to lower the dose to sensitive equipment such as the camera of the imaging detector, thereby also prolonging their functioning life time. In such beams, usually there is a significant fraction of MeV neutron content from the high-energy tail of fission or spallation spectrum of the source. Here, we focus on the application of reactor-based thermal beam lines for fast neutron imaging. This approach may be particularly useful as very few beam lines exist that are specially designed and dedicated to fast neutron imaging (e.g. at the YAYOI fast reactor in Japan [1] or the NECTAR beam line at the 20-MW FRM II reactor in Munich, Germany [2]). To our knowledge, besides our attempts in [3] and [4], no application of fast neutron imaging at a thermal beam line has been reported in the literature.

Experimental setup

The imaging beam line. We have primarily used the RAD beamline of the 10 MW research reactor of the Budapest Neutron Centre (BNC), Hungary, to demonstrate the feasibility of fast neutron radiography and tomography [4]. The beamline is routinely utilized for thermal neutron imaging having a thermal flux of around 4×10^7 n/cm^{-2}s^{-1}. The beam features a relatively high gamma background of 8.5 Gy/h (non-attenuated) and a significant fast neutron contribution of 2.7×10^7 n/cm^{-2}s^{-1} (E > 2.5 MeV). These values have been carefully evaluated by the instrument crew. The fast flux specifically has been measured by the standard procedure, activating a Cd-clad Ni foil utilizing the Ni-58(n,p)Co-58 threshold reaction (~2.5 MeV) as described in [9]. Further details about the beam line are given in [5]. We have used a 10 mm thick borated rubber mat filter (MirroBor), enabling the suppression of practically the entire thermal neutron component of the beam to avoid unnecessary sample activation. We also used a 200 mm thick lead sheet to filter direct in-beam gamma contribution, the efficacy of which as has been carefully determined using the shadow cone technique [3]. Most of the relatively modest gamma background in the experimental hatch is originating from prompt gammas due to activation of surrounding structures by scattered neutrons [3]. Details on how the beam filtering has been optimized to suppress the thermal component and maximize the fast neutron to gamma ratio are given in [3, 4]. The filtering decreases the useful fast neutron flux to around 3.3×10^5 n/cm^{-2}s^{-1}. The primary aperture of the beamline (situated at the boundary of the biological shield and the core reflector) with a diameter of 28 mm results in a calculated L/D of 177 at the position of the detector, which is placed at 4960 mm from the primary aperture, enabling a quasi-parallel beam imaging geometry.

The fast neutron imaging detector and scintillator. An ANDOR Neo 5.5 (2560 × 2160 pix) sCMOS camera with a 50 mm, f/1.2 objective with a total field of view (FOV) of the detector of about 270 × 230 mm^2 at an effective pixel size of about 106 µm was used. Regarding the scintillator screen, it should be pointed out that the most efficient nuclear reaction to convert fast neutrons to a detectable particle is elastic scatter on hydrogen (recoil protons). Therefore, hydrogen rich materials are preferred for fast neutron detection. Specifically, for imaging purposes, two main types of fast neutron scintillators are commonly used as practical choices: organic (plastic) slab scintillators and inorganic (mostly ZnS type) scintillator embedded in a transparent organic matrix used for conversion. We have tested both types of scintillators, a BC400 plastic scintillator slab from St. Gobain [6] and a commercially available scintillator with ZnS:Cu mixed into a polypropylene (PP) matrix [7]. Both types of scintillators have their advantages and disadvantages. The plastic scintillator is transparent to its own light therefore larger thicknesses, up to e.g. a centimeter, can be applied even in slab form. This can improve the conversion efficiency which is important for highly penetrating, few MeV fast neutrons. However, a very large screen thickness has an adverse effect on the spatial resolution (e.g. proper focusing on a thicker screen becomes increasingly a problem). The most severe disadvantage of plastic scintillators compared to ZnS-based scintillators is the roughly factor ten lower light yield. This is especially detrimental for camera-based imaging detectors for which the light collection efficiency is relatively poor due mainly to the small solid angle under which the screen is seen by the objective, amounting to typically a factor of about 1e−4. Therefore, if one uses a scintillator with relatively low light yield in such a setup, neutron collision events can go undetected. This we have pointed out in our earlier work using an 8mm thick BC400 [3]. Furthermore, thick plastic scintillator slabs have a higher gamma sensitivity, while thinner ZnS-based screens are somewhat less sensitive to gammas, on the order of half [3]. On the other hand, a serious drawback of ZnS-based screens is their low transparency due to the polycrystalline ZnS and the granular structure of the screen, which prohibits benefiting from higher thicknesses (it is typically limited to about 2-4 mm), opposed to transparent screens. The diffuse scatter of the scintillation light in ZnS-based screens is the main contributor to the native screen blur. Our

Neutron Radiography - WCNR-11   Materials Research Forum LLC
Materials Research Proceedings **15** (2020) 180-184   https://doi.org/10.21741/9781644900574-28

findings clearly prove that by far the best imaging efficiency can be achieved using a ZnS-based screen for fast neutrons and that the compromise on the spatial resolution is not that severe, as is shown in [4].

Results

We have imaged different objects with increasing complexity as reported in [3,4], with a determined spatial resolution of approximately 1.6 mm. The utility of fast neutron imaging has been demonstrated with the application to more complex and robust objects including a 150 mm diameter viscous coupling used in helicopter blade shafts [4]. Here we report on the latest investigations on a complex, robust object: a lithium-fueled power source for planetary landers

Li-fueled reactor for conceptual planetary lander power system. Another complex object we imaged using fast neutrons was a proof-of-concept Li-fueled batch reactor for powering a planetary lander, that was developed and tested at Pennsylvania State University [8], Fig. 1a. This device would react stored lithium fuel with readily accessible *in situ* oxidizers on planetary surfaces to generate heat and power. This technology has been proposed for Venus surface missions, using the primarily CO_2 atmosphere as an oxidizer, for Lunar and Mars missions (using *in situ* water and/or CO_2), and for missions to Ocean Worlds such as Jupiter's moon Europa (using surface ice as an oxidizer). In a batch configuration, condensed Li-based products (*e.g.*, Li_2C_2, Li_2CO_3, Li_2O, LiOH, LiH…) would be retained in the reactor vessel, avoiding contamination of the surroundings. This technology may enable very high specific-energy power systems because only the low-mass Li fuel, and not the more massive oxidizer, must be transported with the lander [8].

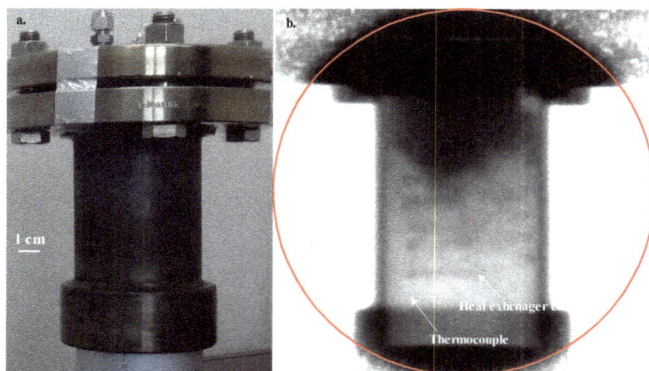

Figure 1: Photo (a) and a fast neutron image (b) of the Li-fueled reactor, post-operation. The flat and dark-field corrected radiographic image reveals a heat exchanger coil, the upper highly oxidized region (dark-gray), and the lower, primarily non-oxidized, elemental Li-rich portion (light gray). The image was taken over four minutes of exposure time. The red line represent roughly the beam size at the detector position.

The reactor vessel is a stainless-steel cylinder (OD ~ 90 mm, ID ~ 75 mm) with a large flange on top. The cylindrical reactor vessel contains a stainless steel heat exchanger coil near its base for reaction heat delivery to a compressed air stream. The reactor was filled ~75% with Li, evacuated to ~200 Pa, and externally heated to the auto-ignition temperature of Li and CO_2 (approximately 410 °C, above the melting temperature of Li at 180 °C). CO_2 was then metered

Neutron Radiography - WCNR-11   Materials Research Forum LLC

Materials Research Proceedings **15** (2020) 180-184   https://doi.org/10.21741/9781644900574-28

into the headspace above the molten Li pool, initiating the oxidation reaction. The reactor test was ended after the CO_2 consumption rate dropped below the intake flow meter measurement range. After this test, the device was imaged by fast-neutrons.

Due to the very high Li content, the reactor would be fully opaque to thermal neutrons. X-ray images would be burdened by metal artifacts. Therefore, fast-neutron imaging offers the best approach to non-destructively imaging the reactor contents. A 3 mm thick ZnS/PP screen was used for the imaging. A fast neutron radiograph is shown in Fig. 1b. It reveals some internal metallic structures, including the heat exchanger coil and thermocouple wires. The reaction products (including C, Li_2O, Li_2CO_3, Li_2C_2 [8]) all have larger attenuation coefficients than elemental Li. As a result, the image is darker near the top, where the oxidation reaction reached high yield and more attenuating products are concentrated. This "crust" of products seems to have prevented CO_2 from reaching the lower portion of the reactor. The lower portion of the reactor is much more transparent, suggesting the presence of unreacted elemental Li.

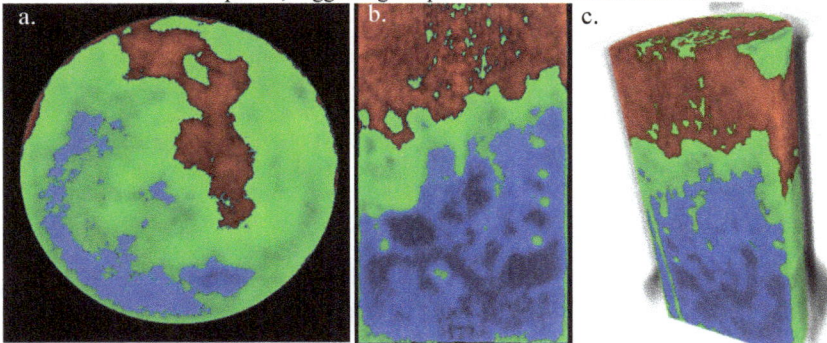

Figure 2: Segmentation (a. - horizontal cut, b.- vertical cut) and volume rendering (c.) of the internal contents of the Li reactor based on the tomographic reconstruction. The segmentation is based on three gray-level intervals: blue represents mainly elemental Li, and green and red represent the reaction products and metallic parts inside the reactor.

We performed tomographic imaging of the reactor, acquiring 376 equally spaced projections as the sample was rotated 360° about its vertical axis, and 3×3 pixel binning applied, resulting in a 317 μm effective voxel size. At each projection angle, five one-minute long exposures of the object were taken and a pixel-wide median of the five images was applied to improve image quality and to suppress noise. The images were filtered for ring artifacts using the combined wavelet and Fourier filter by Munch et al. [10]. After reconstruction by Filtered Back Projection (FBP) the images were treated by an anisotropic diffusion filter [11] to de-noise them with minimal blurring. Figure 2 shows a segmentation of the reactor contents based on the reconstructed images (performed in VG Studio [12]). The segmentation was performed with three grey value intervals corresponding to increasing attenuation of the material (from blue to red). Blue represents mainly elemental Li; green and red mainly represent the reaction products. Note that the metallic parts (heat exchanger coil, thermocouples) have sufficient attenuation to fast neutrons such that they are also segmented into the green band. This could be improved by a more thorough analysis and more sophisticated data segmentation methods, which is planned for future work.

Summary

We have reported our efforts to apply a thermal neutron imaging beam line with a significant fast neutron component for fast neutron imaging. Studies were performed using the RAD beam line at the research reactor of the Budapest Neutron Centre. We have tailored the in-beam filtering of the beam line to provide as high as possible fast neutron flux while suppressing the gamma background and the thermal contribution. We have made a significant effort to optimize the imaging detector to achieve maximal efficiency, and achieved an imaging resolution of approximately 1.6 mm. We demonstrated the utility of fast neutron imaging in the non-destructive analysis of robust and dense objects containing significant portions of both light and heavy elements. Such objects would be impossible to non-destructively image with other approaches. Tomographic images of a $Li-CO_2$ reactor (~90 mm in diameter) were generated and presented.

References

[1] Fujine S., Yoneda K., Yoshii K., Kamata, M., Tamaki, M., Ohkubo, K., Ikeda, Y., Kobayashi, H., Development of imaging techniques for fast neutron radiography in Japan. Nuclear Instruments and Methods in Physics Research A 424 (1999) 190-199. https://doi.org/10.1016/S0168-9002(98)01326-6

[2] T. Bucherl, Ch. Lierse von Gostomski, H. Breitkreutz, M. Jungwirth, F.M. Wagner, NECTAR-A fission neutron radiography and tomography facility, Nuclear Instruments and Methods in Physics Research A, 651, (2011)86–89. https://doi.org/10.1016/j.nima.2011.01.058

[3] Zboray, R., Adams, R., Kis, Z., 2017. Fast neutron radiography and tomography at a 10-MW research reactor beamline. Appl. Radiat. Isot. 119, 43–50. https://doi.org/10.1016/j.apradiso.2016.10.012

[4] Zboray, R., Adams, R., Kis, Z., 2018. Scintillator screen development for fast neutron radiography and tomography and its application at the beamline of the 10MW BNC research reactor, Applied Radiation and Isotopes 140 (2018) 215–223. https://doi.org/10.1016/j.apradiso.2018.07.016

[5] Kis, Z., Szentmiklósi, L., Belgya, T., Balaskó, M., Horváth, L., Maróti, B., 2015. Neutron based imaging and element-mapping at the Budapest Neutron Centre. Phys. Procedia, 69, 40-47. https://doi.org/10.1016/j.phpro.2015.07.005

[6] St.Gobain, 2011. Organic scintillation materials.

[7] RC Tritec AG, T., 2017. Scintillators. ⟨http://www.rctritec.com/en/scintillators.html⟩.

[8] Greer, C. J., Paul, M. V., & Rattner, A. S. (2018). Analysis of lithium-combustion power systems for extreme environment spacecraft. Acta Astronautica, 151, 68-79. https://doi.org/10.1016/j.actaastro.2018.05.039

[9] ASTM, 2002. Standard test method for measuring fast-neutron reaction rates by radioactivation of nickel. E 264-02, ASTM International, United States.

[10] Munch, B., Trtik, P., Marone, F., Stampanoni, M., 2009. Stripe and ring artifact removal with combined wavelet – fourier filtering. Opt. Express 17 (May (10)), 8567–8591. https://doi.org/10.1364/OE.17.008567

[11] Perona, P., Malik, J., 1990. Scale-space and edge detection using anisotropic diffusion. IEEE Trans. Pattern Anal. Mach. Intell. 12, 629–639. https://doi.org/10.1109/34.56205

[12] VolumeGraphics, 2018. VGstudio. ⟨https://www.volumegraphics.com/en/products/vgstudio.html⟩

Neutron Radiography - WCNR-11
Materials Research Proceedings **15** (2020) 185-190

Materials Research Forum LLC
https://doi.org/10.21741/9781644900574-29

3D Reconstruction of the Rotational Axis in Fission Neutron Tomography

Oliver Kalthoff[1, a *], Thomas Bücherl[2,b]

[1]University of Applied Sciences Heilbronn, Max-Planck-Straße 39, 74081 Heilbronn, Germany

[2]ZTWB Radiochemie München, Technical University of Munich, Walther-Meißner-Str. 3, 85748 Garching, Germany

[a]oliver.kalthoff@hs-heilbronn.de, [b]thomas.buecherl@tum.de

Keywords: Fission Neutron Tomography, Image Registration, Rotation Axis

Abstract. In tomography, a misalignment of the rotational axis can introduce blurring. We present intensity-based image registration to calculate the axis' components in three-dimensional space. We have shown that the axis can be deduced from rotating *and* translating image pairs acquired at 0° and 180°. No prior experimental calibration or any a-priori knowledge about the system's mechanical setup is necessary. Three samples of different symmetry and homogeneity were examined to experimentally assess the numerical effects of image registration.

Introduction

In fission neutron tomography the reconstruction quality is affected by the correct alignment of the object and the rotational positioning system relative to the detection system. A lateral shift and tilt of the system's rotation axis (SRA) may cause reconstruction algorithms to obscure or simulate image details. Hence for accurate reconstructions, the spatial location of the SRA is crucial. This especially holds for high throughput experiments or in situations where high image contrast is important.

Various methods have been used to account for the tilting of the SRA. A widely used approach is to use a pair of mirrored radiographs taken from complementary viewing angles, for instance at 0° and 180°. One radiograph is rotated in the detector plane until a similarity measure is optimized with respect to its complement. The rotation angle then represents the tilting of the SRA. A more sophisticated approach has been suggested by [1]. Here, the covariance matrix of an averaged radiograph is calculated from which the (orthonormal) eigenvectors are derived. The rotation axis is identical to the eigenvector corresponding to the largest eigenvalue.

Both approaches make two simplifying assumptions, which do not hold in general. First, a lateral shift of the SRA is assumed to be negligibly small implying that the object's axis is linearly aligned with the SRA. Second, the SRA is supposed to be completely located in the detector plane i.e. perpendicular to the neutron flux.

At TUM's NECTAR facility [2] three samples were investigated with the goal to analytically calculate the SRA's position in three-dimensional space. Since the SRA may be highly sensitive to a particular experimental setup, its position had to be inferable solely from particular radiographs without prior calibration or separate measurements.

The problem has occurred in other fields like X-ray transmission tomography, or X-ray diffraction tomography as well. Gürsoy [3] and earlier Donath [4] assessed artifacts in the reconstructed images, which are evaluated using an optimization framework. Here, the SRA's center is regarded as a parameter, which is iteratively adjusted such that the image's entropy is minimized. Others [5,6] employed feature-based image registration to estimate the lateral shift and tilt of the SRA. Both approaches yield the SRA's projection in the detector plane.

Neutron Radiography - WCNR-11   Materials Research Forum LLC
Materials Research Proceedings 15 (2020) 185-190   https://doi.org/10.21741/9781644900574-29

To the best of our knowledge there is no method to calculate the SRA in three-dimensional space from two-dimensional projections. To achieve this goal, we combined intensity-based image registration with general rotations in three dimensions: While image registration yields the object's *planar* tilting angle and its translation vector, these observables can be used to calculate the SRA's components in three-dimensional space.

The results of the new method are encouraging and show that the quality of the tomographic reconstruction is be enhanced if the SRA is deduced from the measurement itself. We have shown that the approach is self-contained allowing for the re-assessment of completed measurements.

Methods

The golden ratio scheme [11] employs a nonsequential decomposition of the sample rotation angle sequence. If the SRA is perfectly adjusted, each radiograph is identical to its mirrored complement. In practice, this condition is difficult to establish. Instead, the image pairs are not aligned or not registered. This effect is best described by a two-dimensional transformation $\tau = \tau(x; M, b)$, where M and b represent a rotation matrix and a translation vector, respectively. Since M and b are not known in advance, τ has to be deduced from the image pairs. The calculation of M and b is considered an ill-posed problem for which no unique solution exists. Therefore, image registration techniques require a suitable distance measure *and* a regularizer.

Conversely, τ can be viewed as a transformation between a reference and a template coordinate system establishing a mapping of corresponding unit vectors (or any linear combination thereof). The same result could have been obtained by a 180° rotation of unit vectors around the SRA. This rotation is expressed in matrix form from which the SRA's components can be derived.

Image preprocessing. Albeit comprehensive measures to shield the fluroscope detector at the NECTAR facility, any radiograph is subject to imaging noise. This is predominantly reflected in randomly distributed whitish streaks originating from scattered neutrons or secondary gamma radiation. The streaks impose fundamental problems on any image registration technique: Even small disturbances of the input images can cause M and b to become unstable. It is therefore essential to remove the streaks without introducing image blur. A sophisticated method has been suggested by Osterloh et al. [7]. The procedure requires the image's standard deviation as input parameter and operates unattendedly. As a preprocessing step all radiographs were corrected for random streaks.

Image registration. In the following let $R, T \in \mathbb{R}^{m \times n}, m, n \in \mathbb{N}$ be two images acquired at 0° and 180.43°, respectively[1]. R is considered as a (fixed) reference, whereas T is a sliding (moving) template [12]. The goal is to find a spatial transformation $\tau(x; M, b)$ such that $\tau(T; M, b)$ is similar to R. Here, similarity is expressed in terms of a distance measure $d \in \mathbb{R}$. The registration problem can then be stated as follows:

Given two images R and T and a distance measure d, calculate a transformation τ such that

$$d(R, \tau(T; M, b)) \longrightarrow \min. \tag{1}$$

The particular choice of τ depends on the experimental conditions. Three remarks are indicated:

---

[1] Using the golden ratio scan, T cannot be acquired at exactly 180°. Instead, the image closest to 180° was chosen as template. We considered the angular deviation of 0.43° to be negligible compared to other imaging defects originating from beam divergence or blurring.

1. Although beam divergence is inherent to any radiographic experiment, its effects are negligible when both the sample's dimensions and the sample-detector distance are small.
2. The detector's exposure time was set large enough to ensure that the neutron flux can be regarded as constant. Therefore, no intensity mapping of image pairs is necessary.
3. Since the samples are rigid bodies, $M \in \mathbb{R}^{2 \times 2}$ is an orthogonal rotation matrix with $\det(M) = 1$.

Hence, $\tau(x; M, \boldsymbol{b}) = Mx + \boldsymbol{b}$ is a linear or rigid [13] transformation. Several image registration techniques are known to align R and T. Intensity based image registration is of special importance since no landmarks (or fiducials) are necessary. A straightforward approach is based on the minimization of the sum of squared differences (SSD, [8] or [9]). In this case (1) is re-written as

$$d^{SSD}(R, \tau(T; M, \boldsymbol{b})) = \tfrac{1}{2} \| R - \tau(T; M, \boldsymbol{b}) \|^2 = \tfrac{1}{2} \sum_{i,j} (R_{ij} - \tau(T; M, \boldsymbol{b})_{ij})^2 \longrightarrow \min. \qquad (2)$$

The problem is to find a suitable optimization routine to minimize d^{SSD} with respect to M and $\boldsymbol{b}$. A Gauss-Newton regularization method was used. This is advantageous, since the sum of squared errors is quadratic in the parameters to be estimated and no Hessian is required [12]. Alternatively, Levenberg-Marquardt techniques have been suggested in the literature [10].

Rotation around the SRA. A rotation of points around an axis through the origin is given by

$$[R_{\boldsymbol{n}}(\alpha)]_{ij} = (1 - \cos \alpha) n_i n_j + \cos \alpha \, \delta_{ij} + \sin \alpha \, \varepsilon_{ijk} n_k. \qquad (3)$$

where $\boldsymbol{n} = (n_1, n_2, n_3)^T$ is a unit vector pointing into the axis' direction, δ_{ij} is Kronecker's Delta and ε_{ijk} is the Levi-Civita symbol. In our case $\alpha = 180°$, therfore (3) simplifies to

$$R_{\boldsymbol{n}}(180°) = \begin{pmatrix} 2n_1^2 - 1 & 2n_1 n_2 & 2n_1 n_3 \\ 2n_2 n_1 & 2n_2^2 - 1 & 2n_2 n_3 \\ 2n_3 n_1 & 2n_3 n_2 & 2n_3^2 - 1 \end{pmatrix} \qquad (4)$$

Applying (4) to a unit vector of the template frame yields the corresponding unit vector of the reference frame:

$$\boldsymbol{x}_R = R_{\boldsymbol{n}}(180°) \, \boldsymbol{x}_T \qquad (5)$$

where $\boldsymbol{x}_R$ is calculated with respect to M and $\boldsymbol{b}$ and $\boldsymbol{x}_T$ is an arbitrary vector with $\|\boldsymbol{x}_T\| = 1$. Note that only n_1 and n_2 can be calculated from (5). Using spherical coordinates one has

$$\begin{pmatrix} n_1 \\ n_2 \\ n_3 \end{pmatrix} = \begin{pmatrix} \sin \theta \cos \phi \\ \sin \theta \sin \phi \\ \cos \theta \end{pmatrix}, \qquad (6)$$

where θ and ϕ are the polar and azimuthal angles, respectively.

Samples. Three samples of different geometrical shape and homogeneity were chosen to assess the results of the intensity-based registration method: a cylindrically shaped object made of plastic, a turbine blade and a plastic bag randomly filled with electronics components and metal parts. While the first and the second sample exhibit well defined contrasts, the third sample appears blurred and inhomogeneous. For each sample the detector's exposure time was 60s.

Each radiograph was corrected for the dark-current (DC) and open-beam (OB) images according to

$$x_{corr} = \frac{x - x_{DC}}{x_{OB} - x_{DC}}.$$

Results

For all samples intensity-based image registration was applied to the image pairs acquired at 0° (R) and 180° (T). The results are depicted as superimposed images, where T has been shifted and rotated. The following numeric results were obtained:

$$M_{cyl} = \begin{pmatrix} 0.9997 & 0.0232 \\ -0.0232 & 0.9997 \end{pmatrix}, \quad b_{cyl} = \begin{pmatrix} 38 \\ -9 \end{pmatrix}$$

$$M_{blade} = \begin{pmatrix} 0.9998 & 0.0199 \\ -0.0199 & 0.9998 \end{pmatrix}, \quad b_{blade} = \begin{pmatrix} 21 \\ -11 \end{pmatrix}$$

$$M_{bag} = \begin{pmatrix} 0.9998 & 0.0178 \\ -0.0178 & 0.9998 \end{pmatrix}, \quad b_{bag} = \begin{pmatrix} -45 \\ -13 \end{pmatrix}$$

The components of b are given in pixels and are rounded. The polar and azimuthal angles for the samples are (c.f. (6))

$$\theta_{cyl} = 0.6°, \phi_{cyl} = 89.3° \quad \theta_{blade} = 0.6°, \phi_{blade} = 89.4° \quad \theta_{bag} = 0.6°, \phi_{bag} = 89.5°$$

Note that, depending on the direction of the coordinate axes, the values of θ and ϕ have to be interpreted accordingly.

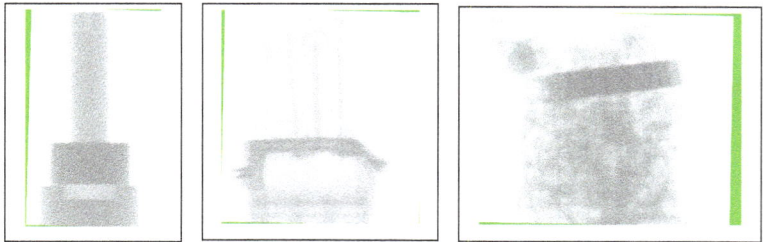

Fig1: Results for mirrored and registrated image pairs. Left: cylinder, middle: turbine blade, right: bag filled with electronics components. The images are displayed in false colors to emphasize the registration effects of rotating and shifting the template images T. Images were cropped and scaled for display.

Discussion

The goal of estimating the SRA from image pairs was achieved. We presented an analytical method to calculate the SRA's components using intensity-based image registration methods and rotation matrices in three-dimensional space. Comparable results were obtained for the rotation matrix, indicating that the SRA is tilted as expected. Since the rotational device was not modified, the SRA is almost identical for all three samples. The SRA's polar angle (0.6°) is in accordance with the empirical value assumed at NECTAR (0.7°). However, the method yielded a different translation vector for the plastic bag. There are two reasons. First, the plastic bag is amorphous and does not exhibit high-contrast structures which are needed for intensity-based registration. Second, image registration is regarded as an ill-posed problem where no unique solution exists. There may be a large number of "local" solutions near the "global" i.e. true minimum of the distance measure. To cure this, manually selected fiducials (or landmarks) can reduce the plurality of solutions. This is in contrast to our intention: The SRA's position had to be allocatable from particular radiographs without prior calibration or additional measurements.

The rotation matrix M and the translation vector b are estimated simultaneously. However, b may be prone to large systematic errors. Recall that the template (or moving) image has to be mirrored prior to image registration. The more the SRA is shifted from the image center, the larger the systematic error in b is: Mirroring anticipates a lateral shift of the template relative to the reference image. Therefore, the template has to be mirrored without a lateral shift. A possible solution is currently being investigated.

It was assumed throughout the experiments, that beam divergence is negligible. If this condition does not hold, image scaling becomes increasingly important. The registration model is still linear, but not any longer rigid in a strict sense. The transformation M is affine and $\det(M) > 0$. The problem is aggravated, if scaling *and* shearing is necessary. This aspect will be assessed for large radioactive waste drums.

Our approach is not restricted to the golden ratio scheme. It can be applied to any acquisition technique as long as the angular difference of the image pairs is (approximately) 180°.

A cautionary note has to be made. Any attempt to retrospectively estimate the SRA from a series of radiographs would be pointless if the true SRA would have been known a priori. Vice versa, it is speculative to assess the correctness of any approach a posteriori if the true SRA is unknown. It would be illustrative to actually compare tomographic reconstruction results based on different methods to estimate the SRA. We feel that this would fit into a review paper.

References

[1] Z. Ji et al., Calibration and Correction Method of the Deflection Angle of Rotation Axis Projection On Neutron Tomography, Physics Procedia 88 (2017) 299-305. https://doi.org/10.1016/j.phpro.2017.06.041

[2] Heinz Maier-Leibnitz Zentrum et al., NECTAR: Radiography and tomography station using fission neutrons, Journal of large-scale research facilities 1 (2015) A19. https://doi.org/10.17815/jlsrf-1-45

[3] D. Gürsoy et al., TomoPy: a framework for the analysis of synchrotron tomographic data, J. Synchrotron Rad. 21 (2014) 1188-1193. https://doi.org/10.1107/S1600577514013939

[4] T. Donath, F. Beckmann, A. Schreyer, Automated determination of the center of rotation in tomography data, J. Opt. Soc. Am. A 23(5) (2006) 1048-1057. https://doi.org/10.1364/JOSAA.23.001048

[5] Y. Yang et al., Registration of the rotation axis in X-ray tomography, J. Synchrotron Rad. 22 (2015), 452–457. https://doi.org/10.1107/S160057751402726X

[6] N. Vo, M. Drakopoulos, R.C. Atwood, C. Reinhard, Reliable method for calculating the center of rotation in parallel-beam tomography, Opt. Express 22 (2014) 19078-19086. https://doi.org/10.1364/OE.22.019078

[7] K. Osterloh, T. Bücherl, U. Zscherpel, U. Ewert, Image recovery by removing stochastic artefacts identified as local asymmetries, Journal of Instrumentation 7(4) (2012) C04018. https://doi.org/10.1088/1748-0221/7/04/C04018

[8] L.G. Brown, A survey of image registration techniques, ACM Computing Surveys 24(4), 325-376. https://doi.org/10.1145/146370.146374

[9] M. Capek, Optimisation strategies applied to global similarity based image registration methods, WSCG '99: 7th International Conference in Central Europe on Computer Graphics (1999) 369-374.

[10] P. Thévenaz, U.E. Ruttimann, M. Unser, A pyramid approach to subpixel registration based on intensity, IEEE Trans. Image Processing 7(1) (1998) 27-41. https://doi.org/10.1109/83.650848

[11] T. Köhler, A projection access scheme for iterative reconstruction based on the golden section, IEEE Symposium Conference Record Nuclear Science 6 (2004), 3961-3965

[12] J. Modersitzky, Numerical methods for image registration, Oxford University Press on Demand, 2004. https://doi.org/10.1093/acprof:oso/9780198528418.003.0008

[13] J. Modersitzky, FAIR: flexible algorithms for image registration, SIAM (2009). https://doi.org/10.1137/1.9780898718843

Software

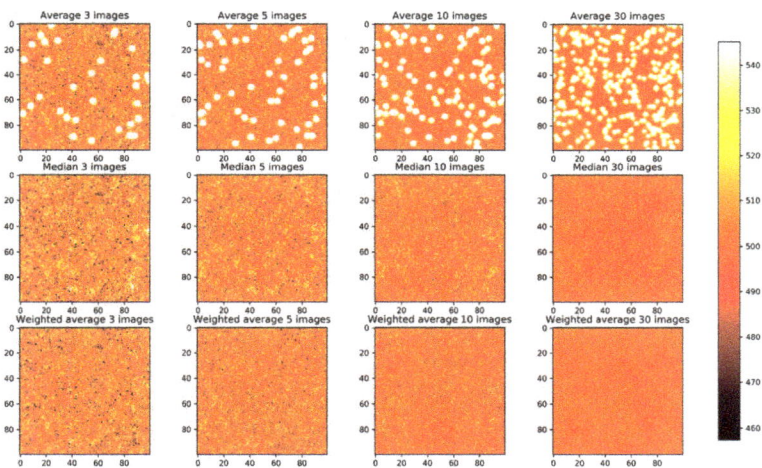

Neutron Radiography - WCNR-11
Materials Research Proceedings 15 (2020) 193-197

Materials Research Forum LLC
https://doi.org/10.21741/9781644900574-30

Methods to Combine Multiple Images to Improve Quality

Anders. P. Kaestner

[1]Laboratory for Neutron Scattering and Imaging, Paul Scherrer Institut, CH5232 Villigen-PSI, Switzerland

anders.kaestner@psi.ch

Keywords: Outliers, Image Combination, Neutron Imaging

Abstract. Noise and artifacts have a negative impact on the image quality and the resulting image analysis. The signal to noise ratio (SNR) caused by the neutron flux and the light conversion efficiency is one component in this. Here, we are more concerned about the effect of outliers, which frequently appear in neutron images. There are two approaches to reduce the impact of outliers (1) by applying spatial outlier rejection filters on each image and (2) by acquiring multiple images which are combined into a single image with a total neutron dose similar to the dose of the image in option (1). Here, we focus on the second option where we will show the importance of the choice of combination method. The impact is demonstrated to show the ability to reject outliers but also that the SNR can be improved. The image combination approach has the advantage that it does not affect neighbor pixels or larger regions. The tested methods are the arithmetic average, median, and a new weighted average. The weighted average shows promising results compared to the other two alternatives both regarding improving SNR and its outlier rejection ability.

Introduction

Neutron imaging is often performed at relatively low neutron counts. The origin of this is that most neutron sources provide a relatively low flux compared to for example X-ray sources. The provided neutrons are used in imaging experiments that require high spatial, or temporal resolution, or even both at the same time. The experiment requirements sometimes result in very low neutron counts per pixel in the acquired images. In addition to the Poisson distributed counting noise from neutrons and photons and binominally distributed noise from the acquisition system], it also very common to observe bright outliers in neutron images. The combination of noise and outliers sometimes degrade the images to the degree that makes any reliable analysis impossible. One solution to mitigate this problem is to increase the signal to noise ratio by increasing the exposure times, but this is mostly not realistic due to limited beamtime or limitations that are given by the observed sample. Therefore, numerical methods and alternative acquisition schemes are needed. The outliers are often suppressed using combinations of median filters and outlier detection [1,2]. The counting noise can be reduced by applying different noise reduction filters. In this paper, a different approach is pursued where multiple images are acquired from the same view of the sample. These images are then to be combined into a single image (figure 1). The total exposure time of the combined image shall be equal to an image in the single shot approach. The advantage of this approach is that it introduces an opportunity to reduce noise and outliers without the need to involve pixel neighbors as would be needed when spatial filters are applied. The focus in this paper is, in particular, on different methods to pixel-wise combine multiple images into a single image. Three different combination approaches (average, median, and weighted average) will be evaluated using phantom images with different levels of Poisson noise and outlier frequency and strength. Finally, the methods are also demonstrated using real neutron images.

Neutron Radiography - WCNR-11   Materials Research Forum LLC
Materials Research Proceedings **15** (2020) 193-197   https://doi.org/10.21741/9781644900574-30

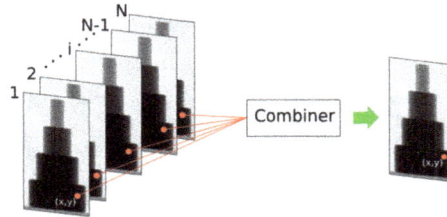

Figure 1: *Multiple images are combined pixel-wise into a single image with improved SNR and ideally fewer outliers.*

Methods

The noise and outlier reduction concept investigated in this paper is based on having multiple observations (images) of the same view. These images are to be combined into a single image which ideally shall have higher SNR and fewer outliers than a single image acquired with the equivalent dose as the collection together. Three different combination operations will be evaluated. These are the arithmetic mean (average), median, and weighted average. These operations are to be applied on the image values at the same pixel position in all images in the collection and will provide a new value that is used in the combined image. The operators are defined as follows. With a collection of N 2D images $f_1 \dots f_N$ the average operator is described as

$$f_{avg}(x,y) = \frac{1}{N}\sum_1^N f_i(x,y) \tag{1}$$

This operator is efficient to improve the SNR for data without outliers; this operator produces a result that corresponds to acquiring a single image with N times longer exposure time. In cases when outliers appear in the data, this operator fails to compute a value corresponding to the expected value. For this reason, the outlier rejecting behavior of the median operator is attractive. This operator is described with

$$f_{med}(x,y) = median(\{f_i(x,y)|i \in [1,N], i \in Z^+\}) \tag{2}$$

The median is a selection operator that never will be able to obtain an averaging effect. Therefore, the theoretical SNR value is always less than the SNR of the average with outlier free data. A method to combine the advantages of the average (improved SNR) and the median (outlier rejection) and a new method using a weighted average operator is proposed to combine the images

$$f_w(x,y) = \sum_{i=1}^N w_i(x,y) \cdot f_i(x,y) \tag{3}$$

This averaging is a generalization of (1) which would have $w = 1/N$. The weights based on estimates of the local standard deviation $\hat{\sigma}_i$ as

$$w_i(x,y) = \frac{\frac{1}{\hat{\sigma}_i(x,y)}}{\sum_{i=1}^N \frac{1}{\hat{\sigma}_i(x,y)}} \tag{4}$$

This type of weight penalizes pixels with high values of $\hat{\sigma}_i$. High values are typical for pixels with outliers in their neighborhood. If all images at a position have approximately the same value of $\hat{\sigma}_i$, they produce weights approaching $w = 1/N$, i.e., the average operation in equation (1). The most efficient way to compute $\hat{\sigma}_i$ is to use convolution with box filter kernels of some given size, typically 3×3 or 5×5 are used. A 5×5 kernel was used in the evaluation.

Evaluation

The performance of the three combination methods was evaluated with a simulation using images with Poisson distributed noise to reflect quantum noise of neutrons arriving at a detector and outliers with different size and strength were added to the noise images. Figure 2 shows examples using increasing number of simulated images in the combination. The blobs visible in the average images show the well-known fact that plain arithmetic averaging is not able reject outliers. In the simulation 25 such image set were created for different signal to noise ratio (SNR) in the different images to study the variation. Two cases were studied: (1) increasing the number of images with constant dose and (2) maintaining the total dose but increasing the number of images in the combination.

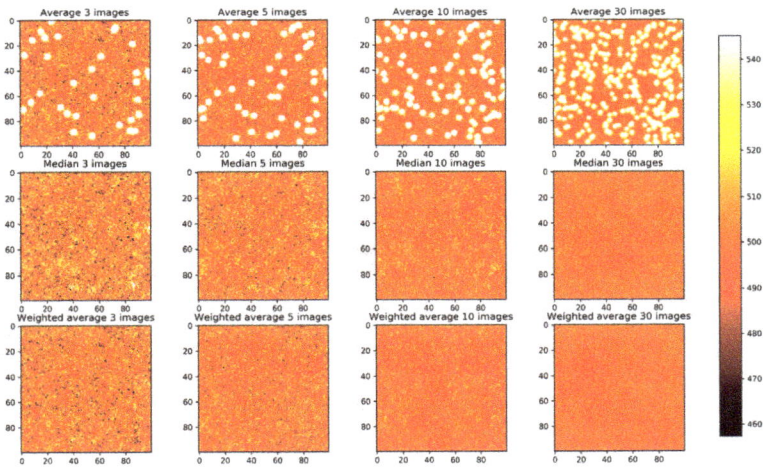

Figure 2: *Examples of combination performance using the same simulated test images with constant average gray level with added spots. All images are shown with the same gray level interval.*

Neutron Radiography - WCNR-11   Materials Research Forum LLC
Materials Research Proceedings **15** (2020) 193-197   https://doi.org/10.21741/9781644900574-30

In both cases, the simulation shows that the weighted average performs best. The SNR of the processed images was computed as SNR=$\sigma_{image}/\mu_{image}$.

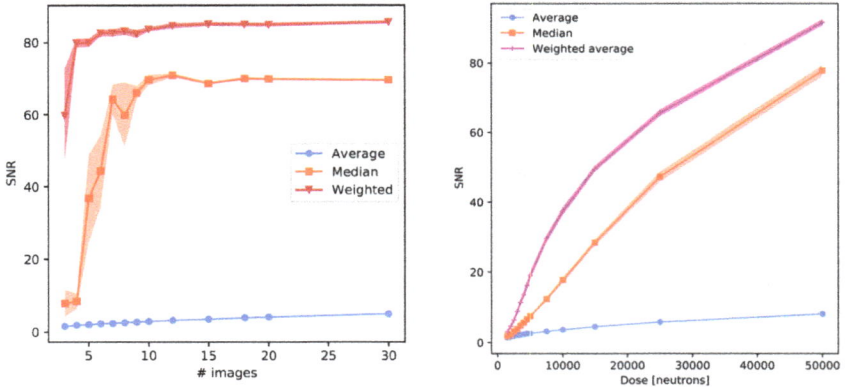

Figure 3: *Results from simulations using images with Poisson distributed noise and outliers added. The curves to the left show the performance using a constant total dose split up on multiple images. The curves to the right show the effect of adding more images to the combination and thus increasing the total dose. The error bands correspond to 95% confidence intervals.*

The results of the numeric experiments are plotted in figure 3. The simulations showed improved SNR and spot rejection ability for median and weighted average compared to the average that cannot reject outliers. The difference between median and weighted average is to be expected as the SNR using median is $\propto \sqrt{2N/\pi}$ [3] for large N compared to $\propto \sqrt{N}$ for averaging methods.

In the example below, the performance of these two methods is demonstrated on measured neutron radiographs. In this example, 3-30 images were combined using the two methods to provide qualitative comparison how each method performs. Both methods can reduce outliers, but the weighted average performs better than the median. There are still some outliers remaining, even when 30 images are combined. These outliers are most likely stuck pixels on the detector and cannot be removed using this method.

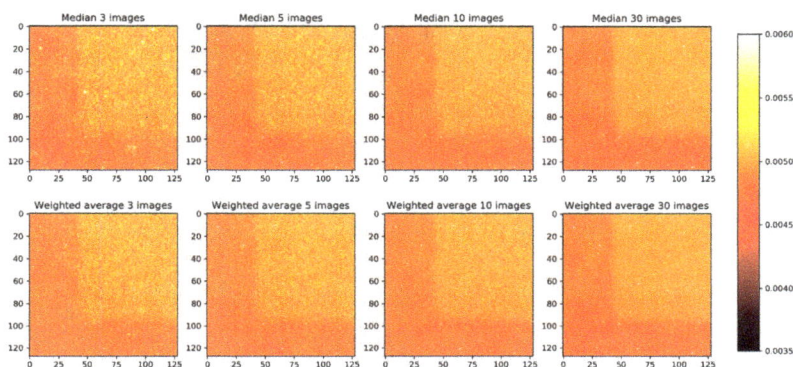

Figure 4: Demonstration of the performance median and weighted average using experimental data.

Summary

This shows the performance of three different methods to combine images into a single image with the aim to reduce noise and outliers. Simulations indicate that a weighted averaging approach based on local standard deviation outperforms median and arithmetic means as a method to combine several images. The weighted average method has shown promising results both in improving SNR and its ability to reduce the number of outliers. It can also be concluded that image combination as a method to reject gamma spot and to improve the signal to noise ratio. The outlier reducing benefit can already be observed with as few as three images, in particular when the weighted average method is used. In general, both median and weighted average perform best using between five and ten images for the same total dose. A higher number of images will only marginally improve image quality. Therefore, the acquisition overhead and increased storage requirements are not motivated. The presented image combination methods are already implemented in our CT reconstruction tool MuhRec [4].

References

[1] Hungler, P.; Bennett, L.; Lewis, W.; Bevan, G.; Metzler, J. Comparison of Image Filters for Low Dose Neutron Imaging. Physics Procedia 2013, 43, 169–178. https://doi.org/10.1016/j.phpro.2013.03.020

[2] Li, H.; Schillinger, B.; Calzada, E.; Yinong, L.; Muehlbauer, M. An adaptive algorithm for gamma spots removal in CCD-based neutron radiography and tomography. Nuclear Instruments & Methods in Physics Research Section A 2006, 564, 405–413. https://doi.org/10.1016/j.nima.2006.04.063

[3] Weisstein, E.W. CRC Concise Encyclopedia of Mathematics, Second Edition; Chapman and Hall/CRC, 2002. https://doi.org/10.1201/9781420035223

[4] Kaestner, A. MuhRec – a new tomography reconstructor. Nuclear Instruments and Methods A 2011, 212 651, 156–160. https://doi.org/10.1016/j.nima.2011.01.129

Neutron Radiography - WCNR-11
Materials Research Proceedings **15** (2020) 198-204

Materials Research Forum LLC
https://doi.org/10.21741/9781644900574-31

Jupyter Notebooks for Neutron Radiography Data Processing and Analysis

Jean-Christophe Bilheux[a] *, Jiao Y. Y. Lin[b] and Hassina Z. Bilheux[c]

[1]Neutron Scattering Division, Oak Ridge National Laboratory, Oak Ridge, TN 37831, USA

[a]bilheuxjm@ornl.gov, [b]linjiao@ornl.gov, [c]bilheuxhn@ornl.gov

(*) corresponding author

Keywords: Notebooks, Jupyter, Python, Neutron, Imaging, Analysis, Normalization

Abstract. Neutron radiography and computed tomography encompass a vibrant range of scientific applications, requiring advanced technique development and cutting-edge data processing and analysis software. We have developed an extensive portfolio of Python-based Jupyter notebooks that are custom-made for a specific experiment and sample geometry. These notebooks do not require any programming skills, although the code is accessible to programming experts if they wish to modify it. The notebooks are available on our analysis servers where the imaging data is also stored, preventing unnecessary and lengthy data transfer. This manuscript gives an overview of our efforts to empower the research community, that uses both the Spallation Neutron Source and High Flux Isotope Reactor imaging capabilities, to process and analyze their data in collaboration with our imaging team.

Introduction

Neutron radiography experiments impact a wide range of scientific areas such as materials science [1], energy [2], physics, engineering [3], archaeometry [4], plant physiology [5], geosciences [6], biology [7, 8], and chemistry. The High Flux Isotope Reactor (HFIR) CG-1D neutron imaging beamline [9] covers a broad range of scientific applications brought by scientists and engineers with different levels of expertise in image data analysis and programming. The facility is accessible at no cost through a peer review process if the visiting research team agrees to publish.

It is very challenging to develop a "one-shoe-fits-all" data processing and analysis software. A limited subset of the CG-1D research community performs data analysis without the help of the imaging team. The beamline is staffed with beamline scientists, data acquisition and software experts. This team works toward the development of intuitive data acquisition, processing and analysis tools that can be used by novice, intermediate and expert researchers. These users interact with our data processing and analysis software through the Python Jupyter notebooks. At this facility, most users rely on tools provided by the imaging team to process their data from normalization to 3-dimensional reconstruction, along with analysis of complex geometry samples. In order to improve the researcher's software experience and to make sure they are provided with adequate tools for their data processing and analysis, we implemented the procedure described undermentioned. Prior to arrival to the facility, we contact the principal investigator (PI) to discuss their data analysis needs as well as the level of software expertise of everyone in the PI's team. We already have many notebooks that can perform the most general needs of the users, but in case a team is coming with the need of a new tool, the pre-beam-experiment contact provides enough time to develop adequate (in terms of both algorithms and usability) tools. Upon arrival of the research team, we demonstrate the usage and provide a full step-by-step tutorial of each software. Finally, we follow up with the researchers after they have

Neutron Radiography - WCNR-11   Materials Research Forum LLC
Materials Research Proceedings **15** (2020) 198-204   https://doi.org/10.21741/9781644900574-31

returned to their home institution, provide further software support, and get their feedback on our procedure and the software. In this article, we detail how we collaborate with researchers who come to use our neutron imaging beamline and demonstrate the advantages of using the Python Jupyter notebooks as a development and deployment environment.

Experimental Planning

While the beamline scientist often interacts with the research team as early as during the beam time proposal writing phase, data software discussions usually start when beam time has been granted. To optimize the visiting research team's experience at the beamline, a few key parameters need to be understood before their experimental beam time. First and foremost, we need to understand how the experiments can provide an answer to their scientific question. This leads to discussions about how the researchers plan to collect their data, what parameters need to be saved in the metadata and later extracted, and what information needs to be extracted from the radiographs and computed tomography (CT) scans. These early communications with the experimental researchers are critical to ensure data processing and analysis is in place either at the time the experiment is conducted or soon after. Second, our researchers have access to our previously developed Jupyter notebooks with tutorials from our imaging website [10]. Finally, the software experts spend time with the researchers at the beamline during the experiments to fully appreciate the complexity of the measurements and to confirm the data analysis strategy previously discussed.

Jupyter Notebooks

As defined by its official web site, the Jupyter Notebooks [11] is a versatile, open source web application that allows to create and share documents that contain live code, equations, visualizations and narrative texts. The notebooks are designed to intuitively develop and run codes. The Jupyter notebooks supports over 40 programming languages [12]. We chose to write code using the Python language which has an easy-to-learn and interpreted object-oriented programming language. The Jupyter interface is straightforward to use and thus, once familiar with the functionality of the interface, one can start coding inside a notebook using a web browser [13], see Fig. 1 (left). A notebook consists of a series of cells that each contains lines of code. Each cell can be run or rerun independently, pending the logic of the programming is respected. The code can be easily modified, such as changing an algorithm in a cell, without impacting the rest of the users.

In order to simplify the utilization of the notebooks, we implemented user interfaces (UI) that are launched from the notebook, as illustrated in Fig. 1 (right). This is possible because the thinlinc [14] connection to the SNS analysis servers is available for our users.

How to Access the Notebooks

Another benefit of using Jupyter notebooks is that they can be run from our analysis servers without having to install software on the researcher's computer. Since the imaging data reside on the servers, this prevents unnecessary transfer of large imaging data sets. Using a web browser and assuming the researcher has an account in our servers, they can log in. Once logged in the Linux server, they then click the **jupyter imaging** options on the "Applications and Software" dropdown menus. Within seconds, the user is switched to a new conda environment [15] which is a directory that contains all the installed python libraries/conda packages necessary to run the notebooks. The portfolio of the python notebooks is automatically imported into the user account, allowing them to modify the notebooks without modifying the original notebooks which are stored elsewhere, and the Jupyter server is started from the notebooks folder location.

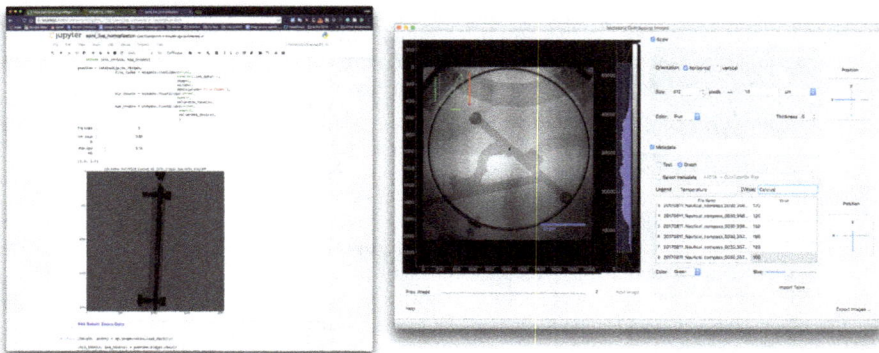

Figure 1. *Left: An example of Python-based Jupyter notebook displayed using a web browser.*
Right: A user interface directly launched from a notebook.

The Jupyter home page shows a long list of various files and folders. The notebooks, files with extension ".ipynb", are the only files most users will interact with. Advanced users can adventure inside the code itself, which can be found in the "_code" folder. Each notebook opens in a new tab and at its top is a link to a step-by-step tutorial (some with video tutorials, see Fig. 2) that explains the purpose of the code and how to run it. A parchment display summarizes the four important rules to remember to successfully run the notebooks, as illustrated in Fig. 2.

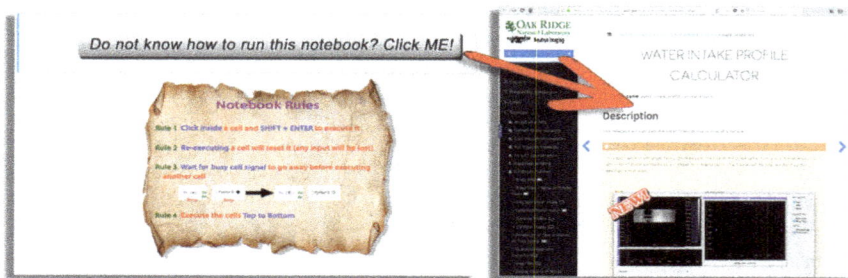

Figure 2. *The top of each notebook provides a direct link (left) to a step-by-step tutorial (right)*
and a list of important rules (left) to execute the notebooks.

Examples of Notebooks

Within the past few years, we have developed a large library of Jupyter notebooks, all of which are available on our servers to all the CG-1D and the future VENUS (an imaging instrument at the Spallation Neutron Source) communities. These notebooks are custom-made for a specific experiment and sample geometry. A few examples are listed below to give a glance at the data processing and analysis software capability.

Neutron Radiography - WCNR-11   Materials Research Forum LLC
Materials Research Proceedings **15** (2020) 198-204   https://doi.org/10.21741/9781644900574-31

-   Example 1: Binning of images based on metadata

This notebook combines radiographs according to the temperature of the furnace saved in the metadata of each radiograph (Fig. 3). The notebook displays the entire temperature history of the experiment with the furnace file information as well as a time lapse of the corresponding radiographs (Fig. 3A). The code allows to select various binning options (by time, by temperature of the furnace, by number of imaging files). The principle of binning is similar to the event filtering for scattering instruments at the SNS [16]. Fig. 3B displays the binning that was applied in this example, and Fig. 3C shows the result of the binning with the created temperature steps.

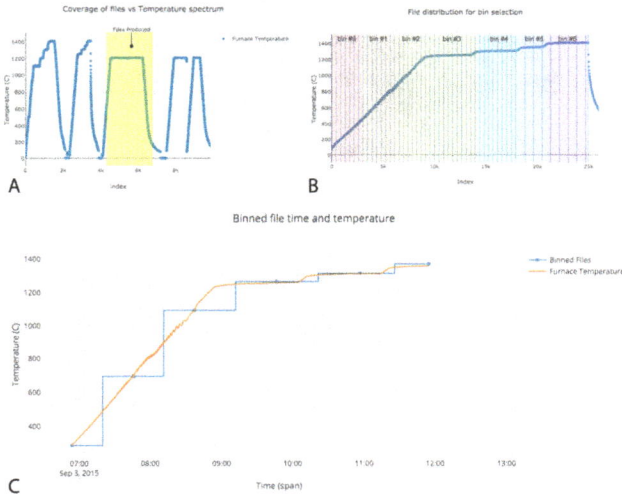

Figure 3. A: Furnace temperature information over entire experiment time (blue) and corresponding selected radiographs (yellow); B: color coded regions to show how the radiographs are binned; C: binning result with average furnace temperature reported for each bin.

-   Example #2: Water uptake in a soil sample

This notebook calculates the water uptake velocity in a soil sample (Fig. 4). The most important notebook requirements for this research team were:

- avoid command line input,
- graphical selection of a region of interest,
- option to integrate region along x or y axis,
- be able to sort files using file name, or time stamp,
- be able to rename files (to improve sorting by file name),
- have the option to easily add algorithm used in the calculation of the front of the water uptake signal,
- export various profiles in ASCII file format.

This example illustrates the complex requirements yet the versatility of a Jupyter notebook. Moreover, in collaboration with the research team, several algorithms that track the water front automatically were implemented and compared with each other. These implementations were straightforward and code-reuse were maximized.

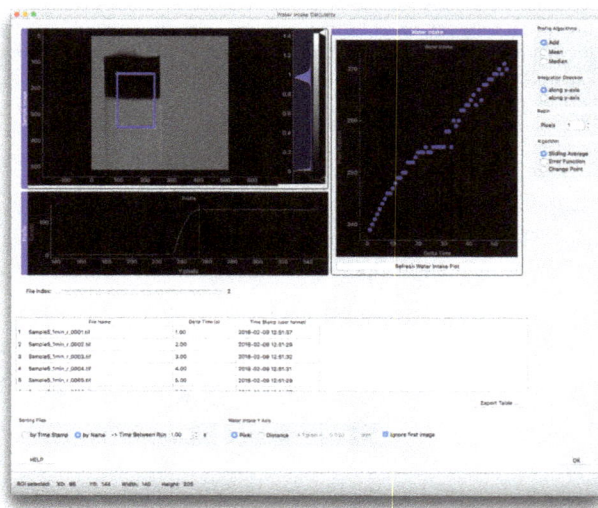

Figure 4. *Water uptake notebook showing the profile of the water front versus time for a given region selected. (courtesy of Prof. Ed Perfect (University of Tennessee, Knoxville)).*

Conclusion

Our team's approach to data processing and analysis is to provide easy-to-use code that is custom-made for a specific experiment and sample geometry, requires little to no programming expertise, and does not require complex software installation on the end-user's computer. Thus, we have successfully developed and deployed Python-based Jupyter notebooks that are available on our analysis servers. With these software capabilities, we can respond to the increasingly challenging and diverse research community that utilizes neutron imaging capabilities at the Spallation Neutron Source and the High Flux Isotope Reactor. The notebooks allow the most advanced users to modify the code directly, while less experienced users can take advantage of the graphical user interface. Step-by-step tutorials of all the notebooks are also available on our neutron imaging website to facilitate the use of the notebooks.

Acknowledgements

The authors would like to thank Prof. Ed Perfect and his team for their valuable input in improving our research community experience using the Jupyter notebooks. This research used resources at the High Flux Isotope Reactor and Spallation Neutron Source, DOE Office of Science User Facilities operated by the Oak Ridge National Laboratory.

This manuscript has been authored [or, co-authored] by UT-Battelle, LLC, under contract DE-AC05-00OR22725 with the US Department of Energy (DOE). The US government retains and the publisher, by accepting the article for publication, acknowledges that the US government retains a nonexclusive, paid-up, irrevocable, worldwide license to publish or reproduce the published form of this manuscript, or allow others to do so, for US government purposes. DOE will provide public access to these results of federally sponsored research in accordance with the DOE Public Access Plan (http://energy.gov/downloads/doe-public-access-plan).

References

[1] R. R. Dehoff, M. Kirka, W. Sames, H. Bilheux, A. Tremsin, L. Lowe, S. Babu. Site specific control of crystallographic grain orientation through electron beam additive manufacturing, Materials Science and Technology 31 (8) (2015) 931-938. https://doi.org/10.1179/1743284714Y.0000000734

[2] Y. Zhang, K. R. Chandran, H. Z. Bilheux, Imaging of the Li spatial distribution within v 2 o 5 cathode in a coin cell by neutron computed tomography, Journal of Power Sources 376 (2018) 125-130. https://doi.org/10.1016/j.jpowsour.2017.11.080

[3] D. J. Duke, C. E. Finney, A. Kastengren, K. Matusik, N. Sovis, L. Santodonato, H. Bilheux, D. Schmidt, C. Powell, T. Toops, High resolution x-ray and neutron computed tomography of an engine combustion network spray g gasoline injector, SAE International Journal of Fuels and Lubricants 10 (2017-01-0824) (2017) 328-343. https://doi.org/10.4271/2017-01-0824

[4] F. Salvemini, S. Olsen, V. Luzin, U. Garbe, J. Davis, T. Knowles, K. Sheedy, Neutron tomographic analysis: Material characterization of silver and electrum coins from the 6th and 5th centuries bce, Materials Characterization 118 (2016) 175-185. https://doi.org/10.1016/j.matchar.2016.05.018

[5] M. Holz, A. Carminati, Y. Kuzyakov, Distribution of root exudates and mucilage in the rhizosphere: combining 14c imaging with neutron radiography, in: EGU General Assembly Conference Abstracts, Vol. 17, 2015.

[6] E. Perfect, C.-L. Cheng, M. Kang, H. Bilheux, J. Lamanna, M. Gragg, D. Wrigth, Neutron imaging of hydrogen-rich fluids in geomaterials and engineered porous media: A review, Earth-Science Reviews 129 (2014) 120-135. https://doi.org/10.1016/j.earscirev.2013.11.012

[7] H. Z. Bilheux, J.-C. Bilheux, W. B. Bailey, W. S. Keener, L. E. Davis, K. W. Herwig, K. Cekanova, Neutron imaging at the Oak Ridge National Laboratory: Application to biological research, in: Biomedical Science and Engineering Center Conference (BSEC), 2014 Annual Oak Ridge National Laboratory, IEEE, 2014, pp. 1-4. https://doi.org/10.1109/BSEC.2014.6867751

[8] H. Z. Bilheux, M. Cekanova, A. A. Vass, T. L. Nichols, J.-C. Bilheux, R. L. Donnell, V. Finochiarro, A novel approach to determine post mortem interval using neutron radiography, Forensic science international 251 (2015) 11-21. https://doi.org/10.1016/j.forsciint.2015.02.017

[9] L. Santodonato, H. Bilheux, B. Bailey, J. Bilheux, P. Nguyen, A. Tremsin, D. Selby, L. Walker, the CG-1D neutron imaging beamline at the Oak Ridge National Laboratory High Flux Isotope reactor, Physics Procedia 69 (2015) 104-108. https://doi.org/10.1016/j.phpro.2015.07.015

[10] Neutron Imaging at ORNL Home Page. https://neutronimaging.pages.ornl.gov.

[11] Jupyter Notebooks. http://jupyter.org.

[12] Python Web Page. https://python.org

[13] Python notebook tutorial.
https://neutronimaging.pages.ornl.gov/tutorial/how_to_run_notebooks/

[14] thinlinc – a remote desktop server. https://www.cendio.com/thinlinc/what-is-thinlinc

[15] Conda environment. https://conda.io/docs/index.html.

[16] Garrett E. Granroth, K. An, H. L. Smith, P. Whitfield, J. C. Neuefeind, J. Lee, W. Zhou, V. N. Sedov, P. F. Peterson, A. Parizzi, H. Skorpenske, S. M. Hartman, A. Huq and D. L. Abernathy. Event-based processing of neutron scattering data at the Spallation Neutron Source, J. Appl. Cryst. (2018 51, 616-629. https://doi.org/10.1107/S1600576718004727

Application

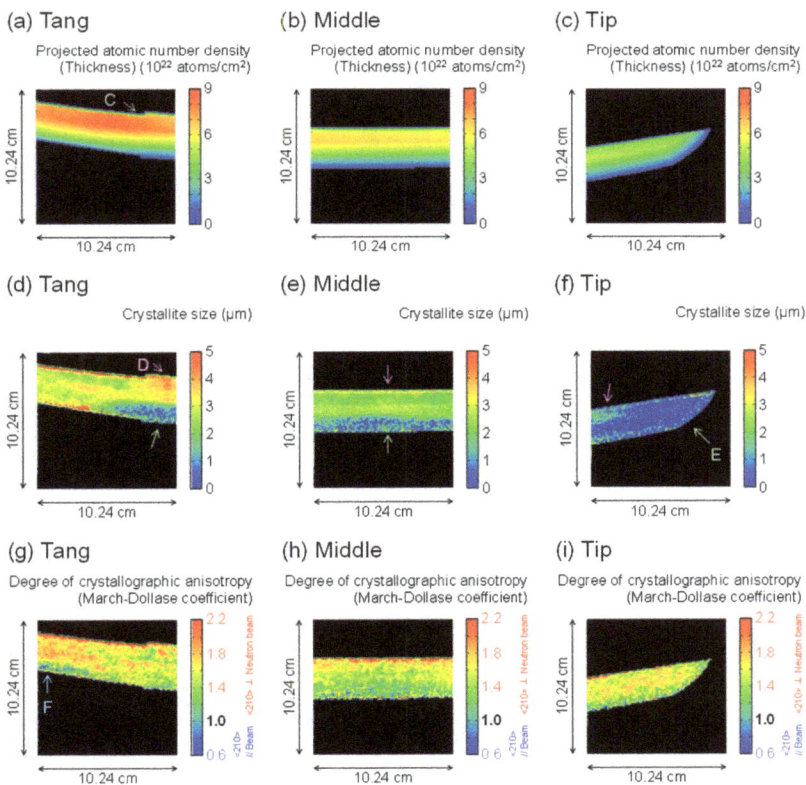

(a) Tang

Projected atomic number density
(Thickness) (10^{22} atoms/cm^2)

(b) Middle

Projected atomic number density
(Thickness) (10^{22} atoms/cm^2)

(c) Tip

Projected atomic number density
(Thickness) (10^{22} atoms/cm^2)

(d) Tang

Crystallite size (µm)

(e) Middle

Crystallite size (µm)

(f) Tip

Crystallite size (µm)

(g) Tang

Degree of crystallographic anisotropy
(March-Dollase coefficient)

(h) Middle

Degree of crystallographic anisotropy
(March-Dollase coefficient)

(i) Tip

Degree of crystallographic anisotropy
(March-Dollase coefficient)

Neutron Radiography - WCNR-11
Materials Research Proceedings 15 (2020) 207-213

Materials Research Forum LLC
https://doi.org/10.21741/9781644900574-32

Pulsed Neutron Imaging Based Crystallographic Structure Study of a Japanese Sword made by Sukemasa in the Muromachi Period

Kenichi Oikawa[1, a *], Yoshiaki Kiyanagi[2, b], Hirotaka Sato[3, c], Kazuma Ohmae[2, d], Anh Hoang Pham[5, e], Kenichi Watanabe[2, f], Yoshihiro Matsumoto[4, g], Takenao Shinohara[1, h], Tetsuya Kai[1, i], Stefanus Harjo[1, j], Masato Ohnuma[3, k], Sigekazu Morito[5, l], Takuya Ohba[5, m], Akira Uritani[2, n], Masakazu Ito[6, o]

[1]J-PARC Center, Japan Atomic Energy Agency, Ibaraki, 319-1195, Japan

[2]Graduate School of Engineering, Nagoya University, Aichi, 464-8603, Japan

[3]Faculty of Engineering, Hokkaido University, Hokkaido 060-8628, Japan

[4]Comprehensive Research Organization for Science and Society, Ibaraki 319-1106, Japan

[5]Interdisciplinary Faculty of Science and Engineering, Shimane University, Shimane 690-8504, Japan

[6]WAKOU MUSEUM, Shimane 692-0011, Japan

[a]kenichi.oikawa@j-parc.jp, [b]kiyanagi@phi.phys.nagoya-u.ac.jp, [c]h.sato@eng.hokudai.ac.jp, [d]oomae.kazuma@f.mbox.nagoya-u.ac.jp, [e]anhpham@riko.shimane-u.ac.jp, [f]k-watanabe@energy.nagoya-u.ac.jp, [g]y_matsumoto@cross.or.jp, [h]takenao.shinohara@j-parc.jp, [i]tetsuya.kai@j-parc.jp, [j]stefanus.harjo@j-parc.jp, [k]ohnuma.masato@eng.hokudai.ac.jp, [l]mosh@riko.shimane-u.ac.jp, [m]ohba@riko.shimane-u.ac.jp, [n]uritani@energy.nagoya-u.ac.jp, [o]masakazu_itoh@cup.ocn.ne.jp

Keywords: Japanese Sword, Quench Hardening, Martensite, Bragg-Edge Transmission, RITS Code

Abstract. Energy-resolved neutron imaging using a pulsed neutron source is capable of visualizing crystallographic information over a large area of a sample by analyzing position-dependent Bragg-edge transmission spectra. We applied this method to a Japanese sword, signed by Sukemasa, to elucidate position dependent crystallographic characteristics, including but not limited to: degree of hardening, crystallite size, degree of preferred orientation. By comparing the degree of hardening to that of a contemporary short Japanese sword (dagger), the Sukemasa showed relatively small changes in the position of Bragg-edge (110) and its broadening. No coarse grain was found within the detector resolution (ca. 1 mm), and the crystallite size of the blade area was analyzed to be almost uniform and less than 1 μm. We thus recognize in a comprehensive manner that the Sukemasa sword was manufactured with great care.

Introduction

Japanese swords are very attractive not only as a work of art but also from a metallurgical point of view. There were several famous traditional styles (Gokaden for instance) of Japanese sword-making in the Koto (old sword) age; A.D. 987–1596. Detailed manufacturing techniques, such as methods of obtaining raw materials [1], an iterative hammering process and the method for combining different steels [2] in the Koto age, are not clear, since they were handed down secretly within each school and have been lost over time. Various kinds of Japanese swords were sliced thickly and the cut surface were studied by conventional methods, such as microscopy and chemical analysis, EPMA, and so on [3]. This analytical approach was possible in the past, but at the present time when Japanese vintage swords have become valuable, it is indispensable to

Neutron Radiography - WCNR-11   Materials Research Forum LLC
Materials Research Proceedings **15** (2020) 207-213   https://doi.org/10.21741/9781644900574-32

establish non-destructive analysis methods to identify some peculiar characteristics related to the sword making procedure. Neutron experiments are a powerful tool to study metallic cultural heritage objects due to their high penetrating power and capability to give micro-structural properties [4,5]. Bragg-edge transmission (BET) imaging, in particular, gives real-space distributions of bulk information in the Japanese sword [6]. In this work, we investigated crystallographic information of a Japanese sword made by Sukemasa in Izumi province (southern part of Osaka prefecture; out of Gokaden) in the first quarter of the 16th century.

Experimental

The Japanese sword Sukemasa (Fig. 1(a); 790 mm in total length and 15 mm in curvature) and a contemporary dagger (300 mm blade length divided into three 100 mm pieces) were used in the present investigation.

The time-of-flight (TOF) neutron transmission imaging experiment was performed at BL22 RADEN [7] in MLF J-PARC with a proton beam power of 150 kW. The downstream sample position with a pinhole optical geometry was used as shown in Fig. 1(b). Two-dimensional TOF BET spectra between 0.5 and 6.5 Å were obtained by use of a gas electron multiplier (GEM) detector [8] with a 0.8×0.8 mm^2 pixel resolution and a 10×10 cm^2 detection area. Three different areas of the sword were measured, as indicated in Fig. 1(a). The BET imaging measurement time was about 8 hours for each sample position and 6 hours without the sample for the detector calibration.

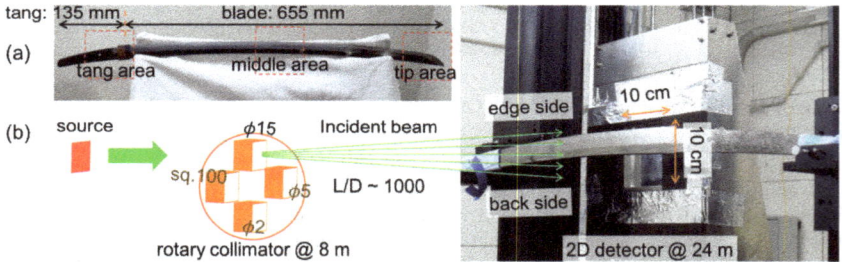

Fig. 1 (a) A picture of the Sukemasa. Three measured areas are indicated by the dashed red boxes. (b) A schematic view and a picture of the BET measurement at RADEN.

Data analysis

Wavelength-range contrast imaging

The wavelength dependent neutron transmission spectrum of a crystalline sample is expressed as follows:

$$Tr(\lambda) = \exp\left(-\sum_p \sigma_{\text{tot},p}(\lambda)\rho_p t_p\right), \qquad (1)$$

where $\sigma_{\text{tot},p}(\lambda)$ is the neutron total cross section, ρ_p is the density, and t_p is the thickness of the crystalline phase p. Here, we tentatively define Tr_{cold} and Tr_{thermal} where the former and latter are the wavelength selective transmission averaged over the wavelengths

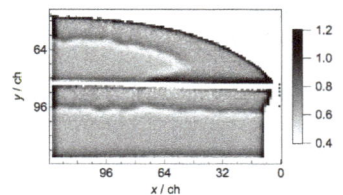

Fig. 2 The wavelength-range contrast image of the tip and middle of the contemporary dagger.

of 4.1–6.4 and 0.5–2.3 Å, respectively. A wavelength-range contrast image is obtained by the ratio, $Ln(Tr_{cold})/Ln(Tr_{thermal})$, where the dependence on $\rho_p t_p$ is canceled. Therefore, the image represents the division of the wavelength-range dependent $\sigma_{tot,p}$, i.e. $\sigma_{cold}/\sigma_{thermal}$.

Fig. 2 shows the wavelength-range contrast image of the contemporary dagger. Figs. 3(a), 3(b) and 3(c) show the wavelength-range contrast images of the tang, middle and tip areas of the Sukemasa, respectively. The quenching boundary of the Sukemasa shows a lower contrast than the contemporary dagger. The magnetic and/or small angle scattering could affect this particular contrast seen in the contemporary dagger, but this is still under investigation.

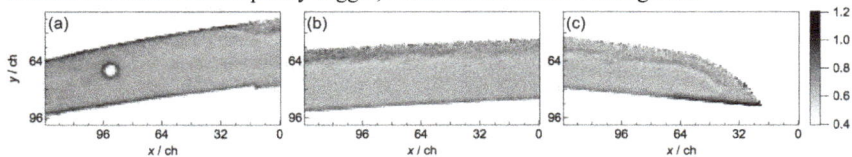

Fig. 3 The wavelength-range contrast images of (a) the tang, (b) middle and (c) tip areas of the Sukemasa. Note that a small circle in the tang area is the Mekugi (fastening pin) hole.

Single edge and full pattern analysis using the RITS code

All BET spectra were analyzed by a Rietveld-like analysis code, RITS [9]. A single-edge analysis was performed to obtain the lattice (110) plane spacing d_{110} and the broadening of the edge width w_{110}, whereas a Rietveld-type analysis was performed to obtain the projected atomic number density, crystallite size and preferred orientation parameter. For all analyses, a single-phase body-centered-cubic structure was assumed.

Fig. 4(a) exemplifies the Rietveld-type fitting pattern for a BET spectrum where wavelengths between 2.0 and 4.5 Å were used. Fig. 4(b) exemplifies the single edge fitting pattern for a (110) Bragg-edge in the BET spectrum for wavelengths between ca. 3.8 and 4.3 Å. To obtain enough data statistics for the analysis, BET spectra for a 2 × 2 pixel (1.6 × 1.6 mm²) area were summed into a single BET spectrum, and the pixel area was stepped at 1 pixel (0.8 mm) intervals in the x- and y-directions.

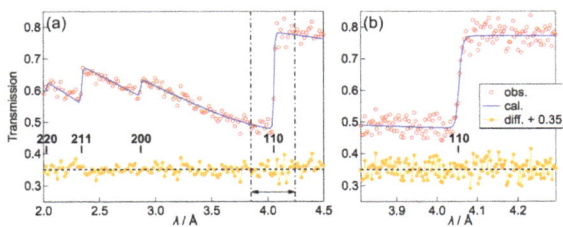

Fig. 4 Examples of (a) Rietveld-type analysis for a full BET spectrum with 100-µs time channels and (b) single-edge analysis for the Bragg-edge (110) with 20-µs time channels. The vertical dashed lines in (a) indicate the range for the single-edge analysis.

Result and discussion

Single-edge analysis

One of the essential processes for the

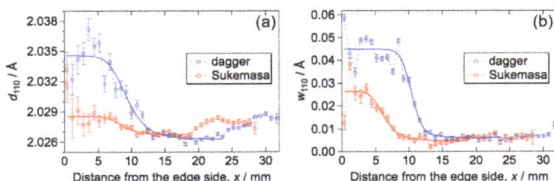

Fig. 5 Distribution of the lattice d-spacing (a) and broadening of the edge width (b) of the (110) plane across the middle area of the Sukemasa and the contemporary dagger.

Neutron Radiography - WCNR-11   Materials Research Forum LLC
Materials Research Proceedings **15** (2020) 207-213   https://doi.org/10.21741/9781644900574-32

fabrication of a Japanese sword is the unique quenching method with an intentional temperature gradient, which produces a hard martensite phase and soft pearlite - ferrite phase along the cutting-edge and the back side of the blade, respectively [10]. As the quenching technique should be different in age and tradition, the martensite characteristics; such as the lattice d-spacing and broadening of the edge width, would be different in the Sukemasa and the contemporary dagger. Single-edge analysis is useful to elucidate these characteristics because of its simple calculation method [11].

Figs. 5(a) and 5(b) show a d_{110} and w_{110} distribution around the center of the middle area of the Sukemasa and the contemporary dagger. To obtain high-statistics data for this analysis, BET spectra for a 50×1 pixel (40×0.8 mm^2) area were summed into a single BET spectrum, and stepped at 1 pixel intervals from the edge side to the back side. The d_{110} and w_{110} of the quenched area of the contemporary dagger is obviously larger than the Sukemasa, but the ratio between the hardened and unhardened regions, however, is rather similar; about 1:2. Interestingly, the variation in d_{110} of the back-side region is nearly the same for each sword. This may be because the carbon content of the material and the firing temperature in the manufacturing process were similar for each of these swords [12]. The verification of this last supposition, however, would require the use of additional, destructive testing methods, which is not possible for the Sukemasa.

Figs. 6 (a-c) and Figs. 7 (a-c) show 2D maps of the lattice spacing d_{110} (Å) and the edge broadening w_{110} (Å), respectively, of the Sukemasa. As shown in Fig. 5(a), the increase in d_{110} is rather small in the hardened area of the Sukemasa, so that the martensite phase distributions are somewhat vague as shown in Figs. 6 (a-c). However, the martensite phase distributions are clearly seen in Figs. 7 (a-c) and appear roughly correlated with the contrast of Figs. 3 (a-c).

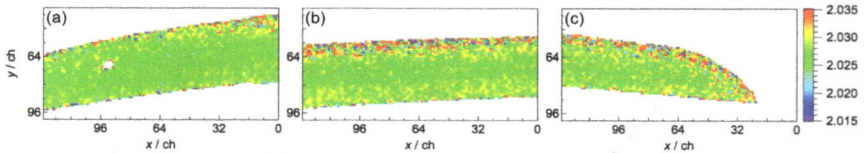

Fig. 6 2D maps of the lattice (110) plane spacing d_{110} (Å) of the Sukemasa.

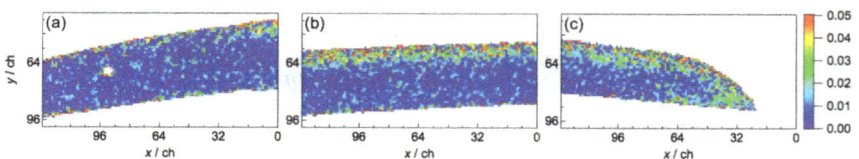

Fig. 7 2D maps of the edge broadening w_{110} (Å) of the Sukemasa.

Full-pattern analysis

In the Japanese sword-making process, the hammering process is repeated several times to remove impurities and ensure uniformity of the blade material. On the other hand, the tang area generally undergoes less processing and different heat treatments compared to the blade area. As shown in Figs. 8 (a-c), the projected density of the sword changes very smoothly, without any wavy/steep fluctuations or voids. In contrast, the tang area exhibits somewhat sudden changes in the crystallite size and preferred orientation parameter as shown in Fig. 9(a) and Fig. 10(a), respectively.

Figs. 9 (a-c) show maps of the crystallite size of the Sukemasa. Sabine's primary extinction function was used to fit the transmission spectra in the RITS code, where the crystallite size can be evaluated by the extinction parameter, s [13]. The crystallite size in the tip and middle areas is small and almost uniform, in contrast to the tang area, which suggests that the hammering and the quenching processes for the blade area were carried out with great care.

Figs. 10 (a-c) show maps of the preferred-orientation parameter, r. The RITS code incorporates a form for the preferred-orientation correction using the March-Dollase (MD) formulation for a cylindrically symmetric distribution [13]. An MD coefficient r provides the degree of the preferred orientation, namely $r = 1$ for a random textured sample and $r < 1$ (<hkl> parallel to the incident beam) or $r > 1$ (<hkl> perpendicular to the incident beam) for a textured sample, where <hkl> represents the preferred-orientation vector. For the present measurement, the preferred orientation <110> for ferrite/martensite gave better fitting results than those using other major orientations like <100> or <111>. However, it gave poor fitting results in regions with $r > 2$. We tried to fit using other <hkl> close to <110> and found the preferred orientation <750> gave good overall fitting results over the whole range of r found in the sample. The maps indicate that the <750> vector is perpendicular to the incident beam in the present sample setting. The preferred-orientation parameter in most of the blade area, except for the hardened edge, is not so large ($1 < r < 1.8$) and the position dependence is small compared with the tang area.

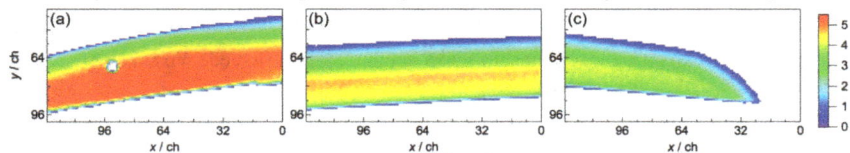

Fig. 8 2D maps of the projected density ρt ($10^{22} cm^{-2}$) of the Sukemasa. The projected densities for the red area in (a) exceed the maximum scale of the current plots.

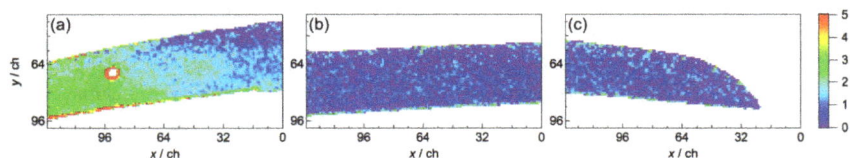

Fig. 9 2D maps of the crystallite size s (μm) of the Sukemasa.

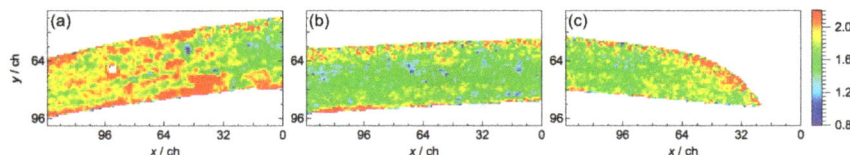

Fig. 10 2D maps of the preferred-orientation parameter r of the Sukemasa. The values of r in the red areas of (a) exceed the maximum scale of the current plots.

Summary

Crystallographic information of an antique Japanese sword made by Sukemasa was studied using an energy-resolved neutron imaging method. We found the following.

(1) The wavelength-range contrast images showed increased contrast in the quenching area and its boundary. These contrast images correlate well to the edge-broadening maps rather than the lattice-plane-spacing maps of the Sukemasa.

(2) From an observed increase of the lattice plane spacing and broadening of the edge width, we confirmed that the martensite phase is concentrated in the hardened cutting edge.

(3) Crystallite size seems to be small and almost uniform on the tip and middle areas. On the other hand, the tang area is clearly coarser than the other areas.

(4) The preferred orientation of the tang area seems to be the most anisotropic and the preferred orientation of the hardened edge seems to be somewhat stronger than the rest of the blade.

Together with the evaluation using neutron tomography [14], we will further investigate the detailed information on the micro-structural properties of the Japanese sword Sukemasa.

Acknowledgement

Neutron experiments at the MLF J-PARC were performed under Proposal Nos. 2017A0099 and 2016B0163. This work partially includes the result of 'Collaborative Important Researches' organized by JAEA, QST and U. Tokyo. The authors would like to thank Japanese swordsmith T. Sasaki for providing test pieces of the dagger sample.

References

[1]   A. Amata, T. Tsuchiko, Tetsu to Nihontō, first ed., Keiyūsha, Tōkyō, 2004 (in Japanese).

[2]   T. Saito, Kinzoku ga kataru Nihonshi : senka, nihontō, teppō, first ed., Yoshikawa Kōbunkan, Tōkyō, 2012 (in Japanese).

[3]   T. Takahashi, T. Murakami, S. Okada, N. Fujii, Discovering New Aspects in a Japanese Sword, Tetsu-to-Hagane 71 (1985) 108-114 (in Japanese). https://doi.org/10.2355/tetsutohagane1955.71.15_1818

[4]   F. Grazzi, L. Bartoli, F. Civita, R. Franci, A. Paradowska, A. Scherillo, M. Zoppi, From Koto age to modern times: Quantitative characterization of Japanese swords with Time of Flight Neutron Diffraction, J. Anal. At. Spectrom. 26 (2011) 1030-1039. https://doi.org/10.1039/c0ja00238k

[5]   F. Salvemini, F. Grazzi, S. Peetermans, F. Civita, R. Franci, S. Hartmann, E. Lehmann, M. Zoppi, Quantitative characterization of Japanese ancient swords through energy-resolved neutron imaging, J. Analytical Atomic Spectrometry 27 (2012) 1494-1501. https://doi.org/10.1039/c2ja30035d

[6]   Y. Shiota, H. Hasemi, Y. Kiyanagi, Crystallographic analysis of a Japanese sword by using Bragg edge transmission spectroscopy, Phys. Procedia 88 (2017) 128-133. https://doi.org/10.1016/j.phpro.2017.06.017

[7]   T. Shinohara, T. Kai, K. Oikawa, M. Segawa, M. Harada, T. Nakatani, M. Ooi, K. Aizawa, H. Sato, T. Kamiyama, H. Yokota, T. Sera, K. Mochiki, Y. Kiyanagi, Final design of the Energy-Resolved Neutron Imaging System "RADEN" at J-PARC, J. Phys. Conf. Ser. 746 (2016) 012007. https://doi.org/10.1088/1742-6596/746/1/012007

[8]   S. Uno, T. Uchida, M. Sekimoto, T. Murakami, K. Miyama, M. Shoji, E. Nakano, T. Koike, K. Morita, H. Sato, T. Kamiyama, Y. Kiyanagi, Two-dimensional neutron detector with GEM and its applications, Phys. Procedia 26 (2012) 142-152. https://doi.org/10.1016/j.phpro.2012.03.019

[9]   H. Sato, T. Kamiyama, Y. Kiyanagi, A Rietveld-type analysis code for pulsed neutron Bragg-edge transmission imaging and quantitative evaluation of texture and microstructure of a welded α-iron plate, Mater. Trans. 52 (2011) 1294-1302. https://doi.org/10.2320/matertrans.M2010328

[10]   B. Dodd, The making of old Japanese swords, J Mech Work Tech 2 (1978) 75-84. https://doi.org/10.1016/0378-3804(78)90016-5

[11]   H. Sato, T. Sato, Y. Shiota, T. Kamiyama, A.S. Tremsin, M. Ohnuma, Y. Kiyanagi, Relation between Vickers hardness and Bragg-edge broadening in quenched steel rods observed by pulsed neutron transmission imaging, Mater. Trans. 56 (2015) 1147-1152. https://doi.org/10.2320/matertrans.M2015049

[12]   T. Inoue, Tatara and the Japanese sword: the science and technology, Acta Mech 214 (2010) 17-30. https://doi.org/10.1007/s00707-010-0308-7

[13]   Y. Kiyanagi, H. Sato, T. Kamiyama, T. Shinohara, A new imaging method using pulsed neutron sources for visualizing structural and dynamical information, J. Phys.: Conf. Ser. 340 (2012) 012010. https://doi.org/10.1088/1742-6596/340/1/012010

[14]   Y. Matsumoto, K. Watanabe, K. Ohmae, A. Uritani, Y. Kiyanagi, H. Sato, M. Ohnuma, A.H. Pham, S. Morito, T. Ohba, K. Oikawa, T. Shinohara, T. Kai, S. Harjo, M. Ito, This proceeding.

Neutron Radiography - WCNR-11
Materials Research Proceedings 15 (2020) 214-220

Materials Research Forum LLC
https://doi.org/10.21741/9781644900574-33

Crystallographic Microstructure Study of a Japanese Sword made by Noritsuna in the Muromachi Period by Pulsed Neutron Bragg-Edge Transmission Imaging

Hirotaka Sato[1,a*], Yoshiaki Kiyanagi[2,b], Kenichi Oikawa[3,c], Kazuma Ohmae[2,d], Anh Hoang Pham[4,e], Kenichi Watanabe[2,f], Yoshihiro Matsumoto[5,g], Takenao Shinohara[3,h], Tetsuya Kai[3,i], Stefanus Harjo[3,j], Masato Ohnuma[1,k], Shigekazu Morito[4,l], Takuya Ohba[4,m], Akira Uritani[2,n] and Masakazu Itoh[6,o]

[1]Faculty of Engineering, Hokkaido University, Hokkaido 060-8628, Japan

[2]Graduate School of Engineering, Nagoya University, Aichi 464-8603, Japan

[3]J-PARC Center, Japan Atomic Energy Agency (JAEA), Ibaraki 319-1195, Japan

[4]Interdisciplinary Faculty of Sci. and Engineering, Shimane University, Shimane 690-8504 Japan

[5]Comprehensive Research Organization for Science and Society, Ibaraki 319-1106, Japan

[6]Wakou Museum, Shimane 692-0011, Japan

[a]h.sato@eng.hokudai.ac.jp, [b]kiyanagi@phi.phys.nagoya-u.ac.jp, [c]kenichi.oikawa@j-parc.jp, [d]oomae.kazuma@f.mbox.nagoya-u.ac.jp, [e]anhpham@riko.shimane-u.ac.jp, [f]k-watanabe@energy.nagoya-u.ac.jp, [g]y_matsumoto@cross.or.jp, [h]takenao.shinohara@j-parc.jp, [i]tetsuya.kai@j-parc.jp, [j]stefanus.harjo@j-parc.jp, [k]ohnuma.masato@eng.hokudai.ac.jp, [l]mosh@riko.shimane-u.ac.jp, [m]ohba@riko.shimane-u.ac.jp, [n]uritani@energy.nagoya-u.ac.jp, [o]masakazu_itoh@cup.ocn.ne.jp

Keywords: Japanese Sword, Noritsuna, Bizen, Muromachi Period, Martensite Phase, Crystallographic Texture, Microstructure, Crystalline Grain, Neutron Imaging, Bragg Edge

Abstract. Large-area real-space distribution of crystallographic microstructural information in a Japanese sword made by Noritsuna at Bizen in A.D. 1405 was non-destructively investigated by Bragg-edge neutron transmission imaging using the RADEN instrument at BL22 of MLF (Materials and Life Science Experimental Facility) in J-PARC and the data analysis software RITS (Rietveld Imaging of Transmission Spectra), as one of the series of a systematic research project. As a result, unique properties of the Noritsuna sword were revealed as follows. Hard martensite which d-spacing is close to that of a modern quenched steel exists at the cutting edge, but the area is smaller than that of a modern sword. Coarse grains exist near the notch at the back of the tang. Fine and coarse crystallite-size steels are separately distributed. The texture is not so strong, and the preferred orientation <210> is perpendicular to the normal direction of the sword plate except for the front region of the tang region.

Introduction

Japanese sword is one of the most interesting cultural-heritage artifacts. In fact, a sword typically consists of multiple-phase structures (e.g., ferrite, cementite and martensite), it was crafted in many steps (e.g., selective quenching (hardening) and iterative hammering etc.), and the manufacturing process varied depending on the era and the geographical location. However, being a tradition orally transmitted, there are many unknown information due to lack of written records, in particular, about ancient swords. Therefore, systematic metallurgical characterizations are indispensable for investigating the history of Japanese swords. For this reason, a systematic research project was launched by the authors of this paper. This project aims to non-destructively characterize Japanese swords made by swordsmiths in various places and eras, by using neutron diffraction (S. Harjo, K. Oikawa *et al.*), neutron tomography (Y. Matsumoto, K. Watanabe *et al.* [1]) and Bragg-edge neutron transmission imaging. From viewpoints of both metallurgical characterization inside a bulk and valuable art/heritage preservation, we consider the best method is a neutron beam experiment like previous works performed by neutron diffraction [2], tomography [3] and Bragg-edge transmission imaging [4,5].

In cooperation with Wakou Museum, we systematically investigated four Japanese swords by Bragg-edge neutron transmission imaging; a sword made by Morikage at Bizen in 3rd quarter of 14th century (by A. H. Pham *et al.*), a sword made by Noritsuna at Bizen in 1st quarter of 15th century (this work), a sword made by Sukemasa at Izumi in 1st quarter of 16th century (by K. Oikawa *et al.* [6]) and a modern sword made by Masamitsu in A.D. 1969 (by K. Ohmae *et al.* [7]). This paper reports characterization results of one of the series, the Noritsuna sword.

Specimen: Japanese sword made by Noritsuna at Bizen in A.D. 1405

The Japanese sword investigated in this work was made by swordsmith Noritsuna at Bizen in the Muromachi period (A.D. 1336-1573). There were five famous traditional sword-making styles (schools) "Gokaden", i.e., Soshu, Mino, Yamashiro, Yamato and Bizen, in the old-sword (Koto) age in A.D. 987-1596. The investigated sword was made at one of the Gokaden, Bizen.

Three regions in the sword, the tip region (Tip), the tang region (Tang) and the region between Tip and Tang (Middle), were measured by wavelength-resolved neutron radiography.

Pulsed neutron imaging experiment and Bragg-edge transmission spectrum analysis method

Time-of-flight (TOF) neutron radiography experiment was performed at the RADEN instrument [8] at BL22 of MLF (Materials and Life Science Experimental Facility) in J-PARC which was the energy-resolved neutron imaging instrument connected to a megawatt-class pulsed spallation neutron source. The power of the 3 GeV proton synchrotron was 150 kW during this experiment. The neutron moderator was a decoupled-type supercritical para-H_2 moderator. The sword and a detector were placed at 24 m from the neutron source. The collimator ratio L/D was 1000 since a pinhole placed at 8 m from the neutron source was set in the diameter of 15 mm. The wavelength resolution $\Delta\lambda/\lambda$ at $\lambda = 0.4$ nm was 0.2%. The used neutron TOF-imaging detector, nGEM, was a GEM (gas electron multiplier) type detector [9] which had the pixel size of 800 μm and the detection area of 10.24 cm × 10.24 cm. The measurement time were 7.1 hours for Tip, 8.4 hours for Middle, 8.2 hours for Tang and 6.0 hours for open beam. The measurement time was relatively long due to limit of the maximum counting rate of the detector.

The used data analysis software was the latest version of RITS (Rietveld Imaging of Transmission Spectra) [10-12] for analysis of Bragg-edge neutron transmission spectra at each pixel. Since Bragg-edges in wavelength-dependent neutron transmission data (spectrum) are caused by diffraction phenomenon of neutrons, we can get crystallographic microstructural information of a specimen by analyzing the Bragg-edge transmission spectrum. The single Bragg-edge wavelength-position/broadening analysis mode of RITS [11], which we can exploit to obtain the information of the average value and the full width at half maximum (FWHM) of distribution of crystal lattice plane spacing (*d*-spacing), was used for mapping martensite phase in steel, similarly to a previous work [11]. On the other hand, the Rietveld-type (wide wavelength bandwidth) Bragg-edge transmission spectrum analysis mode of RITS [10] was used for evaluating the information of projected atomic number density (effective thickness), preferred orientation (the most probable crystal orientation) along the beam direction, degree of crystallographic texture (microstructure with anisotropic crystal orientation distribution) evolution and crystallite size. The initial Bragg-edge broadening profile due to only the instrumental resolution was determined by measuring a standard α-iron (ferrite).

Results of single Bragg-edge analysis for imaging of martensite phase in the sword

Figs. 1 (a) and (b) show two types of spectrum near {110} Bragg-edge of α-iron and the profile fitting results obtained by the single Bragg-edge analysis mode of RITS [11], at the blade body region (Position A, see Fig. 2) and the cutting-edge region (Position B, see Fig. 2). It was confirmed that Bragg-edge of the cutting-edge region was broadened due to presence of martensite. In fact, it is well known that the cutting edge is hardened by quenching resulting in the formation of a martensite structure. The measured Bragg-edge broadening reflected it, and this was consistent with a previous work [5].

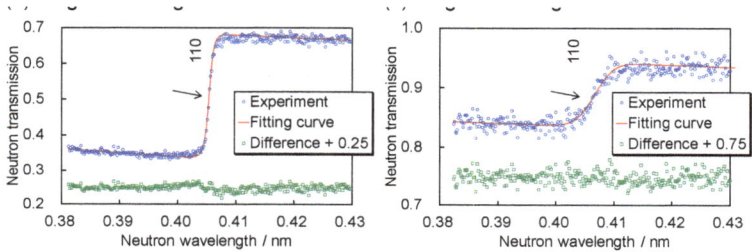

Fig. 1: Results of single Bragg-edge (BE) fitting analysis at (a) the blade body region and (b) the cutting-edge region.

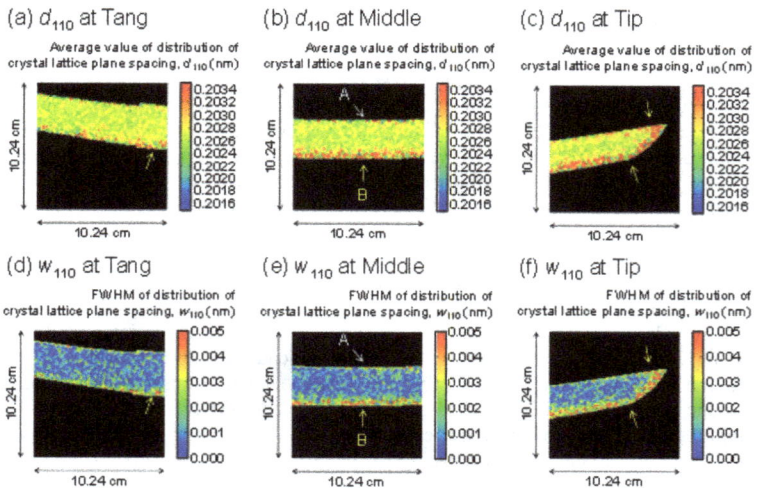

Fig. 2: Imaging results of average value (d_{110}) and FWHM (w_{110}) of the distribution of {110} crystal lattice plane spacing at Tang, Middle and Tip, obtained by single Bragg-edge profile fitting analysis.

Figs. 2 (a)-(c) show maps of the average value of d-spacing distribution (d_{110}), and Figs. 2 (d)-(f) show results of FWHM of d-spacing distribution (w_{110}), at each region (Tang, Middle and Tip). High d_{110} and w_{110} indicate existence of martensite phase in steel, and low d_{110} and w_{110} indicate existence of ferrite phase in steel [11]. The absolute values of d_{110} and w_{110} evaluated in this sword are close to those evaluated in a previous work [11]. Namely, the imaging results show that martensite phase (quenched and hardened region) exists near the cutting edge (*yellow arrows* in Fig. 2). In addition, an interesting aspect of the Noritsuna sword is the width (area) of the martensite region. The width was

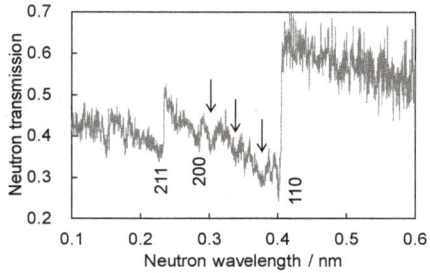

Fig. 3: Coarse-grain type transmission spectrum observed near notch at the back of Tang (Position C).

smaller than that of a modern sword [7]. This means two possibilities; modern sword-quenching process was more well-controlled than ancient one because of the advance in metallurgical knowledge, or the hardened cutting-edge was lost by re-sharpening of the sword during its service. In any case, although the martensite width is small, the hardness seems not to be low because both d_{110} and w_{110} are close to those of a modern quenched 0.45mass%C-steel [11].

Results of Rietveld-type analysis for imaging of microstructural information in the sword
First of all, we discover a relatively coarse-grained region in the Noritsuna sword. Fig. 3 shows a

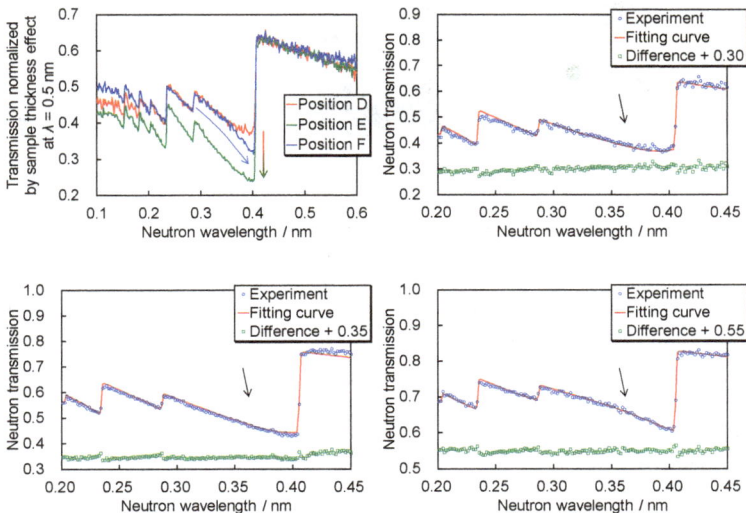

Fig. 4: (a) Change of neutron transmission spectra among Positions D, E and F. Results of Rietveld-type profile fitting analysis for transmission spectra measured at Positions D (b), E (c) and F (d).

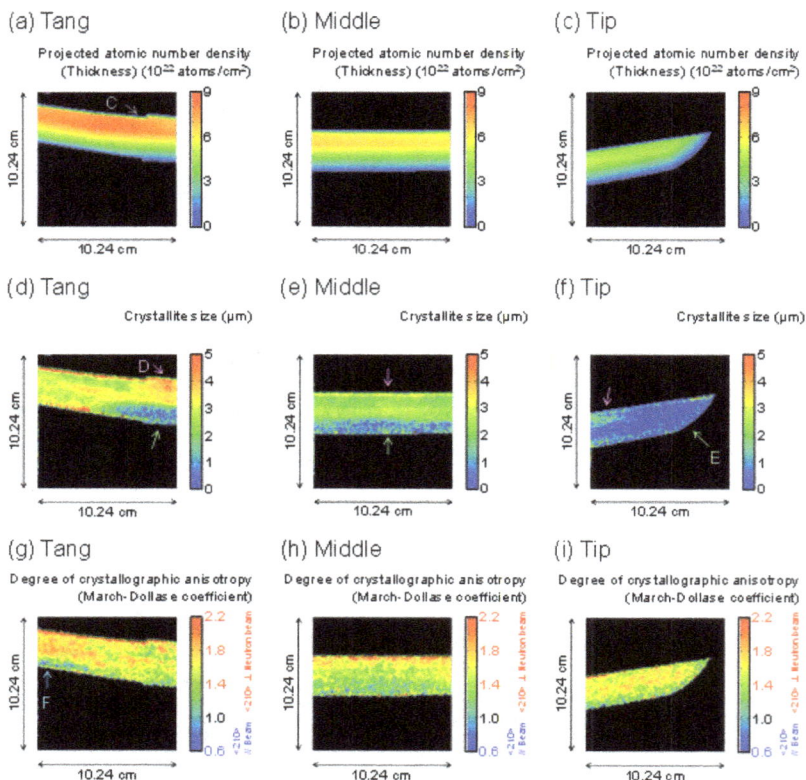

Fig. 5: Imaging of projected atomic number density, crystallite size and degree of crystallographic texture evolution.

neutron transmission spectrum measured near the notch at the back of Tang (Position C, see Fig. 5 (a)). Some transmission dips due to relatively-coarse grains [13] were observed here.

Fig. 4 shows neutron transmission spectra with the fitting curves obtained by RITS, at Positions D, E and F (see Fig. 5). Fig. 4 indicates that intensity or shape of transmission spectrum are changed by crystallite size or preferred orientation [10]. The details are discussed as follows.

Fig. 5 shows imaging results of (a)-(c) projected atomic number density, (d)-(f) crystallite size and (g)-(i) degree of crystallographic texture evolution (the March-Dollase coefficient) at Tang, Middle and Tip, obtained by RITS; the detailed information of each parameter was explained in Ref. [10]. From imaging results of crystallite size, it was found that this sword consisted of two types of iron; one was fine-microstructure steel which composed the cutting-edge side and the tip region (*green* arrows in Figs. 5 (d)-(f)), and the other was coarse-microstructure steel which composed the inner body and the tang region (*pink* arrows in Figs. 5 (d)-(f)).

We also found that the main preferred orientation over the whole region in the sword was <210> which was perpendicular to neutron transmission direction (ND), through the Bragg-edge texture analysis using the March-Dollase function of RITS [10]. However, the preferred orientation <210> becomes to be parallel to ND near the front of Tang (Position F, see Fig. 5 (g)). In any case, degree of texture of this sword was generally weak (see Figs. 4 and 5 (g)-(i)).

Conclusion

We found some interesting information of crystallographic microstructure in a Japanese sword made by Noritsuna, by using Bragg-edge neutron transmission imaging.

(1) The cutting edge consists of *hard* quenched steel (martensite phase). The width is smaller than that of a modern sword [7], but the hardness seems not to be low because the *d*-spacing distribution of martensite in this sword is close to that of a modern quenched steel [11].

(2) This sword includes coarse grains only near the notch at the back of the tang region.

(3) Both fine-crystallites steel which mainly composes the tip region and coarse-crystallites steel which mainly composes the tang region are separately distributed.

(4) The texture is not so strong. The preferred orientation <210> is perpendicular to ND over the whole region, but becomes to be parallel to ND only around the front of the tang region.

Acknowledgements

The neutron experiment at J-PARC MLF BL22 "RADEN" was performed under a user program (Proposal No. 2017A0099). This work partially includes the results of "Collaborative Important Researches" organized by JAEA, QST and The University of Tokyo.

References

[1] Y. Matsumoto, K. Watanabe *et al.*, Mater. Res. Proc. (these proceedings).

[2] F. Grazzi, L. Bartoli, F. Civita, R. Franci, A. Paradowska, A. Scherillo and M. Zoppi, J. Anal. At. Spectrom. 26 (2011) 1030-1039. https://doi.org/10.1039/c0ja00238k

[3] F. Salvemini, F. Grazzi, S. Peetermans, F. Civita, R. Franci, S. Hartmann, E. Lehmann and M. Zoppi, J. Anal. At. Spectrom. 27 (2012) 1494-1501. https://doi.org/10.1039/c2ja30035d

[4] K. Kino, N. Ayukawa, Y. Kiyanagi, T. Uchida, S. Uno, F. Grazzi and A. Scherillo, Phys. Procedia 43 (2013) 360-364. https://doi.org/10.1016/j.phpro.2013.03.043

[5] Y. Shiota, H. Hasemi and Y. Kiyanagi, Phys. Procedia 88 (2017) 128-133. https://doi.org/10.1016/j.phpro.2017.06.017

[6] K. Oikawa *et al.*, Mater. Res. Proc. (this proceedings).

[7] K. Ohmae *et al.*, Mater. Res. Proc. (this proceedings).

[8] T. Shinohara, T. Kai, K. Oikawa, M. Segawa, M. Harada, T. Nakatani, M. Ooi, K. Aizawa, H. Sato, T. Kamiyama, H. Yokota, T. Sera, K. Mochiki and Y. Kiyanagi, J. Phys. Conf. Ser. 746 (2016) 012007. https://doi.org/10.1088/1742-6596/746/1/012007

[9] S. Uno, T. Uchida, M. Sekimoto, T. Murakami, K. Miyama, M. Shoji, E. Nakano, T. Koike, K. Morita, H. Satoh, T. Kamiyama and Y. Kiyanagi, Phys. Procedia 26 (2012) 142-152. https://doi.org/10.1016/j.phpro.2012.03.019

[10] H. Sato, T. Kamiyama and Y. Kiyanagi, Mater. Trans. 52 (2011) 1294-1302. https://doi.org/10.2320/matertrans.M2010328

Neutron Radiography - WCNR-11   Materials Research Forum LLC
Materials Research Proceedings **15** (2020) 214-220   https://doi.org/10.21741/9781644900574-33

[11] H. Sato, T. Sato, Y. Shiota, T. Kamiyama, A. S. Tremsin, M. Ohnuma and Y. Kiyanagi, Mater. Trans. 56 (2015) 1147-1152. https://doi.org/10.2320/matertrans.M2015049

[12] H. Sato, K. Watanabe, K. Kiyokawa, R. Kiyanagi, K. Y. Hara, T. Kamiyama, M. Furusaka, T. Shinohara and Y. Kiyanagi, Phys. Procedia 88 (2017) 322-330. https://doi.org/10.1016/j.phpro.2017.06.044

[13] W. Kockelmann, G. Frei, E. H. Lehmann, P. Vontobel and J. R. Santisteban, Nucl. Instrum. Methods A 578 (2007) 421-434. https://doi.org/10.1016/j.nima.2007.05.207

Neutron Radiography - WCNR-11
Materials Research Proceedings **15** (2020) 221-226

Materials Research Forum LLC
https://doi.org/10.21741/9781644900574-34

Comparative Study of Ancient and Modern Japanese Swords using Neutron Tomography

Yoshihiro Matsumoto[1, a *], Kenichi Watanabe[2,b], Kazuma Ohmae[2,c], Akira Uritani[2,d], Yoshiaki Kiyanagi[2,e], Hirotaka Sato[3,f], Masato Ohnuma[3,g], Anh Hoang Pham[4,h], Shigekazu Morito[4,i], Takuya Ohba[4,j], Kenichi Oikawa[5,k], Takenao Shinohara[5,l], Tetsuya Kai[5,m] Stefanus Harjo[5,n] and Masakazu Ito[6,o]

[1]Comprehensive Research Organization for Science and Society, Ibaraki 319-1106, Japan

[2]Graduate School of Engineering, Nagoya University, Aichi, 464-8603, Japan

[3]Faculty of Engineering, Hokkaido University, Hokkaido 060-8628, Japan

[4]Interdisciplinary Faculty of Science and Engineering, Shimane University, Shimane 690-8504, Japan

[5]J-PARC Center, Japan Atomic Energy Agency, Ibaraki, 319-1195, Japan

[6]WAKOU MUSEUM, Shimane 692-0011, Japan

[a]y_matsumoto@cross.or.jp, [b]k-watanabe@energy.nagoya-u.ac.jp, [c]oomae.kazuma@f.mbox.nagoya-u.ac.jp, [d]uritani@energy.nagoya-u.ac.jp, [e]kiyanagi@phi.phys.nagoya-u.ac.jp, [f]h.sato@eng.hokudai.ac.jp, [g]ohnuma.masato@eng.hokudai.ac.jp, [h]anhpham@riko.shimane-u.ac.jp, [i]mosh@riko.shimane-u.ac.jp, [j]ohba@riko.shimane-u.ac.jp, [k]kenichi.oikawa@j-parc.jp, [l]takenao.shinohara@j-parc.jp, [m]tetsuya.kai@j-parc.jp, [n]stefanus.harjo@j-parc.jp, [o]masakazu_itoh@cup.ocn.ne.jp

Keywords: Japanese Sword, Neutron Tomography, RADEN

Abstract. We have performed neutron tomography using two ancient Japanese swords (designated Morikage and Sukemasa) and one modern Japanese sword (Masamitsu) at RADEN in the J-PARC Materials and Life Science Experimental Facility. For the ancient Japanese sword Morikage, it is found that the martensite iron is distributed in the region of about 3 mm from the cutting-edge and the ferrite iron is distributed in the inner region of the blade. The martensite iron area surrounds the inner ferrite iron area. For the ancient Japanese sword Sukemasa, the martensite iron is distributed only in a very narrow region at the cutting-edge and the homogeneous ferrite iron area is dominantly distributed in the inner region of the blade. In contrast to the ancient Japanese swords, the distribution of the martensite iron is about 8 mm from the cutting-edge and is similar to the wave pattern visible on the blade surface for the modern Japanese sword Masamitsu. Additionally for Masamitsu, a region where the neutron transmittance slightly increases was found at the interface between the martensite area and inner ferrite iron area. Line-like structures due to inclusions produced in the manufacturing process were also found in the blade. These results indicate that the manufacturing processes and raw materials of the Japanese swords are significantly different depending on the era and place of manufacture. Going forward, it is necessary to compare systematically the internal structure of more samples in order to clarify historical changes in Japanese sword making, and the non-destructive approach using neutron tomography is one of the powerful tools to elucidate them.

Neutron Radiography - WCNR-11   Materials Research Forum LLC
Materials Research Proceedings 15 (2020) 221-226   https://doi.org/10.21741/9781644900574-34

Introduction

Japanese swords have superior characteristics in strength and toughness, giving them important value not only as historical works of art but also from a metallurgical point of view [1,2]. It is known that Japanese swords differ in the raw materials and manufacturing processes depending on the era and place of manufacture. Most modern Japanese swords use a special steel called Tamahagane that is smelted from iron sand and charcoal in the traditional steel-making system Tatara as the raw materials, and they are produced based on the manufacturing processes developed after the Edo period (17th - 19th century). On the other hand, with regard to ancient Japanese swords manufactured from the Kamakura period to the Muromachi period (12th - 16th century), there are still many unknown points both in the manufacturing processes and raw materials used due to lack of historical records. Comparing ancient Japanese swords with modern Japanese swords from the viewpoint of materials science can provide an important foothold for understanding the historical changes of the Japanese sword. In fact, investigations on Japanese swords have been conducted using various methods and valuable knowledge has been obtained [3-6]. Therefore, based on the above background, we have investigated the three-

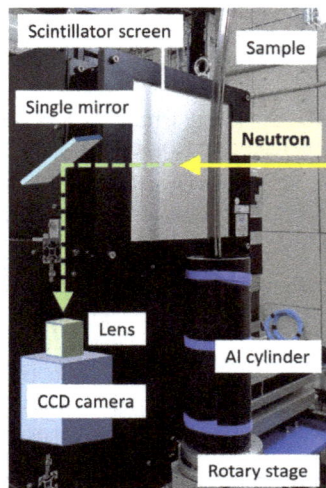

Fig. 1. Illustraion of the experimental setups of neutron tomography.

dimensional internal structure and crystallographic structure of ancient and modern Japanese swords non-destructively using neutron tomography and Bragg-edge transmission (BET) imaging [7]. In this paper, we report the results of the neutron tomography measurement.

Experimental

Characteristics of the Japanese swords used in the present study are listed in Table 1. Morikage was manufactured in the Bizen district, the location of one of the five major sword schools, or "Gokaden", in the early Muromachi period. Sukemasa was manufactured in the Izumi district during the late Muromachi period. These swords are categorized as the ancient Japanese sword, or "Koto". Masamitsu was manufactured in Fukuoka prefecture in A.D. 1969 and is categorized to the modern Japanese sword, or "Gendaito".

Neutron tomography measurements were carried out at the pulsed neutron imaging instrument RADEN located at BL22 of the J-PARC Materials and Life Science Experimental Facility (MLF) [8]. The layout of the detection system used in the present study is illustrated in Fig. 1. This system consists of a camera-type detector and a rotary stage. Neutrons which penetrate the sample are converted to green light by the ^{6}LiF + ZnS scintillator screen (RC Tritec Ltd.) of area 300×300 mm^2 and 0.1 mm thickness and focused on the camera sensor by a single mirror. In the present study, we adopted a large area CCD camera, the ANDOR iKon-L (Andor Technology Ltd.). This CCD camera provides 16 bit images, and its sensor (2048×2048 active pixels) is cooled to -100 °C with a Peltier module and coolant circulation to suppress dark current noise. These components are installed in the detector box covered with thin black aluminum plates, thick lead plates and boron rubber sheets to reduce undesirable noise due to scattered neutrons, γ-ray irradiation and light leaks. Each Japanese sword was adequately

Neutron Radiography - WCNR-11   Materials Research Forum LLC
Materials Research Proceedings **15** (2020) 221-226   https://doi.org/10.21741/9781644900574-34

protected so as not to destroy its cultural and artistic values, and then held vertically in the aluminum cylinder mounted on the rotary stage as shown in Figure 1. This aluminum cylinder was covered with boron rubber sheets to avoid activation of the sample outside the neutron irradiation area. As for neutron tomography measurements, both the camera-type detector and the rotary stage can be controlled simultaneously using the device control software framework IROHA2 of J-PARC MLF [9]. J-PARC consists of the multi-purpose facilities, and neutron intensity changes during data acquisition due to accelerator trips, proton beam sharing and so on. Thus, the transmission image data is saved and the rotary stage turns to the next condition only when the number of proton pulses injected to the neutron target satisfies a set point estimated from the camera exposure time. In this way, a projection image that is not influenced by fluctuation of the pulsed neutron source can be acquired automatically. Each projection image was normalized by the neutron beam intensity after subtracting dark current noise. The computed reconstruction of the two-dimensional images (tomograms) of the sample by the filtered-back-projection method was performed using the visualization software VG studio MAX (Volume Graphics GmbH). All projection images were measured with a field of view of 150×150 mm^2, L/D of 400 and camera exposure time of 100 sec using a white neutron beam. As the proton beam power in the present study was about 150 kW, neutron fluence per projection would be about 2.9×10^8 n/cm^2 as estimated from the time-averaged neutron flux of 5.8×10^7 n/sec/cm^2/MW expected from the design parameters of RADEN [7]. Distance from the rotational center of the sample to the scintillator screen was set to 70 mm, giving a geometric blurring of approximately 0.18 mm at the scintillator screen.

Result and Discussion

Fig. 2 shows the tomograms from around the middle area of the swords. The XZ-plane and the XY-plane correspond to a parallel plane and a perpendicular plane with respect to the longitudinal direction of the blade. The projection images used for the reconstruction were obtained by rotating the sample from -180° to +180° by 1° steps for Morikage and Masamitsu and from 0° to 171° by 1° steps for Sukemasa, respectively. Since we had to reduce the number of the projection images significantly from the usual value of 500-800 [10] due to the limited available measurement time and an unexpected neutron beam stoppage, the spatial resolution for this neutron tomography should be less than 0.5 mm judging from our previous study [11]. As for Morikage, it is found that there is a difference in the grayscale between the region of about 3 mm from the cutting-edge and the inner region of the blade in the XZ-plane tomogram. It is known that the martensite iron formed by quench hardening is distributed near the cutting-edge and ferrite iron is distributed in the inner region of the blade depending on the manufacturing processes of the swords [2,5]. Since the similar crystallographic structure was also obtained from our BET imaging measurement conducted to Morikage, the observed difference is thought to correlate with the distribution of the martensite and ferrite irons. From the XY-plane tomograms, it can be seen that the martensite iron area surrounds the ferrite iron area. The concave structures appearing on both sides of the blade are cutting structures for achieving weight reduction and improved rigidity of the sword, and they reach into the ferrite iron area. In contrast, the difference in the grayscale is only barely visible near the cutting-edge in both the XZ-plane and XY-plane tomograms for Sukemasa. The grayscale in the inner region of the blade is nearly constant, so it is considered that the homogeneous ferrite iron is dominant in this region. From the BET imaging measurement [12], the crystallite size distribution is considerably uniform except in a very narrow region from the cutting-edge. The manufacturing processes and raw materials used for Morikage and Sukemasa seem to be different even though they were produced in the same Muromachi period. As for Masamitsu, the martensite iron is distributed near the cutting-edge, as in the ancient Japanese swords, judging from the XZ-plane and XY-plane tomograms, but its range is significantly different. The martensite iron appears to spread more than 8 mm from the cutting-edge, in agreement with the results of the BET imaging measurement [12], and its distribution closely resembles the wave-pattern visible on the blade surface, called the "Hamon". From this, it is considered that the quench hardening was done to the deep part of the blade to produce a sharp Hamon pattern. Furthermore, the grayscale of the interface between the martensite iron area and the inner ferrite iron area seems to be slightly darker, which means the neutron transmission is higher compared to the surrounding areas. At the present time, it is inferred that a component with small neutron cross section or low atomic density is distributed in the interface region, though a more detailed analysis is needed. On the other hand, bright line-like structures were found in the blade surrounded by the yellow circle especially from the XZ-plane tomogram. This is considered to be due to inclusions containing components having a large neutron cross section (for example, borax used in the forging process). Thus, the manufacturing processes, raw materials and residual materials of the modern Japanese sword Masamitsu are seen to be significantly different compared to the ancient Japanese swords. The grayscale contrast observed in the tomogram depends on not only the neutron absorption cross section but also the various factors such as neutron scattering, diffraction effects and so on. It will be necessary to investigate what kind of factors cause the contrast and feed back to the tomographic analysis of the Japanese sword.

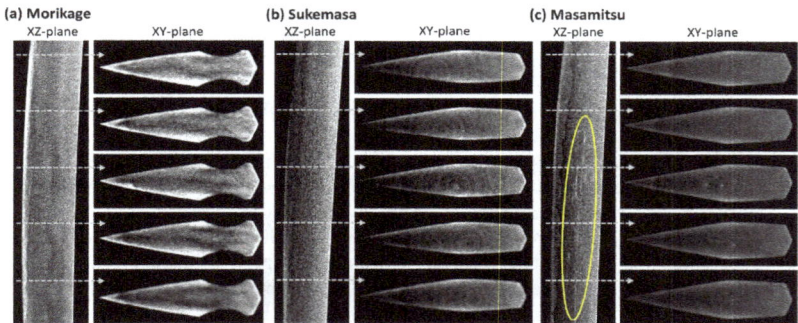

Fig. 2. The XZ-plane and XY-plane computed tomograms around the middle area of the blades of the ancient Japanese swords (a) Morikage and (b) Sukemasa and the modern Japanese sword (c) Masamitsu.

Summary

The three-dimensional internal structure of the ancient Japanese swords (Morikage and Sukemasa) and the modern Japanese sword (Masamitsu) were investigated non-destructively using neutron tomography. In both the ancient and modern Japanese swords, the martensite iron formed by quench hardening is distributed near the cutting-edge and the ferrite iron is distributed in the inner region of the blade. However, the range of the martensite iron is very different for each sword. In the case of the ancient Japanese swords, the martensite iron is distributed only to a few millimeters from the cutting-edge, while in the case of the modern Japanese sword, it extends to a wide area of about 8 mm from the cutting-edge. This should be due to the quench hardening treatment to the deep part of the modern blade done to produce the clear Hamon pattern. In addition, it is found that there are inclusions and residual materials that may be produced in the manufacturing processes of the modern Japanese sword, which cannot be confirmed with the ancient Japanese swords. We thus succeeded in obtaining some new

Neutron Radiography - WCNR-11
Materials Research Proceedings **15** (2020) 221-226

Materials Research Forum LLC
https://doi.org/10.21741/9781644900574-34

knowledge about the manufacturing processes and raw materials of the Japanese swords. In the future, we will systematically investigate many samples covering different manufacturing eras and locations to clarify historical changes in Japanese sword making.

Acknowledgement

Neutron tomography experiments at the J-PARC MLF were performed under Proposal Nos. 2017A0099 and 2016B0163. This work partially includes the results of 'Collaborative Important Researches' organized by JAEA, QST and U. Tokyo.

References

[1] M.R. Notis, The history of the metallographic study of the Japanese sword, Materials Characterization 45 (2000) 253-258. https://doi.org/10.1016/S1044-5803(01)00101-2

[2] T. Inoue, The Japanese sword; the material, manufacturing and computer simulation of quenching process, Materials Science Research International 3 (1997) 193-203. https://doi.org/10.2472/jsms.46.12Appendix_193

[3] F. Grazzi, L. Bartoli, F. Civita, A.M. Paradowska, A. Scherillo, M. Zoppi, Non destructive characterization of phase distribution and residual atrain/stress map of two ancient (Koto) age Japanese swords, Materials Science Forum 652 (2010) 167-173. https://doi.org/10.4028/www.scientific.net/MSF.652.167

[4] F. Grazzi, L. Bartoli, F. Civita, R. Franci, A. Paradowska, A. Scherillo, M. Zoppi, From Koto age to modern times: Quantitative characterization of Japanese swords with time of flight neutron diffraction, J. Anal. At. Spectrom. 26 (2011) 1030-1039. https://doi.org/10.1039/c0ja00238k

[5] M. Yaso, T. Takaiwa, Y. Minagi, T. Kanaizumi, K. Kubota, T. Hayashi, S. Morito, T. Ohba, Study of Japanese sword from a viewpoint of steel strength, J. Alloys Compd. 577 (2013) S690-S694. https://doi.org/10.1016/j.jallcom.2012.06.141

[6] F. Salvemini, F. Grazzi, N. Kardjilov, I. Manke, F. Civita, M. Zoppi, Neutron computed laminography on ancient metal artefacts, Anal. Methods 7 (2015) 271-278. https://doi.org/10.1039/C4AY02014F

[7] Y. Kiyanagi, H. Sato, T. Kamiyama, T. Shinohara, A new imaging method using pulsed neutron sources for visualizing structural and dynamical information, J. Phys. Conf. Ser. (2012) 340 021010. https://doi.org/10.1088/1742-6596/340/1/012010

[8] T. Shinohara, T. Kai, K. Oikawa, M. Segawa, M. Harada, T. Nakatani, M. Ooi, K. Aizawa, H. Sato, T. Kamiyama, H. Yokota, T. Sera, K. Mochiki, Y. Kiyanagi, Final design of the Energy-Resolved Neutron Imaging System "RADEN" at J-PARC, J. Phys. Conf. Ser. 746 (2016) 012007. https://doi.org/10.1088/1742-6596/746/1/012007

[9] T. Nakatani, Y. Imamura, T. Ito, T. Otomo, The control software framework of the web base, J. Phys. Conf. Ser. 8 (2015) 036013. https://doi.org/10.7566/JPSCP.8.036013

[10] A.C. Kak, M. Slaney, Principles of computerized tomographic imaging, Classics in Applied Mathematics, SIAM, Philadelphia, USA, 2001, pp. 49-112. https://doi.org/10.1137/1.9780898719277

[11] Y. Matsumoto, M. Segawa, T. Kai, T. Shinohara, T. Nakatani, K. Oikawa, K. Hiroi, Y.H. Su, H. Hayashida, J.D. Parker, S.Y. Zhang, Y. Kiyanagi, Recent progress of radiography and

Neutron Radiography - WCNR-11   Materials Research Forum LLC
Materials Research Proceedings 15 (2020) 221-226   https://doi.org/10.21741/9781644900574-34

tomography at the energy-resolved neutron imaging system RADEN, Phys. Procedia 88 (2017) 162-166. https://doi.org/10.1016/j.phpro.2017.06.022

[12] K. Oikawa, Y. Kiyanagi, H. Sato, K. Ohmae, A.H. Pham, K. Watanabe, Y. Matsumoto, T. Shinohara, T. Kai, S. Harjo, M. Ohnuma, S. Morito, T. Ohba, A. Uritani, M. Ito, Crystallographic structure study of a Japanese sword made by Sukemasa in the Muromachi period using pulsed neutron imaging, Mater. Res. Proc. (this proceedings)

[13] K. Ohmae, Y. Kiyanagi, H. Sato, K. Oikawa, A.H. Pham, K. Watanabe, Y. Matsumoto, T. Shinohara, T. Kai, S. Harjo, M. Ohnuma, S. Morito, T. Ohba, A. Uritani, M. Ito, Crystallographic structure study of a modern Japanese Masamistu using pulsed neutron imaging, Mater. Res. Proc. (these proceedings)

Neutron Radiography - WCNR-11
Materials Research Proceedings **15** (2020) 227-232

Materials Research Forum LLC
https://doi.org/10.21741/9781644900574-35

Crystallographic Structure Study of a Japanese Sword Masamitsu made in the 1969 using Pulsed Neutron Imaging

Kazuma Ohmae[1, a *], Yoshiaki Kiyanagi[1,b], Hirotaka Sato[2,c], Kenichi Oikawa[3,d], Anh Hoang Pham[4,e], Kenichi Watanabe[1,f], Yoshihiro Matsumoto[5,g], Takenao Shinohara[3,h], Tetsuya Kai[3,i], Stefanus Harjo[3,j], Masato Ohnuma[2,k], Shigekazu Morito[4,l], Takuya Ohba[4,m], Akira Uritani[1,n], Masakazu Ito[6,o]

[1]Graduate School of Engineering, Nagoya University, Aichi, 464-8603, Japan

[2]Faculty of Engineering, Hokkaido University, Hokkaido 060-8628, Japan

[3]J-PARC Center, Japan Atomic Energy Agency, Ibaraki, 319-1195, Japan

[4]Interdisciplinary Faculty of Science and Engineering, Shimane University, Shimane 690-8504, Japan

[5]Comprehensive Research Organization for Science and Society, Ibaraki 319-1106, Japan

[6]WAKOU MUSEUM, Shimane 692-0011, Japan

[a]oomae.kazuma@f.mbox.nagoya-u.ac.jp, [b]kiyanagi@phi.phys.nagoya-u.ac.jp, [c]h.sato@eng.hokudai.ac.jp, [d]kenichi.oikawa@j-parc.jp, [e]anhpham@riko.shimane-u.ac.jp, [f]k-watanabe@energy.nagoya.ac.jp, [g]y_matsumoto@cross.or.jp, [h]takenao.shinohara@j-parc.jp, [i]tetsuya.kai@j-parc.jp, [j]stefanus.harjo@j-parc.jp, [k]ohnuma.masato@eng.hokudai.ac.jp, [l]mosh@riko.shimane-u.ac.jp, [m]ohba@riko.shimane-u.ac.jp, [n]uritani@energy.nagoya-u.ac.jp, [o]masakazu_itoh@cup.ocn.ne.jp

Keywords: Japanese Sword, Masamitsu, Martensite, Crystallographic Structure, Pulsed Neutron Imaging, Bragg-Edge Transmission, RITS Code

Abstract. Energy-resolved neutron transmission imaging technique using an accelerator-driven pulsed neutron source and a time-resolved two-dimensional detector can obtain the Bragg-edge transmission spectrum at each pixel. In this technique, crystallographic information can be visualized over a wide area of a sample. We used this method to investigate the crystallographic information of a modern Japanese sword produced by Masamitsu in 1969. As a result, shifting and broadening of the (110) Bragg-edge was confirmed at the cutting-edge side. It is evidence that the martensite phase exists at the cutting-edge side as a result of the quenching process. By comparing with results of the old Japanese swords, it is found that the martensite region in the modern Japanese sword is wider or deeper than one in the old swords. In addition, we can see strongly preferred orientation of the crystal grains at the boundary between the martensite phase on the cutting-edge side and the ferrite phase on the back-edge side.

Introduction

The metallographic characteristics are essential information to clarify making processes of the Japanese swords, because these processes were not documented well in the past. The Japanese swords have been produced in various eras and areas since ancient times. In order to discuss characteristics of the Japanese swords depending on areas and eras, we need to analyze many Japanese swords. It is necessary to analyze them non-destructively because the old Japanese swords have historical value. Neutron imaging is a powerful tool to study metallic cultural heritages because of non-destructive, high penetrating power and capability to get crystallographic information [1-2]. By analyzing the position dependent Bragg-edge spectra,

Neutron Radiography - WCNR-11   Materials Research Forum LLC
Materials Research Proceedings **15** (2020) 227-232   https://doi.org/10.21741/9781644900574-35

quantitative visualization of the crystallographic information of the Japanese swords can be achieved [3].

We plan to investigate historical Japanese swords. The comparison of modern and historical swards will lead to a better understanding of the various crystallographic characteristics. In this work, we investigated crystallographic information of the modern Japanese sword made by Masamitsu in 1969.

Experimental procedure

In this work, we measured a modern Japanese sword in order to compare with the old Japanese swords. The Japanese swords are generally made of **a carbon steel** and was quenched and tempered. Overview of the modern Japanese sword made by Masamitsu in 1969 is shown in Fig. 1. This experiment was performed at BL22 RADEN in J-PARC MLF [4]. The proton beam power in the present work was 150 kW. The neutron moderator was a decoupled-type supercritical para-H_2 moderator. As a time-resolved two-dimensional detector, we adopted a neutron sensitive gas electron multiplier (n-GEM) detector [5] with a 0.8×0.8 mm^2 pixel resolution and a 10.24×10.24 cm^2 detection area (128ch $\times$ 128ch). The sword and a detector were placed at 24 m from the neutron source. The collimator ratio L/D

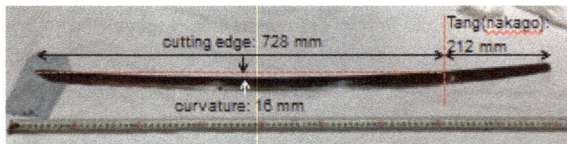

Fig. 1 A photograph of the Masamitsu sword (1969).

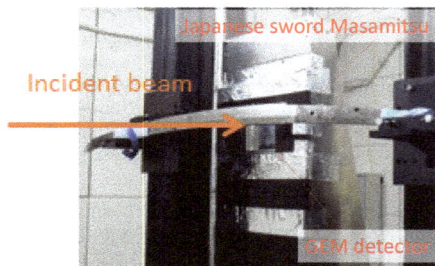

Fig.2 The experimental setup of the present work at RADEN.

was 1000 since a pinhole placed at 8 m from the neutron source was set in the diameter of 15 mm. The wavelength resolution $\Delta\lambda/\lambda$ at $\lambda = 0.4$ nm was 0.2%.

The experimental setup used is shown in Fig. 2. As shown in Fig. 3, the Japanese sword was measured in three different areas, Tang, Middle and Tip areas. Measurement time was 7.2 hours in the Tang area, 8.6 hours in the Middle area, 7.2 hours in the Tip area and 6 hours for the direct beam.

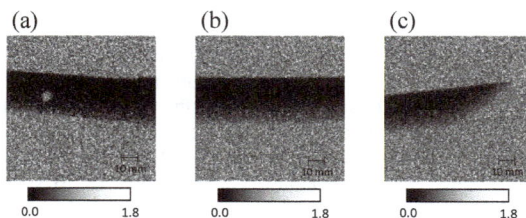

Fig. 3 Transmission images of the (a)Tang area, (b)Middle area and (c) Tip area.

Neutron Radiography - WCNR-11　　　　　　　　　　　　　　　　Materials Research Forum LLC
Materials Research Proceedings **15** (2020) 227-232　　　　　https://doi.org/10.21741/9781644900574-35

Bragg-edge analysis

By using a time-resolved two-dimensional neutron detector at a pulsed neutron source, we can acquire an energy- or wavelength-dependent transmission spectrum at each pixel in the detector with the time-of-flight (TOF) technique. From the neutron transmission spectra of a polycrystalline metallic material, the Bragg-edge structures shown in Fig. 4 can be obtained. Since the Bragg-edge shape appears as a result of neutron diffraction in a polycrystalline or a powder sample, this spectrum contains crystal structure information of the sample. We can extract some information, such as the projected atomic number density, lattice constant, preferential orientation and crystallite size, from the Bragg edge spectrum.

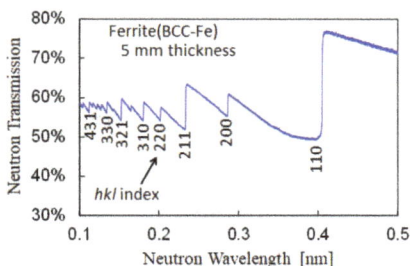

Fig.4 The Bragg-edge transmission spectrum of a ferrite plate (BCC-Fe, 5 mm

The Bragg-edge spectra were analyzed by the neutron Bragg-edge analysis code RITS [6]. The RITS code has two analysis modes, a single-Bragg-edge analysis mode and a Rietveld-type (wide wavelength bandwidth) analysis mode. The single-Bragg-edge analysis mode was used to extract the precise Bragg-edge position and the Bragg-edge broadening by fitting the experimental spectrum to Jorgensen function. The Rietveld-type analysis, in which the whole Bragg-edge spectrum shape is determined through the non-linear least square fitting procedure, was used to extract the projected atomic number density, preferential orientation and crystallite size [6]. In order to improve data statistics, neutron counts in 2×2 pixels (1.6×1.6 mm^2) were summed up into one pixel. As a result, 64×64 pixels two-dimensional Bragg-edge spectra were obtained. We can visualize the crystallographic information of the modern Japanese sword (Masamitsu, 1969) by mapping those refined parameters.

Results and Discussion
Single-Bragg-edge analysis

In the Japanese sword that is generally made of a carbon steel, the cutting-edge side is usually quenched to make the martensite phase to improve the hardness. The martensite is formed in carbon steels by rapid cooling or quenching of the austenite of iron, which allows for higher carbon content than the ferrite. In the quenching process, carbon atoms have no time to diffuse out of the crystal structure. Consequently, the face-centered cubic austenite transforms into a strained body-centered tetragonal martensite phase. The crystal lattice of martensite is distorted from the pure body-centered cubic structure. A clear difference in the Bragg-edge-position shift (d_{110}) and its broadening (w_{110}) between the ferrite and the martensite phase can be seen in the Bragg-edge spectra [7]. In general, for the martensite phase, the Bragg-edge position is shifted to the longer wavelength side and the width is broader than that of the ferrite phase. This tendency was also confirmed in the Japanese sword.

Fig. 5 (a-c) and Fig. 6 (a-c) shows the 2D maps of the lattice spacing d_{110} and the Bragg-edge broadening w_{110}. Because the Bragg edge of the {110} crystal lattice plane is the largest, it was used in this analysis. As shown in Fig. 5 (a-c), d_{110} increases near the cutting-edge. However, the boundary region between the martensite and the ferrite phase seems to have slightly smaller d_{110} values compared with the normal ferrite region. Fig. 6 (a-c) shows that w_{110} also increases on the cutting-edge side. The width of a region showing large d_{110} and w_{110} is about 8 mm. The wider region on the cutting-edge side was quenched compared with the old Japanese swords [8,9]. It is considered that quenching condition of the modern Japanese swords might differ from that of the old sword. Almost the whole region in Fig. 5 (a) shows non-martensite feature. This is because the Tang part was not quenched.

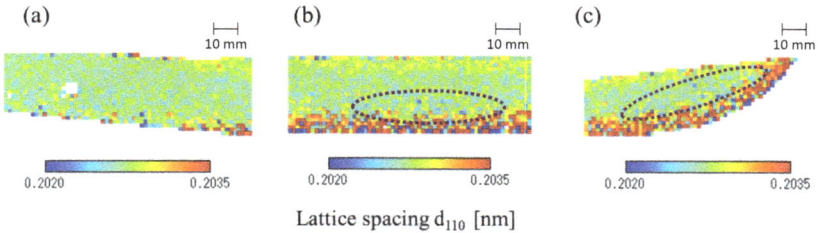

Lattice spacing d_{110} [nm]

Fig.5 2D maps of the lattice spacing d_{110} of the Masamitsu sword. (a)Tang area, (b)Middle area and (c) Tip area. The regions surrounded by dashed circles show lower values of the spacing d_{110}.

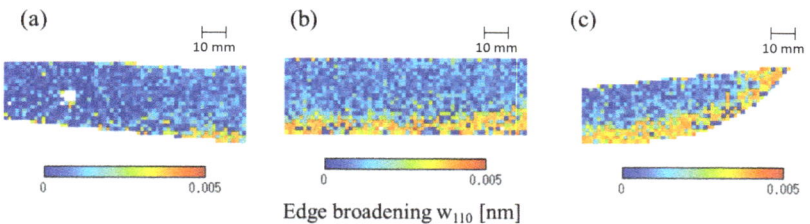

Edge broadening w_{110} [nm]

Fig.6 2D maps of the edge broadening w_{110} of the Masamitsu sword.

Rietveld-type analysis

Fig. 7 (a-c) shows the 2D maps of the projected atomic number density (ρt) of the Masamitsu. The projected atomic number density corresponds to the thickness of the sample. It shows that the physical shape of the Masamitsu changes smoothly. This map reflects the actual sword thickness.

Fig. 8 (a-c) shows the 2D maps of the crystallite size (s) of the Masamitsu. The crystallite size is quantified by Sabine's primary extinction function [6]. This result shows that the crystallite size is sufficiently small over the whole region in the sword. It is considered that coarsening of crystal grain was not occurred in the heating processes.

Fig. 9 (a-c) shows the 2D maps of the preferred-orientation parameter (r). The parameter r is quantified by March-Dollase function [6]. If the orientation distribution of crystal grains is completely random, r should be 1. If the preferred-orientation vector <hkl> is parallel to the neutron beam axis, r becomes less than 1. On the other hand, if the <hkl> plane is perpendicular to the beam axis, r becomes greater than 1. In this case, the <210> vector showed the best fitting results. Fig. 9 (c) shows that r is nearly 1 for the whole region at the Tip area. However, the boundary between the martensite and ferrite region shows greater r values in the Tang and Middle areas (Fig. 9 (a-b)). In order to discuss the difference in preferred-orientation feature between the tip and other regions, we have to investigate in more detail. The spatial distribution of the preferred-orientation might suggest some differences in sword fabrication processes.

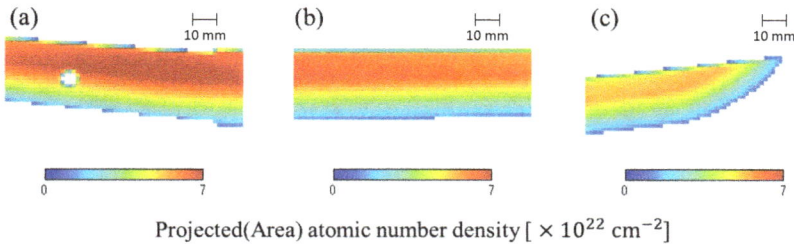

Projected(Area) atomic number density $[\times 10^{22} \text{ cm}^{-2}]$

Fig. 7 2D maps of the projected atomic number density ρt of the Masamitsu sword.

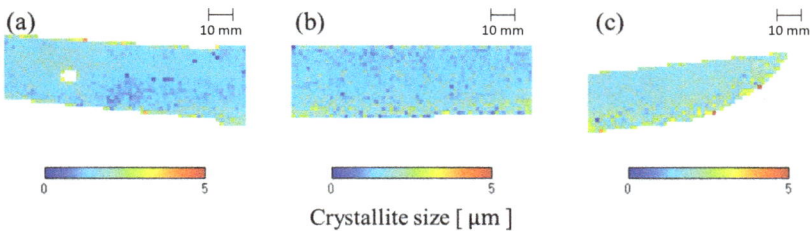

Crystallite size [μm]

Fig.8 2D maps of the crystallite size s of the Masamitsu sword.

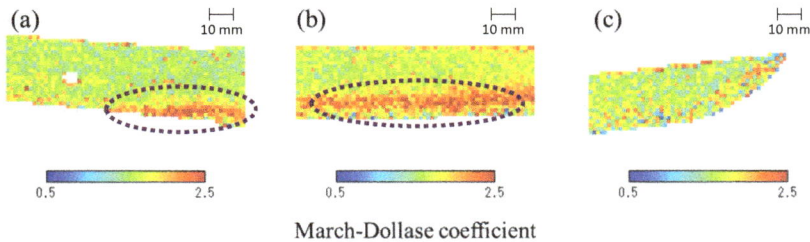

March-Dollase coefficient

Fig.9 2D maps of the preferred-orientation parameter r of the Masamitsu sword. The regions surrounded by dashed circles show higher values of the preferred-orientation parameter r.

Conclusions

We conducted the neutron Bragg-edge imaging to extract crystallographic information of a modern Japanese sword Masamitsu made in 1969. Valuable information of the crystallographic microstructure was revealed as follows:

(1) The cutting-edge side has larger lattice spacing and broader Bragg-edge width than that of the other side. This means that the martensite phase was created in the cutting-edge region. The width of the martensite region in the modern Japanese sword is wider than that of the old ones. It is considered that quenching condition of the modern Japanese swords might differ from that of the old sword.

Neutron Radiography - WCNR-11   Materials Research Forum LLC
Materials Research Proceedings **15** (2020) 227-232   https://doi.org/10.21741/9781644900574-35

(2) Crystallite size seems to be well small over the whole region in the sword. It is considered that coarsening of crystal grain was not occurred in the heating processes.

(3) At the Middle and Tang areas, the preferred orientation near the boundary between the martensite and ferrite phase seems to be stronger than that of the other part. The spatial distribution of the preferred-orientation might suggest some differences in sword fabrication processes.

This work is step towards a better understanding of fabrication processes of historical Japanese swords. Some differences between the modern and old Japanese swords were confirmed. We will further investigate the detailed information on the micro-structural properties of the Japanese swords.

Acknowledgements

The neutron experiments at the J-PARC MLF BL22 "RADEN" performed under Proposal No. 2017A0099. This work partially includes the result of 'Collaborative Important Researches' organized by JAEA, QST and U. Tokyo.

References

[1] F. Salvemini, F. Grazzi, S. Peetermans, *et al.*, Quantitative characterization of Japanese ancient swords through energy-resolved neutron imaging. J. Analytical Atomic Spectrometry **2012**, 27, 1494-1501. https://doi.org/10.1039/c2ja30035d

[2] A. Fedrigo, M. Strobl, A.R. Williams, *et al.*, Neutron imaging study of 'pattern-welded' swords from the Viking Age, Archaeol. Anthropol. Sci., **2018**, 10, 1249 – 1263. https://doi.org/10.1007/s12520-016-0454-5

[3] Y.Shiota, H. Hasemi, Y. Kiyanagi, Crystallographic analysis of a Japanese sword by using Bragg edge transmission spectroscopy, Phys. Procedia **2017**, 88, 128–133. https://doi.org/10.1016/j.phpro.2017.06.017

[4] T. Shinohara, T. Kai, K. Oikawa, *et al.*, Final design of the Energy-Resolved Neutron Imaging System "RADEN" at J-PARC. J. Phys. Conf. Ser. **2016**, 746, 012007. https://doi.org/10.1088/1742-6596/746/1/012007

[5] M. Shoji, S. Uno, T. Uchida, *et al.*, Development of GEM-based detector for thermal neutron, J. Inst., **2012**, 7, C05003. https://doi.org/10.1088/1748-0221/7/05/C05003

[6] H. Sato, T. Kamiyama, Y. Kiyanagi, A Rietveld-type analysis code for pulsed neutron Bragg-edge transmission imaging and quantitative evaluation of texture and microstructure of a welded α-iron plate. Mater. Trans. **2011**, 52, 1294–1302. https://doi.org/10.2320/matertrans.M2010328

[7] H. Sato, T. Sato, Y. Shiota, *et al.*, Relation between Vickers Hardness and Bragg-Edge Broadening in Quenched Steel Rods Observed by Pulsed Neutron Transmission Imaging. Mater. Trans. **2015**, 56, 1147-1152. https://doi.org/10.2320/matertrans.M2015049

[8] K. Oikawa, Y. Kiyanagi, H. Sato, *et al.*, Pulsed neutron imaging based crystallographic structure study of a Japanese sword made by Sukemasa in the Muromachi period, Mater. Res. Proc. (this proceedings).

[9] H. Sato, Y. Kiyanagi, K. Oikawa, *et al.*, Crystallographic microstructure study of a Japanese sword made by Noritsuna in the Muromachi period by pulsed neutron Bragg-edge transmission imaging, Mater. Res. Proc. (this proceedings).

Neutron Radiography - WCNR-11   Materials Research Forum LLC
Materials Research Proceedings **15** (2020) 233-238   https://doi.org/10.21741/9781644900574-36

A Neutron Tomographic Analysis of Plated Silver Coins from Ancient Greece Official or Illegal?

Scott Olsen[,a1][*], Filomena Silvemini[,b1], Vladimir Luzin[c1,3], Ulf Garbe[d1],
Max Avdeev[e1], Joel Davis[f1] and Ken Sheedy[,g2]

[1]ANSTO New Illawarra Rd Lucas Heights, NSW 2234, Australia

[2]Macquarie University, Macquarie NSW 2109 Australia

[3]The University of Newcastle, Callaghan, New South Wales 2308, Australia

[a]sol@ansto.gov.au, [b]filomenas@ansto.gov.au, [c]vll@ansto.gov.au, [d]ulg@ansto.gov.au
[e]max@ansto.gov.au. [f]jda@ansto.gov.au [g]ken.sheedy@mq.edu.gov.au

Keywords: Neutron Tomography, Numismatics, Archaeometallugy

Abstract. This study focuses on a neutron tomographic analysis conducted on a set of plated silver coins minted in the city-state of Athens and in the Greek colonies of Kroton and Metapontum (South Italy or Magna Graecia) during the 6th and 5th centuries BC. The investigation aims to define the plating method by characterising the morphological and structural features of the specimens, i.e. the volume fraction of metallic and non-metallic components, and thickness maps of the plating and porosity. The status of these coins is uncertain: were they official issues authorized by state-authorities during periods of trouble (and silver shortages in the public treasury) or the product of ancient or modern counterfeiters?

Introduction

In the 6th century BC different techniques of coin manufacture were employed by mints in mainland Greece and in the Greek colonies in Southern Italy. In Greece these techniques were evidently derived from the Lydians and consisted in striking a piece of cast metal of predetermined weight (a 'blank' or flan) between two engraved dies made of hardened bronze [1] [2]. Colonies in Magna Graecia, however, uniquely developed another set of minting techniques to produce what today is called *incuse* coinage [3]. One of the most distinctive feature of these incuse issues lies in the fact that the reverse type is the same as that on the obverse but is rendered as a 'negative' or 'incuse' image sunk into the flan. The study of this technique is part of a dedicated on-going project [4] [5] [6]. It is evident that plated coins begin to appear at the very earliest stage in the history of coinage, and then become common place under the Romans [7] [8] [9] [10]. Plating, of course, involved an important modification of the mint's usual production processes – as has been explored by La Niece [11]. Were these plated coins issued by ancient state-authorities or the product of illegal counterfeiters? In an attempt to gain a better understanding of the technology of plated coins, numismatic and historical studies were combined together with metallurgical research based on neutron methods.

Materials and Methods

The numismatic collection of the Australian Centre for Ancient Numismatic Studies (ACANS) at Macquarie University in Sydney (AU) includes silver coins minted in cities of the Greek mainland and Magna Graecia. It includes three specimens, one each from Athens, Metapontum and Kroton that are plated. On the coin from Metapontum (inv. 07GS527), mineralisation is visible on the silver surface as green-blue deposits. These deposits are typically associated with the corrosion of copper; their presence could be explained either as contamination from the burial contest or as a sign of plating. The coin from Kroton (inv. 16A25) showed the presence of a multilayered coating on the protruding ridge. Finally, the coin from Athens (inv. 14A09), one of the first issues of coinage struck by that city, had been cut in half perhaps once it had been discovered to be plated; the cut clearly shows the copper core concealed underneath the silver plating (Table 1).

In order to properly describe the plated coin, and to clarify the manufacturing technique, the morphology and bulk structure of each was investigated by means of neutron tomography. This method is a valuable analytical tool to extract and quantify information such as morphology, porosity inclusions and the presence of composite structures. Most importantly neutron tomography is non-destructive. As demonstrated in [4], by detecting and evaluating structural features, relevant information on the manufacturing process can be inferred.

Table 1: Inventory number, provenance and physical details are listed for the three coins. [5] [6]

Inv.	07GS527	16A25	14A09
City	Metapontum	Kroton	Athens
Year	510-470 BC	510-480 BC	525-515 BC
Diameter	24 mm	~20 mm	~17 mm
Weight	8.1 g	7.4 g	6.7 g

The neutron tomography analysis was performed on DINGO, the ACNS[1] neutron imaging instrument located on a thermal beam tangentially facing the 20MW OPAL research reactor at ANSTO[2], Sydney [12].

The measurements were conducted in the high-resolution acquisition mode (with the ratio of collimator-detector length L to inlet collimator diameter D equals to 1000) with a pixel size of 27 μm by setting a 55x55 mm^2 field of view with 100 mm lens coupled with a 50μm thick scintillation screen. In this configuration, up to 4 samples can be mounted in an aluminium cylindrical holder and measured at the same time. The coins were spaced with aluminium foil. Projections were acquired with an equiangular step of 0.25°over 360°with an exposure time of 50 seconds each, resulting in a total scan time of about 20 hours. The data set were reconstructed with Octopus package [13], while AVIZO [14] was used for visualization and analyses.

Results and discussion

Results from the tomographic analysis confirmed that all selected specimens are plated (Figure 1). The virtual cross sections through the tomographic reconstructions show a homogeneous copper core in all coins. A silver layer surrounds the core with porosities localised at their interface (Figure 2). The amount and size of pores were quantified (Table 2). Analysis also showed variation in the thickness and distribution of the silver coating (Figure 3) as well as the copper and silver volume fractions in the samples (Table 3). This analysis is part of a broader program of investigation that includes neutron diffraction, neutron texture analysis, and SEM-EDS analyses. The study is still in progress; here we present a selection of interesting results

---

[1] Australian Centre for Neutron Scattering
[2] Australian Nuclear Science and Technology Organisation

Neutron Radiography - WCNR-11   Materials Research Forum LLC
Materials Research Proceedings **15** (2020) 233-238   https://doi.org/10.21741/9781644900574-36

based on neutron tomographic analysis and refer the reader to other publications for more a detailed coverage of earlier research [5] [6].

Sample 07GS527 (Metapontum). The neutron tomographic reconstruction clearly shows a fine layer of silver wrapped around a copper base. The coating is evenly distributed with an average thickness of 0.28 mm and constitutes 32.9 % of the coin volume. The structure of the lamina suggests that a silver foil was bonded to the copper core by a diffusion (metallurgical) bonding process. During the process the core was wrapped in pure silver foil and rapidly heated until it was close to the melting point of silver for a short time. Good control of temperature was required to prevent the core melting and the copper from being too rapidly diffused into the silver. Complementary investigation on the High Resolution Powder Diffractometer ECHIDNA [15] at ANSTO confirmed the presence of copper and silver by obtaining full diffraction patterns. No evidence of any other metal constituents, including low temperature melts that could have acted as soft soldering materials, were identified in the diffraction pattern [5]. A low amount (0.1 vol. %) of small porosities (average volume of 0.01 mm^3) are localised at the rim of the coin; this suggests that they were probably created during the plating process through the imperfect adhesion of the silver foil to the core on the coin periphery.

Sample 16A25 (Kroton). The peculiar feature detected in this coin is a multilayered coating constituted by three layers of silver over the copper core. The overall plating, quantified to be around 36.8 vol. %, has a variable thickness, ranging from a minimum of 0.06 to a maximum 3.75 mm in the protruding ridge, with an average value of 0.46 mm. The gaps detected between the silver foils resulted in a larger amount of pores (0.8 vol. %) and on average bigger porosities (volume of 0.2 mm^3) in comparison to the sample from Metapontum (07GS527). It is probable that the coin was recoated as a result of a manufacturing defect. Since no trace of solder was identified, it is assumed that, again the diffusion bonding technique was adopted.

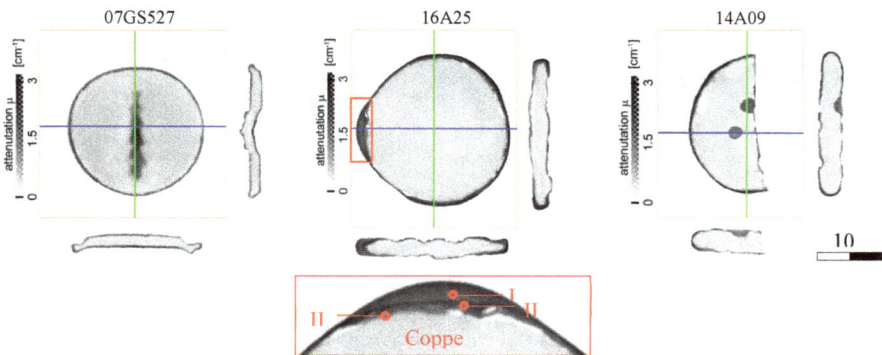

Figure 1: Maps of the planes cropping the reconstructed volume of each coin are shown together with the corresponding cross sections. The scale bar at the left of each figure indicates the colour code for the attenuation coefficient. A detailed view of the multi-layered plating is reported for coin 16A25 at the bottom.

Sample 14A09 (Athens). Some observations can be drawn about the plating of the Athenian coin based on the tomographic reconstruction. The average thickness of the foil is similar to the sample from Metapontum (0.34 mm) but less evenly distributed; it constitutes 19 vol. % of the total volume, a very small amount in comparison to the previous samples. Most of the coin is made of a homogenous copper core that appears heavily corroded at the interface with the plating, and on the exposed cut surface. At the interface between silver coating and copper core, pores are visible. These can be also attributed to the diffusion of corrosion products. The level of porosity is the highest among the three coins discussed in this paper (1.7 vol. %). On average the size of the pores is similar to those in coin **16A25 (Kroton)** but with a higher variability (up to a volume of 1.92 mm^3). Two undefined inserts are visible in **14A09**. These two elements, featuring a drop-like shape, were identified as indentation from the plating into the core. They might be interpreted as plugs applied to repair two holes in the original plating, holes that would reveal the copper core and shown the possessor that it was not of the expected metal value (Figure 2).

■ plating   ■ core   ■ corrosion   ■ plug   10 mm

Figure 2: Left, the structural components are separated and rendered in different colour for each sample; the colour code is expressed by the legend on the bottom. Right, the maps show the distribution of porosities. A scale bar on the right side of each figure indicates the coding colour-volume.

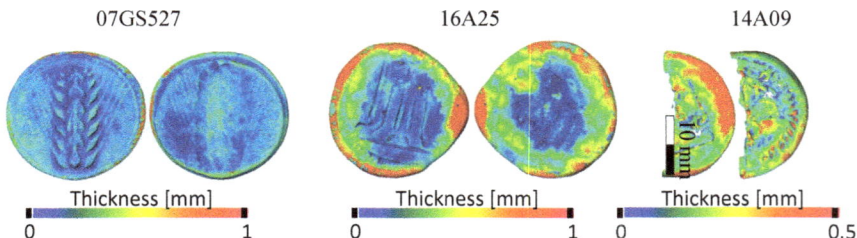

Figure 3: The thickness of the silver plating is reported in false colour for the obverse and revere of each coin. A scale bar indicates the coding colour-thickness.

All 3 coins were fabricated using the diffusion bonding technique, but neutron tomography has revealed noticeable variations in the quality of the coins in terms of the silver to copper ratio and porosity.

Table 2: Statistical analysis of porosities

	Feret shape	Equivalent diameter [mm]			Ave. width	Ave. length	Aspect ratio	Volume [mm3]		
		min	max	average	[mm]	[mm]		min	max	average
07GS527	2.79	0.10	0.55	0.19	0.21	0.75	0.28	0.00	0.09	0.01
16A25	2.26	0.10	1.05	0.23	0.21	0.60	0.35	0.00	0.61	0.02
14A09	1.80	0.03	1.54	0.12	0.14	0.31	0.46	0.00	1.92	0.02

Table 3: Results of the plating and porosity analysis of the 3 silver plated coins.

			07GS527	16A25	14A09
Plating feature	Min/max thickness	mm	0.13 – 1.49	0.06-3.75	0.06 – 0.80
	Mean		0.28	0.46	0.34
	Standard Deviation		0.11	0.36	0.19
Volume	Total	cm^{-3}	0.42	0.83	0.73
	Porosity	%	0.1	0.8	1.7
	Silver		32.9	36.8	19
	Copper		67	62.4	70.5
	Plugs & Mineralisation		-	-	8.8
Mass	Silver	g	1.5	3.2	1.5
	Copper		6.6	4.2	5.2

The Metapontum coin **07GS527** is a technically accomplished product with a very uniform silver layer, minor porosity and a silver-copper ratio of 0.2 (Table 3). The coin from Kroton **16A25** was replated and its silver-copper ratio is the highest (0.4) among the investigated samples. This was due to the additional plating on the ridge as shown in figure 3. On this basis the coin might be an authentic specimen recoated by the official mint of Kroton, it is very unlikely to be a modern forgery as the plating is clear. Further investigation by means of neutron diffraction methods, already in progress, might provide further clues. The last coin, that from Athens (**14A09**), which was minted according the methods typical of the Greek mainland (and not by the incuse coin processes), has a ratio of silver over copper of 0.3. Once again there is the suggestion that the makers touched up the coin as evidenced by the two plugs seen in figure 2. Is it possible then that recoating was a regular part of the official plating process or is this evidence of an illegal counterfeit?

Successful plating (which completely concealed the underlying base metal core) required a high degree of technical skill, and in the examples we have studied we appear to have evidence that first attempts were not always convincing. Should we conclude that a low level of expertise is indicative of a forger? Can we assume that an official mint could be expected to have artisans competent in plating and therefore if the job was 'botched' it is likely that we are looking at a counterfeit coin? Plated coins, however, were never a regular part of the job of an official mint – and until recently it was widely believed that official mints did not produce plated or counterfeit coins [16]. Indeed there is still a general view among numismatists and ancient historians that all plated coins inevitably are fakes [16]. The critical question is 'who is producing the plated coins'? La Niece has made the claim that official mints used manufacturing techniques suitable for mass production of coins but the forger could afford to use relatively labour-intensive methods [11]. But is this juxtaposition true? Ancient city-states only resorted to plating coins when they were in difficulties. Herodotus (Book 3.56) tells a story, for example, that the Samians coated lead with electrum to produce coins to bribe the Spartans to raise the siege of the Samos. It then seems possible in such circumstances where the survival of the city depended on acceptance of the plated coin that special effort would have gone into making sure that this money was believable.

Conclusion

In exploring the plating technique of a set of coins from the ACANS collection at Macquarie University in Sydney (AU), neutron tomography provided critical information on macrostructure and morphology. In all cases the silver plating was applied by diffusion bounding to a copper core, but noticeable differences in the coating structure, and in the silver-copper ratio suggest different procedures, variations in effort required to complete the work, and finally variation in the expertise available to undertake plating.

Our studies of these coins, which in two cases show evidence of recoating, rather than pointing to unofficial counterfeiting, in fact point to the amount of effort that was always needed to make sure the plated coins were acceptable. Ancient states would not have issued plated coins on a regular basis. They were a last resort in times of crisis. It is quite possible that these mints did not have workers trained to produce competent plated coins.

References

[1]   D. Sellwood, "Some experiments in Greek minting technique," NC, vol. 7, no. 3, p. 217–231, 1963.

[2]   D. M. Schaps, "The Invention of Coinage and the Monetization of Ancient Greece," 2004. https://doi.org/10.3998/mpub.17760

[3]   N. Rutter, Historia Numorum. Italy., London, 2001.

[4]   F. Salvemini, S. R. Olsen, V. Luzin, U. Garbe, J. Davis, T. Knowles and K. Sheedy, "Neutron tomographic analysis: Material characterization of silver and electrum coins from the 6th and 5th centuries BCE.," Materials Characterization , vol. 118, pp. 175-185, 2016. https://doi.org/10.1016/j.matchar.2016.05.018

[5]   F. Salvemini, K. Sheedy, S. R. Olsen, M. Avdeev, J. Davis and V. Luzin, "A multi-technique investigation of the incuse coinage of Magna Graecia," Journal of Archaeological Science: Report, pp. 748-755, 2018. https://doi.org/10.1016/j.jasrep.2018.06.025

[6]   K. A. Sheedy, P. Munroe, F. Salvemini and V. Luzin, "An incuse stater from the series 'Sirinos/Pyxoes'," Journal of the Numismatic Association of Australia, vol. 26, pp. 36-52, 2015.

[7]   R. Wallace, "The Production and Exchange of Early Anatolian Electrum Coinages," Revue des Etudes Anciennes , pp. 87-94, 1989. https://doi.org/10.3406/rea.1989.4367

[8]   G. M. Ingo, S. Balbi, T. de Caro, I. Fragala and C. Riccucci, "Microchemical investigation of Greek and Roman silver and gold plated coins: coating techniques and corrosion mechanisms," Applied Physics A, vol. 84, no. 4, pp. 623-629, 2006. https://doi.org/10.1007/s00339-006-3536-x

[9]   A. Deraisme, L. Beck, F. Pilon and J. Barrando, "A study of the silvering process of the Gallo-Roman coins forged during the third century ad.," Archaeometry, vol. 48, no. 3, pp. 469-480, 2006. https://doi.org/10.1111/j.1475-4754.2006.00267.x

[10]   K. Anheuser and . P. Northover , "Silver Plating on Roman and Celtic Coins from Britain – a Technical Study," British Numismatic Journa, vol. 64, pp. 22-32.

[11]   S. La Niece, "Technology of Silver-Plated Coin Forgeries," Metallurgy in Numismatics, vol. 3, pp. 227-239, 1993.

[12]   U. Garbe, T. Randall, C. Davidson, G. Pangelis and S. Kennedy, "A New Neutron Radiography / Tomography / Imaging Station DINGO at OPAL," Physics Procedia, vol. 69, pp. 27-32, 2015. https://doi.org/10.1016/j.phpro.2015.07.003

[13]   M. Dierick, B. Masschaele and L. Van Hoorebeke, "Octopus, a fast and user-friendly tomographic reconstruction package developed in LabView®," Measurement Science and Technology, vol. 15, no. 7, 2004. https://doi.org/10.1088/0957-0233/15/7/020

[14]   "FEI," [Online]. Available: https://www.fei.com/software/amira-avizo/] .

[15]   K.-D. Liss, B. Hunter, M. Hagen, T. Noakes and S. Kennedy, "Echidna—the new high-resolution powder diffractometer being built at OPAL," Physica B: Condensed Matter, vol. 385–386, no. 2, pp. 1010-1012, 2006. https://doi.org/10.1016/j.physb.2006.05.322

[16]   P. Van Alfen, "Problems in ancient imitative and counterfeit coinage," in Making, moving and managing : the new world of ancient economies, 323-31 BC, Z. Archibald and el al., Eds., p. 354.

Neutron Radiography - WCNR-11
Materials Research Proceedings 15 (2020) 239-243

Materials Research Forum LLC
https://doi.org/10.21741/9781644900574-37

The 15th-18th Terracotta Doll Investigation Using a Compact Neutron Tomography System at Thai Research Reactor

Sarinrat Wonglee [a *], Sasiphan Khaweerat [b], Thiansin Liamsuwan [c], Jatechan Channuie [d], Roppon Picha [e] and Weerawat Pornroongruengchok [f]

Thailand Institute of Nuclear Technology (Public Organization),
9/9 Moo 7, Tambon Saimoon, Amphur Ongkharak, Nakhon Nayok 26120, Thailand

[a *]sarinratwl@gmail.com, [b]sasiphank@tint.or.th, [c]thiansinl@tint.or.th, [d]jatechan.c@gmail.com, [e]aeroppon@gmail.com, [f]weerawat@tint.or.th

Keywords: Neutron tomography, Thai Research Reactor, Medium Neutron Flux, Cultural Heritage, Nondestructive Technique

Abstract. It is well known that neutron imaging is a powerful nondestructive technique in archaeological studies, especially for visualization of organic contents or low-density parts inside antiques. In Thailand, the neutron imaging system has been developed to perform neutron tomography (NT) for archeological studies since 2015. A compact NT system, which is composed of a CCD camera coupled with a LiF/ZnS fluorescence screen and an in-house developed rotation stage, was used to investigate the internal structure of an object. The experiment was set up at Thai Research Reactor (TRR-1/M1) of Thailand Institute of Nuclear Technology (Public Organization) with the power of 1.2 MWth. The neutron intensity at the radiographic position was about 10^6 n.cm^{-2}s^{-1}. In this work, an ancient sample of interest, namely, the 15th-18th century terracotta doll was investigated to perform the developed NT system. The resulting 2D neutron image showed a crack at the neck and a small gravel inside the body. Then, the projections were reconstructed by means of the Octopus Imaging software. Even with the compact NT system (L/D: 50), the 3D neutron image of the ancient doll was successfully reconstructed. The image revealed some hidden organic materials coated on the neck of the doll. Moreover, the elemental composition of the terracotta doll was analyzed by using X-ray fluorescence technique. The result could further inform the historical records of the ancient doll.

Introduction

Thailand has a long history and is rich in archaeological evidences. The use of advanced science and technology in cultural heritage are important to reveal the history of the country. Neutron imaging is one of nondestructive tools for internal structure investigation. Light elements or organic compositions can be recorded by neutron imaging which usually are difficult to be imaged by X-ray or gamma ray. Nuclear Research and Development Division of Thailand Institute of Nuclear Technology (Public Organization) has established the NT system at Thai Research Reactor TRR-1/M1 for multipurpose services including archeological study. To investigate an archeological sample, the 15th-18th century terracotta doll was preliminary imaged by using the upgrade NT system. This research aims to enhance the potential of NT system in Thailand which is continued from the previous works [1-2].

Methods

The terracotta doll was made in the period of 15th - 18th century (Ayutthaya Kingdom). The creating purposes were probably the worship of god as well as the kid's toys. The doll was made from red clay and its shape is representative of a human being with a hair bun. In this work, the

Neutron Radiography - WCNR-11
Materials Research Proceedings **15** (2020) 239-243

Materials Research Forum LLC
https://doi.org/10.21741/9781644900574-37

terracotta doll was purchased from an antique store in Bangkok and was stylistic date by archaeologist. The sample size is about 9 cm in height and 6 cm in width. The appearance of the doll is shown in Fig. 1. By shaking it, we found that there is something inside the body of the doll and it is movable.

Fig. 1 The 15th-18th century terracotta doll.

The NT system was setup at the south beam tube of Thai Research Reactor. The schematic diagram of the developed NT system is shown as Fig. 2. The thermal neutron flux at radiographic position was about 10^6 n.cm^{-2}s^{-1} with the reactor power of 1.2 MWth. The L/D ratio of the imaging system was approximately 50. The compact NT system [2] was composed of a 2048 × 2048 pixels-CCD camera [3] coupled with the 20 cm × 20 cm-size of LiF/ZnS scintillation screen and the in-house developed rotation stage. From our previous work [2], the resolution of the 2D neutron image was about 370 micrometers for plastic materials, interpreted from the smallest size of plastic wire which was well separated from the background on the image. For initial experiments, using the compact NT system, 101 neutron projections were acquired over 180-degree angles with the angular step of 1.8 degree. The exposure time per step was 30 seconds. The neutron projections were then reconstructed by using Octopus Imaging software. Parallel beam condition for tomographic reconstruction as well as dark beam and open beam data were used to normalize background.

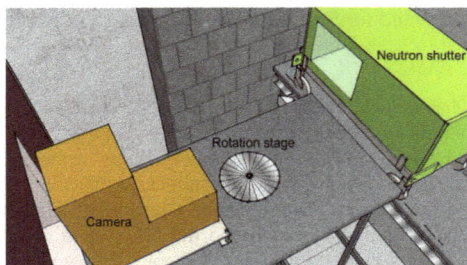

Fig. 2 The schematic diagram of the developed neutron tomography system.

Fig. 3 The elemental analysis by using a handheld XRF spectrometer.

In addition, the elemental composition of the terracotta doll was carried out nondestructively by using X-ray fluorescence (XRF) technique. The handheld XRF spectrometer, Thermo Scientific Niton^{TM} XL3t GOLDD+ was used to analyze the terracotta doll with the "test all geo" analyzing mode and 8 mm-X-ray beam diameter. Five measurements were done for head and body parts as shown in Fig. 3. Then, the average elemental concentration was determined.

Result and discussion
The NT was successfully performed at Thai Research Reactor with a compact system. The total exposure time for NT was about 1 hour. The obtained 2D neutron image of the terracotta doll clearly revealed a crack at the neck, a gravel inside the body hollow and pores which could not been observed by naked eyes, as shown in Fig. 4. The pores size located in the body wall are approximately 3 - 4 mm. The smaller pore size could not be observed in this experiment, however, the image sensitivity can be improved by increasing the exposure time and the number of projections. Then, a stack of tomographic images was created by using Octopus Reconstruction and Visualization functions as shown in Fig. 5 and Fig. 6, respectively. Fig. 5 shows the reconstructed neutron images of the terracotta doll sections at X, Y and Z-axis. By combining neutron projections from the different angles, the high neutron attenuation material appeared more clearly as white color areas. Fig. 6 shows the 3D neutron images with various views by using surface rendering, volume rendering and cutting functions.

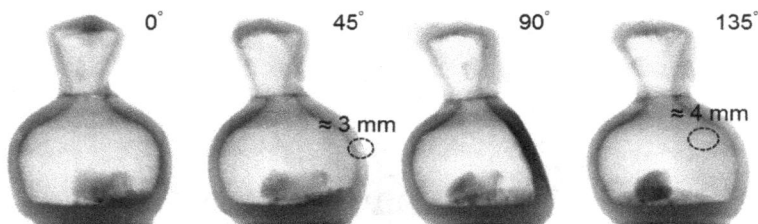

Fig. 4 The neutron projections at each observation angle.

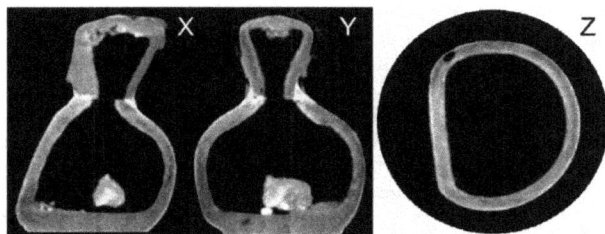

Fig. 5 The reconstructed neutron images of the terracotta doll section at X, Y and Z-axis.

The benefit of tomographic reconstructions provides more information such as a precise pore position could be readout. Furthermore, by using the Octopus Visualization function, light material and/or an organic based coating found around the neck can be observed more obviously. The light material also appeared at the inner wall of the head and inside the gravel. Fig. 7 shows light materials in the terracotta doll as the pseudo-yellow color areas.

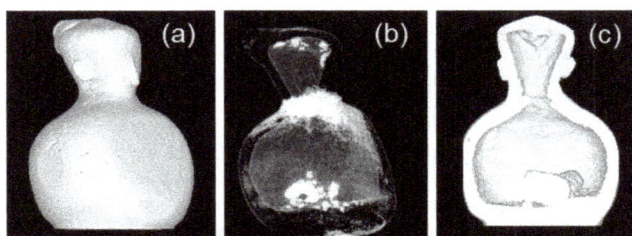

Fig. 6 The 3D neutron reconructed images of the terracotta doll with suface rendering (a), volume rendering (b) and cutting functions (c).

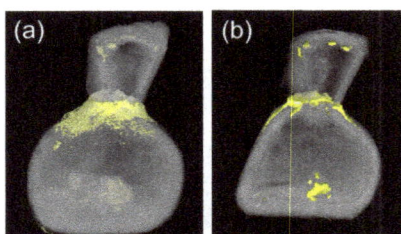

Fig. 7 The light material coated around the neck (a) and located in the gravel (b).

The elemental analysis by XRF showed that the terracotta doll composed of 10 major elements which were Si, Fe, Al, Mg, K, Ca, Ti, Mn, S and P. Besides, the elemental concentrations are shown in Table 1. The revealed content data indicated the difference of the elemental concentrations between the head and the body especially for Fe, K, Ca, Mn and S. From the information of the 15th-18th terracotta doll which was obtained from the nondestructive investigations, it could be concluded that the doll composed of the head and the body which came from different origins but were connected together with organic adhesive.

Table 1 The elemental concentrations in the terracotta doll obtained from XRF.

Element	Si	Fe	Al	Mg	K	Ca	Ti	Mn	S	P
Body	23.21 ±0.96	8.22 ±0.89	5.50 ±0.38	1.01 ±0.14	0.87 ±0.09	0.78 ±0.13	0.73 ±0.05	0.51 ±0.12	0.34 ±0.10	0.20 ±0.03
Head	24.83 ±2.61	4.65 ±0.47	6.31 ±1.03	Not detected	1.35 ±0.14	1.26 ±0.25	0.74 ±0.07	0.08 ±0.02	0.90 ±0.29	0.30 ±0.05

Conclusions

The preliminary 3D neutron image of the 15th-18th terracotta doll could be visualized by using the developed NT system at Thai research reactor. The internal structure details were successfully investigated. In this case, the hidden information of the terracotta doll was disclosed by the contribution of NT and XRF. To increase the functionality of the NT system, further development is required, for example, the rotation stage should be improved to support various types of samples with more stability and also the exposure condition should be optimized for enhancing image quality and getting better resolution. The improvement will support a variety of applications to meet national requirements at last.

References

[1] S. Khaweerata, W. Ratanatongchaia, S.Wonglee and B. Schillinger, The Early Stage of Neutron Tomography for Cultural Heritage Study in Thailand, J. Phys. Con. Ser. 88 (2017) 123-127. https://doi.org/10.1016/j.phpro.2017.06.016

[2] S. Wonglee, S. Khaweerat, J. Channuie, R. Picha, T. Liamsuwan and W. Ratanatongchai, Development of Neutron Imaging System for Neutron Tomography at Thai Research Reactor TRR-1/M1. J. Phys. Con. Ser. 901 (2017) 012149. https://doi.org/10.1088/1742-6596/901/1/012149

[3] A. W. Hewata 2015, Inexpensive Neutron Imaging Cameras using CCDs for Astronomy, Phys. Procedia 69 (2014) 185-8. https://doi.org/10.1016/j.phpro.2015.07.026

Neutron Radiography - WCNR-11
Materials Research Proceedings 15 (2020) 244-249

Materials Research Forum LLC
https://doi.org/10.21741/9781644900574-38

The first Record of Plicidentine in Varanopseidae (Synapsida, Pelycosauria)

Michael Laaß[1, a *] and Burkhard Schillinger[2,b]

[1,2] Technical University of Munich, Heinz Maier-Leibnitz Centre and Faculty of Physics E21, Lichtenbergstraße 1, D-85747 Garching, Germany

[a]michael.laass@gmx.de, [b]burkhard.schillinger@frm2.tum.de

* corresponding author

Keywords: Neutron tomography, Plicidentine, Pelycosauria, Synapsida, Varanosaurus

Abstract. Infolded dentine (plicidentine) around the pulp cavities of the tooth roots is an ancient dental feature of fishes (sarcopterygians and actinopterygians) and several basal tetrapod groups. But, plicidentine is completely unknown in synapsids except some sphenacodontid pelycosaurs. An investigation of a skull of *Varanosaurus acutirostris* (Synapsida, Pelycosauria, Varanopsidae) by means of neutron tomography showed that plicidentine is also present in another group of pelycosaurs and was obviously wider distributed among basal synapsids than previously thought. Furthermore, the presence of plicidentine in this taxon with relatively short tooth roots supports the hypothesis that plicidentine played a functional role in strengthening tooth attachment.

Introduction

Usually, vertebrate teeth are constructed by three kind of hard-tissue. The main body of the tooth root and the crown consists of dentine. The dentine body of the tooth crown is covered by a cap of enamel. In contrast, the dentine, which forms the tooth root, is surrounded by a thin layer of cementum. In some vertebrate groups the dentine layer may be infolded around the pulp cavity at the base of the tooth root. This special kind of dentine is named plicidentine. According to the definition of E.E. Maxwell et al. [1] plicidentine includes both simple as well as complex infolding of dentine structures.

Plicidentine is known from several basal vertebrate groups: actinopterygian and sarcopterygian fishes, some basal tetrapods, amphibians, parareptiles, captorhinids, reptiles and ichthyosaurs (see [1, 2] and references therein). As summarized by K.S. Brink et al. [2] plicidentine has been interpreted as a phylogenetically constrained feature of major vertebrate groups, as a structure to strengthen tooth attachment or a combination of both.

Remarkably, up to 2014 plicidentine was completely unknown in synapsids. Sectioning of the tooth roots of some sphenacodontid pelycosaurs showed that plicidentine was also present in basal synapsids ([2, 3]). However, several phylogenetic analyses revealed that sphenacodontians represent the most derived pelycosaurian clade [4-6]. This result as well as the fact that plicidentine is a typical feature of anamniote and basal tetrapod groups suggests that it was probably also present in more basal pelycosaur groups. As described in the following sections our investigation of a skull of *Varanosaurus acutirostris* showed that this is the case.

Material and Methods

Material. Subject of our investigation was the holotype of the skull of *Varanosaurus acutirostris* (Synapsida, Pelycosauria, Varanopseidae) from the collection of the Bayerische Staatssammlung in Munich (inventory number BSPG 1901 XV 20, Fig. 1). The specimen derives from the

Neutron Radiography - WCNR-11   Materials Research Forum LLC
Materials Research Proceedings **15** (2020) 244-249   https://doi.org/10.21741/9781644900574-38

Craddock Bonebed, Craddock ranch north of Seymour, Baylor County, Texas, USA. This horizon belongs to the Arroyo Formation, Clear Fork Group, Lower Permian [6].

The specimen has been firstly described by F. Broili [7]. Further anatomical observations were published by F. Broili [8] and D.M.S. Watson [9]. A modern description and reconstruction of the skull have been provided by D.S. Berman et al. [6]. Except of a slight lateral compression, the skull is well preserved. It only lacks small parts of the basicranium and the caudal part of the mandible.

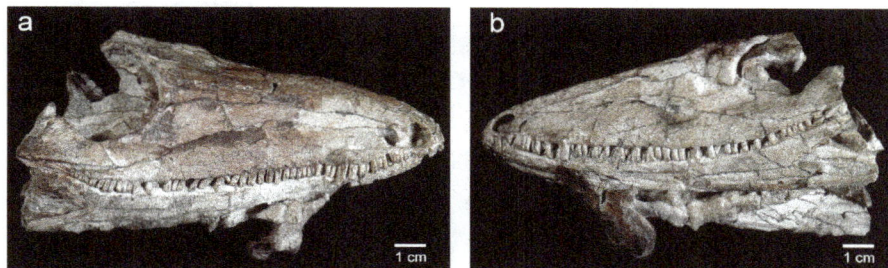

Fig. 1. Varanosaurus acutirostris (coll. number BSPHM 1901 XV 20) (a) right lateral view; (b) left lateral view.

Neutron tomography and three-dimensional reconstruction of the tomographic volume. This work is based upon experiments performed at the ANTARES instrument operated by FRM II at the Heinz Maier-Leibnitz Zentrum (MLZ), Garching, Germany. A description of the beamline is provided by M. Schulz et al. [10]. The radiographic images of the specimen were produced by using cold neutron radiation in the range between 5 Å and 1 Å. The field of view was 150 x 150 mm. The resolution of the camera was 2048 x 2048 pixels. The effective (projected) pixel size on the scintillation screen was 75 μm x 75 μm. The tomographic volume was reconstructed by means of the software Octopus 8.5 (Inside Matters; http://www.octopusreconstruction.com/en/home) using the "filtered backprojection algorithm". Processing of the raw projection data acquired parallel beam geometry. As a result a stack of horizontal tomographic slices, which represents the whole tomographic volume, was generated.

3D modelling and visualisation of cranial structures. The tomographic volume produced by the software Octopus was visualized in Fig. 2 by means of the software VGStudioMaxx. VGStudioMaxx also provided the tools for preparation of the virtual sections through the skull as shown in Figs. 2 c, e-g.

The 3D models of the teeth shown in Fig. 3 were generated following the method of M. Laaß [11] and M. Laaß et al. [12]. Structures of interest such as the teeth were vectorized slice by slice with the segmentation tool of the software AMIRA 5.4 (FEI Visualization Sciences Group; http://www.vsg3d.com/). After smoothening and polygon reduction to 10 percent the regions of interest were saved as Wavefront files (Object files, OBJ). After this the object files were imported into the software SimLab Composer 2014 SP1 (Simlab Soft; http://www.simlab-soft.com/). Then the teeth objects were colored and exported as 3D-PDF.

Results

The tomographic investigation of the specimen of *Varanosaurus acutirostris* revealed the presence of plicidentine in the tooth roots for the first time in a varanopseid pelycosaur (Figs. 2,

3). The infolded dentine was invisible from outside, because the tooth roots are almost completely covered by the jaw bones. Plicidentine is present both in the caniniform and postcaniniform teeth. The infolding is relatively simple, i.e. at the base of the tooth roots 4 to 5 folds can be observed, which become weaker towards the tooth crown (Fig 3).

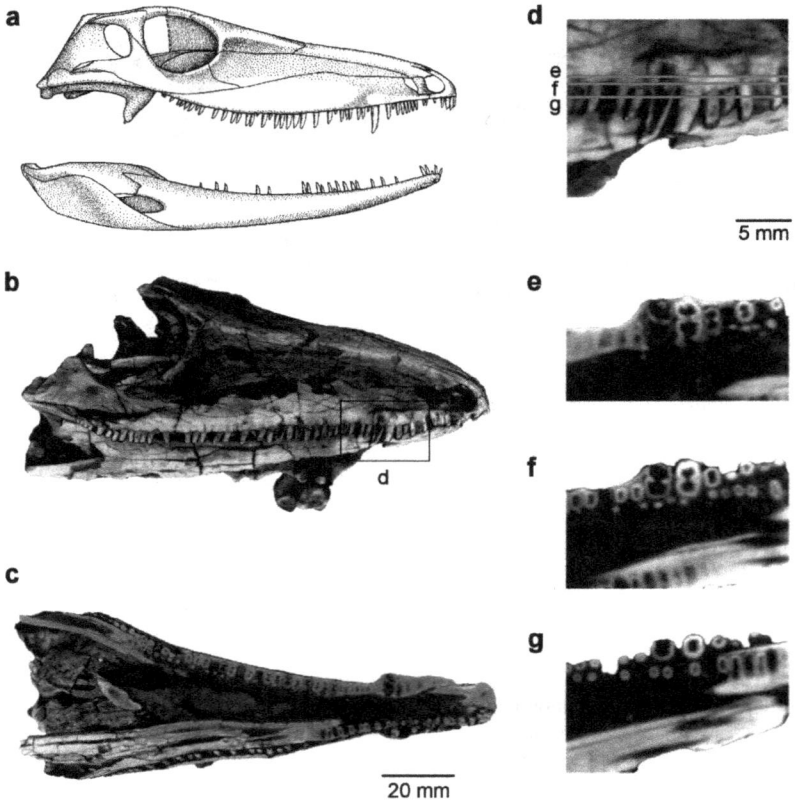

Fig. 2. Varanosaurus acutirostris (inventory number BSPHM 1901 XV 20) (a) reconstruction of the skull of Varanosaurus acutirostris from Berman et al. (1995); (b) Virtual 3D model of the skull in right lateral view; Note that the box marks the position of the image detail shown in (d); (c) coronal section through the skull at the level of the teeth; (d) lateral detailed view on the jaws. Note that the red lines mark the positions of the coronal sections shown in (e), (f), and (g); (e), (f), (g) coronal sections through the teeth at different levels of the roots.

If plicidentine indeed played a functional role to increase the area of attachment of tooth roots in the jaws to withstand the forces from feeding as proposed by H. Preuschoft et al. [13], J.D. Scanlon and M.S.Y. Lee [14] and M.J. MacDougall et al. [15], it can be expected that plicidentine is predominantly present in taxa with relatively shallow tooth roots in comparison to the crowns.

Therefore, the ratios of tooth root lengths vs. total tooth lengths are of special interest and were determined (see table 1) following the method of K.S. Brink et al. [2]. All measurements were taken from virtual sections through the 3D model of the skull.

Table 1. Measurements of the teeth of Varanosaurus acutirostris (BSPHM 1901 XV 20).

Teeth	Total tooth length [mm]	Average crown length [mm]	Average root length [mm]	Root/total tooth length [%]
Postcaniniform teeth	5.7	3.9	1.8	30.1
Caniniform teeth	7.7	5.5	2.3	31.3

As a first result it can be stated that there are no significant differences between the ratios of tooth root length vs. total tooth length of caniniform and postcaniniform teeth. Second, the average ratio of tooth root length vs. total tooth length is ca. 30 % in *Varanosaurus acutirostris*, which shows that *V. acutirostris* possessed relatively shallow tooth roots.

Discussion and conclusions

Our measurements of the root vs. total tooth lengths in *Varanosaurus acutirostris* support the hypothesis that plicidentine seems to be restricted to taxa with relatively shallow tooth roots and probably played a functional role to strengthen tooth attachment. For comparison, K.S. Brink et al. [2] observed plicidentine in four sphenacodontid pelycosaurs (*Ianthodon schultzei, Sphenacodon sp., Dimetrodon limbatus* and *Secodontosaurus obtusidens*) with ratios of root vs. total tooth length smaller than 41%. Interestingly, plicidentine was absent in the spenacodontid pelycosaur *Dimetrodon grandis*, which possesses relatively long tooth roots (50-57%) in relation to the total tooth length [2]. The ratio of tooth root length vs. total tooth length of ca. 30 % in *Varanosaurus acutirostris* fits well to the range of pelycosaur taxa with shallow tooth roots.

The first record of plicidentine in varanopseid pelycosaurs also shows that plicidentine was more widespread among basal synapsids that previously thought, which might be an indication that plicidentine was an ancient feature in early synapsids inherited from basal tetrapods. However, further investigations of other pelycosaur groups are necessary to clarify whether plicidentine evolved several times independently in early synapsid evolution or not.

Finally, the results show that non-destructive technologies such as neutron imaging are a very useful tool for investigation of non-mammalian synapsids.

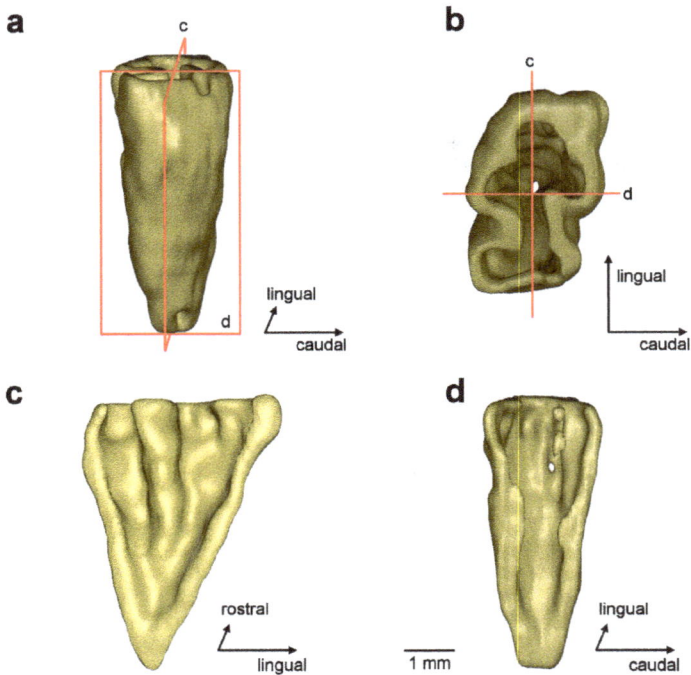

Fig. 3. Virtual 3D model of the left canine of Varanosaurus acutirostris (coll. number BSPHM 1901 XV 20) (a) lateral view; (b) view into the pulp cavity; Note the infolded dentine at the level of the tooth root; (c) and (d) sections through the tooth; Note that the red lines mark the positions of the sections in (a) and (b).

Acknowledgements

The authors gratefully acknowledge FRM II for financial support and the possibility to perform the neutron scattering measurements at the Heinz Maier-Leibnitz Zentrum (MLZ), Garching, Germany. We are also thankful to Oliver Rauhut, Munich, for access to material. Thomas Kuner, Heidelberg, provided the hard- and software for 3D modelling. The anonymous referees are thanked for their valuable comments on an earlier version of this paper.

References

[1] E.E. Maxwell, M.W. Caldwell, D.O. Lamoureux, The structure and phylogenetic distribution of amniote plicidentine. J. Vert. Paleontol. 31 (2011a) 553–561. https://doi.org/10.1080/02724634.2011.557117

[2] K.S. Brink, A.R.H. LeBlanc, R.R. Reisz, First record of plicidentine in Synapsida and patterns of tooth root shape change in Early Permian sphenacodontians. Naturwissenschaften (2014). https://doi.org/10.1007/s00114-014-1228-5

[3] K.S. Brink, R.R. Reisz, Hidden dental diversity in the oldest terrestrial apex predator *Dimetrodon*. Nat. Commun. 5 (2014). https://doi.org/10.1038/ncomms4269

[4] T.S. Kemp, Interrelationships of the Synapsida, in: M. J. Benton (Ed.) The Phylogeny and Classification of the Tetrapods (2), Clarendon Press, Oxford, 1988, pp. 1-22

[5] T.S. Kemp, The origin and evolution of mammals, Oxford Univ. Press, New York, 2005.

[6] D.S. Berman, R.R. Reisz, J.R. Bolt, D. Scott, The cranial anatomy and relationships of the synapsid *Varanosaurus* (Eupelycosauria: Ophiacodontidae) from the Early Permian of Texas and Oklahoma. Ann. Carnegie Mus. 58(2) (1995) 99-138.

[7] F. Broili, Permische Stegocephalen und Reptilien aus Texas. Palaeontogr. 51 (1904) 1-120.

[8] F. Broili, Über den Schädelbau von *Varanosaurus acutirostris*. Centralblatt für Mineralogie, Geologie, Paläontologie, Jahrbuch 1 (1914) 26-29.

[9] D.M.S. Watson, Notes on *Varanosaurus acutirostris* Broili. Ann. Mag. Nat. Hist. 13 (1914) 297-310. https://doi.org/10.1080/00222931408693483

[10] M. Schulz, B. Schillinger, E. Calzada, D. Bausenwein, P. Schmakat, T. Reimann, P. Böni, The new neutron imaging beam line ANTARES at FRM II. Restaur. Archäol (2015) 8.

[11] M. Laaß, Bone conduction hearing and seismic sensitivity of the Late Permian anomodont *Kawingasaurus fossilis*. J. Morphol. 276(2) (2014) 121-143. https://doi.org/10.1002/jmor.20325

[12] M. Laaß, B. Schillinger, I. Werneburg, Neutron tomography and X-ray tomography as tools for the morphological investigation of non-mammalian synapsids. Phys. Procedia 88 (2017) 100-108. https://doi.org/10.1016/j.phpro.2017.06.013

[13] H. Preuschoft, W.-E. Reif, C. Loitsch, E. Tepe, The function of labyrinthodont teeth: big teeth in small jaws, in: N. Schmidt-Kittler, K. Vogel (Eds.), Constructional Morphology and Evolution, Springer-Verlag, Berlin-Heidelberg, 1991, pp. 151–171. https://doi.org/10.1007/978-3-642-76156-0_12

[14] J.D. Scanlon, M.S.Y. Lee, Varanoid-like dentition in primitive snakes (Madtsoiidae). J. Herpetol. 36 (2002) 100–106. https://doi.org/10.1670/0022-1511(2002)036[0100:VLDIPS]2.0.CO;2

[15] M.J. MacDougall, A.R.H. LeBlanc, R.R. Reisz, Plicidentine in the Early Permian Parareptile *Colobomycter pholeter*, and its phylogenetic and functional significance among coeval members of the clade. PLoS ONE 9(5) (2014) e96559. doi:10.1371/journal.pone.0096559. https://doi.org/10.1371/journal.pone.0096559

Neutron Radiography - WCNR-11
Materials Research Proceedings 15 (2020) 250-255

Materials Research Forum LLC
https://doi.org/10.21741/9781644900574-39

Digitally Excavating the Hidden Secrets of an Egyptian Animal Mummy: a Comparative Neutron and X-ray CT Study

Carla A Raymond[1, a], Joseph J Bevitt[2, b]

[1] Department of Earth and Planetary Sciences, Macquarie University, NSW (AU)

[2] Australian Nuclear Science and Technology Organization (ANSTO), Lucas Heights, NSW (AU)

[a]carla.raymond@hdr.mq.edu.au, [b]jbv@ansto.gov.au

Keywords: Neutron Tomography, Mummification, Segmentation, Non-Invasive, Non-Destructive, Cultural Heritage

Abstract. Here we present further analysis and interpretation of our recently published work (Raymond et al. 2019) on a mummified cat (IA.2402) on loan from the Australian Institute of Archaeology (AIA) in Melbourne, Australia. This was the first published case to implement X-ray and neutron CT to votive animal mummies, and is the first in a series of similar studies undertaken at ANSTO. The application of neutron CT to this type of artefact was ideal to non-destructively study mummification techniques and learn about its hidden contents. Using a combination of X-ray and neutron CT provided valuable insight, both individually and collectively, revealing: a partial animal skeleton, an amulet, several layers of coarse and fine textile, and folded padding. Combining both techniques also allowed for complementary study of bones, soft tissue, and textile components. Use of multiple segmentation tools in 3D reconstruction and visualization software VG Studio Max 3.0 enabled detailed digital excavation of the sample, allowing for identification of species, age at death, and how textiles were used to shape and wrap the mummy. Results revealed the animal remains belong to a small, juvenile feline.

Introduction

The ancient Egyptians are known for mummification of both humans and animals alike, however, the process of mummification is not well documented. The mummification of animals as votive offerings was a industrial scale process, and reached a peak popularity between the Late period to Roman Period (664 BC – 395 AD). The current understanding of mummification comes from a small selection of tomb paintings at Thebes and Giza that illustrate the process (Ikram 2011), and the recollections of contemporary sources – Herodotus (5th century BC) and Porphyry (3rd century AD). With the increased interest in all thing Ancient Egypt in the 1800s, "Egyptomania" took the Victorian world by storm and so began a number of unwrapping parties (Shaw 2004). These events were for not for research, rather for finding treasures and amulets in the wrapping, and the prestige that was associated with such an event (Ikram & Dodson 1988; David 1997; Smith 2016). In recent years, Distinguished Professor Salima Ikram, from the American University in Cairo, and her team have been working on understanding the mummification process by experimentation on modern animal cadavers (Ikram 2011). This notion of experimental mummification will help in the understanding of the process, and has already shed much light on the conditions required, ie. quantities of natron, and frequency for change of natron salts for optimal desiccation. In conjunction with this method of study, it is also important that the archaeological community finds other appropriate, and minimally to non-destructive techniques to study and understand ancient mummification methods. Neutron tomography was chosen for this case study of a votive animal mummy, as a comparative and complementary technique to X-ray CT. This project formed part of the authors Master of Research thesis (Raymond, unpublished, 2017), and recent *Archaeometry* publication (Raymond et al. 2019).

The Sample - Mummified "Cat" *IA.2402*

The mummified votive offering IA.2402 was thought to contain the skeleton of a kitten, due to the appearance (Figure 1) and size of the bundle (29.2 cm x 4.8 cm x 7.8 cm). Mummy bundles that contains complete adult skeletons tend to be between 36 cm to 88 cm tall. This mummy has been in the AIA collection since the 1950s, however how it came to be in the collection is less clear. Acquired by Sir Charles Nicholson in 1856-7 on an expedition to Egypt, the cat has no recorded provenance or age. The mummy bundle is quite unusual, particularly the painted green and red markings across the body and head of the mummy (Figure 1), and no similar painted examples have been found in online catalogues at this time. The external wrapping exhibits a corkscrew pattern, downward towards the base, and is secured with the brown resinous material. A detailed object description, images and acquisition story can be found in Raymond et al 2019.

Experiment

The mummy was scanned using the neutron tomography beamline (DINGO) at the Australian Nuclear Science and Technology Organisation (ANSTO) in Lucas Heights, Australia. A series of 720 radiographs at 95µm were collected over 180°, using a 20 cm x 20 cm ^{6}LiF ZnS scintillator (0.050 mm), and ANDO iKON-L CCD camera. Total scan time was 6 hours. Image processing was done using Image J (NIH) and Octopus

Figure 1. Mummified cat (IA.2402) with green and red painted markings and brown resinous material.

8.2(Octopus Imaging). The 3D volume was rendered and visualised using VG Studio Max 3.0 (Volume Graphics GmbH).

X-ray scans were acquired by Prof John Magnussen and his team at Macquarie Medical Imaging in Sydney, Australia, using both a Newtom 5G (Newtom, Italy) cone-beam (CBCT) scanner, and a GE HD750 (General Electric, Milwaukee, USA) dual-energy (DECT) multi-detector (MDCT) scanner. Scans were acquired in a matter of seconds and processed using InteleViewer 4-11-1-P130 for various filters (bone and soft tissue), and reconstructed in 3D using RadiAnt DICOM Viewer 3.4.2. For more detailed parameters, please refer to Raymond et al. (2019).

Results and discussion

The neutron and X-ray scans both revealed that there were remains inside the mummified bundle and enabled further investigation into species and age at death. The X-ray data had greater contrast for studying the bones and other highly absorbing materials on (Figure 2b) and within the wrapping (Figure 3). The first component that stands out is the tiny (4.5 mm x 3.9 mm), bright object in Figure 3a, beside the paws of the animal. This is interpreted to be an amulet of religious or ceremonial significance as it was a common practice to include such talismans in the mummification process. However, there are no definitive features discernible at this resolution, thus it cannot yet be confirmed. The high contrast of the skeletal remains (two legs and a disarticulated tail) enabled a detailed study of the epiphyses (growth plates) and any indications

Neutron Radiography - WCNR-11   Materials Research Forum LLC
Materials Research Proceedings **15** (2020) 250-255   https://doi.org/10.21741/9781644900574-39

of trauma that may have caused the animals demise. Segmentation of the skeleton was achieved using thresholding, setting an upper and lower limit for grayscale displayed. It was determined from the epiphyses that this animal was approximately 11 months old at death, confirming the initial hypothesis, however there were no indications of fatal trauma to the remains to infer cause of death.

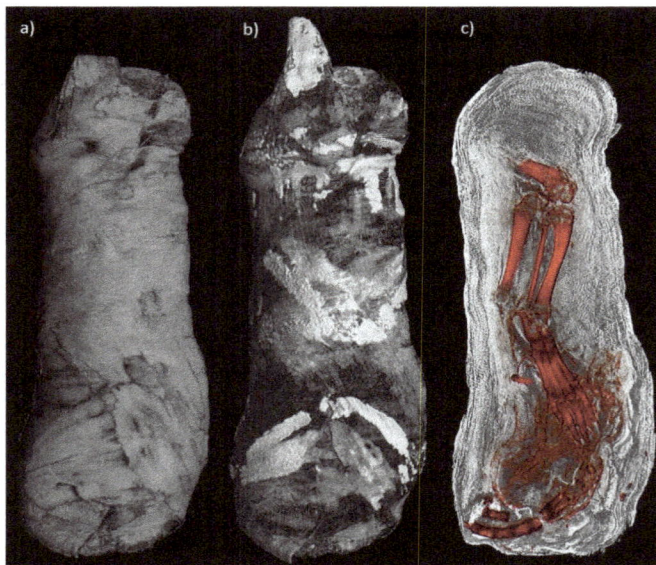

Figure 2. 3D Reconstructions of a) neutron and b) x-ray CT scans; c) X-ray reconstruction (red overlay) on neutron reconstruction, to illustrate the orientation of skeletal remains within the wrappings. These visualisations were made with VG Studio Max 3.0.

Further segmentation of the X-ray results also allowed for speciation of the remains; firstly by visualising the whole tail, it was possible to count the number of caudal vertebrae present. A total of 23 vertebrae were identified (Figure 4), which classified the remains as feline. In addition, by using manual segmentation, it was possible to identify a small bone called the calcaneus (highlighted in green in Figure 5b and c), the shape of which is distinctive of feline species (Van Neer et al. 2014). This manual segmentation process is intensive and slow, involving a selection tool across all three orthogonal planes. Once the region of interest (ROI) was well segmented, it was extracted, and the shape was clearly matched to the species *Felis silvestris*. The segmented ROI of the skeleton was later made into an STL. file, and 3D printed for educational uses at the Macquarie University Museum of Ancient Cultures, and the AIA. This has been a valuable outcome of the project, as it has made the findings accessible to the greater public, and to young students who can now discover the wider implications of when science and archaeology are combined in a complimentary way.

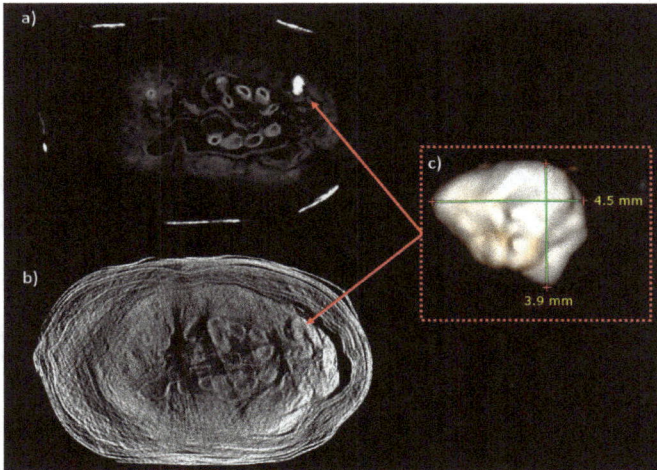

Figure 4. Coronal slices through the mummy revealed a highly x-ray attenuating object close to the paws, seen clearly in x-ray slice (a), however not so in the neutron results (b). c) Is a 3D visualisation of the object, made with RadiAnt DICOM Viewer 3.4.2.

The reconstructed neutron CT results were incredibly informative about the way the bundle had *been* wrapped and shaped using folded textiles. In Figure 2a the corkscrew nature of the external bandages is clearly illustrated, with the end of the bandage flicking back upward. This wrapping direction is not particularly clear on the X-ray CT reconstruction (Figure 2b), beneath the pigment markings. The neutron CT scans also revealed that the "head" and ears of the mummy were shaped using wads of folded textiles, seen above the dashed yellow line in Figure 5a and 2c. In the X-ray reconstruction, this detail was limited to areas of higher and lower density (Figure 5b).

The internal wrapping style and detail is significantly clearer in the neutron reconstruction; there are two clear layers of different grade textile, the outer is much coarser than the inner bandages (Figure 5a and d). This discovery is in agreement with the findings of Ginsburg (1999), who unwrapped a collection of small cat mummies from Saqqara which exhibited two distinct layers, the inner being tight and fine quality, and the outer being a coarser shroud.

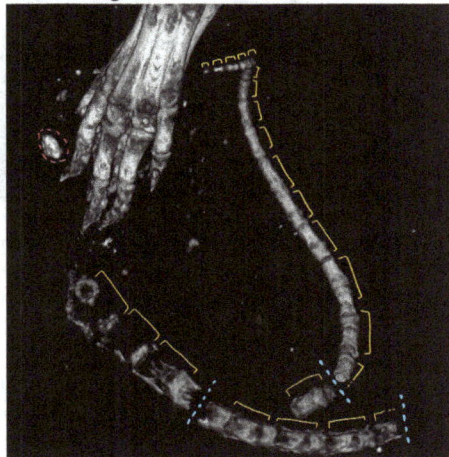

Figure 3. Segmented skeleton showing disarticulated tail (breaks marked in blue) with 23 caudal vertebrae (indicated in yellow), metatarsals, and amulet (red).

In this case, the layer closest to the skeleton is a very fine threaded textile, as the individual threads are difficult to isolate in reconstruction (Figure 5d). The outer layer exhibits a much coarser, loosely weaved textile, to the extent that the individual threads can be identified (Figure 5d). The findings of Ginsburg, while undeniably informative and valuable, were destructive, however this case study has demonstrated that the same level of information can be gleaned by entirely non-invasive methods.

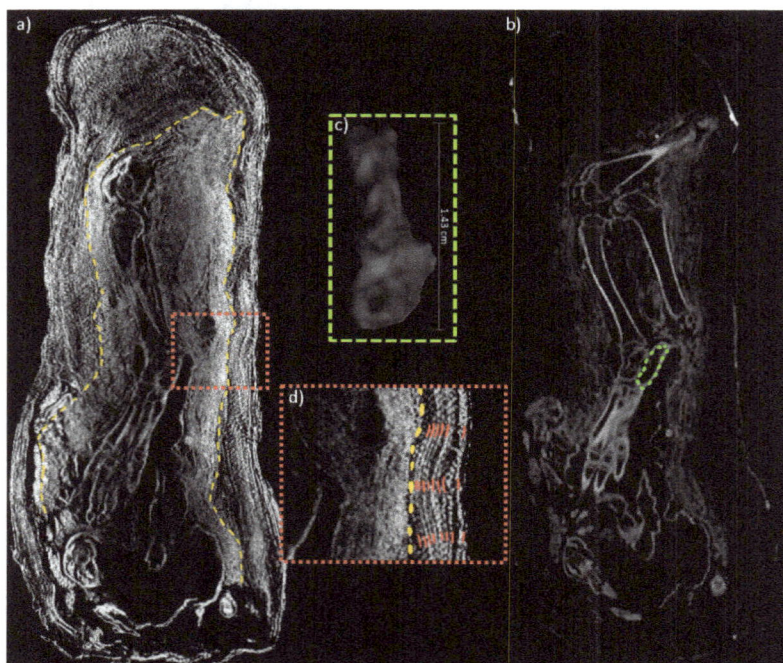

Figure 5. a) Neutron CT slice shows layers of textile wrapping; b) X-ray CT slice shows higher grayscale contrast; c) Segmented and visualised calcaneus bone [green area in b)]; d) Close up showing layered wrapping of varying tightness and coarseness [red box on b)].

Conclusions

Both X-ray and neutron CT scans provided invaluable insight into the contents and manufacture of this mummified votive offering, in a non-invasive way. The reconstructed volumes were most useful and informative when "digital excavation tools" were employed, as this enabled in-depth analysis of the remains. Using assisted and manual segmentation tools we were able to undertake a detailed study of this specimen including identification of the specimen age from epiphyses (equating to approximately 11 months old), as well as identification of species (*Felis silvestris*). We were also able to identify a possible amulet. The segmented ROIs further enabled production of outreach materials and 3D printed educational tools, which have broadened the general audience for CT imaging studies. Most importantly, the neutron CT reconstruction allowed for detailed and non-destructive investigation of wrapping and shaping techniques, showing two distinct wrapping layers and giving an unprecedented glimpse into ancient mummification practices. Overall, the combination of the two imaging techniques proved invaluable, as together they presented a comprehensive and complimentary study of all facets of the artefact.

Acknowledgments
Many thanks extended to our amazing team: A. Prof Ronika Power, Mr Anthony Lanati, Dr Tyr Fothergill, Dr Yann Tristant, Prof John Magnussen, and Prof. Simon Clark for their time and input into this case study. We also want to acknowledge the generous loan of the cat mummy by the Australian Institute of Archaeology and particularly Dr Christopher Davey for trusting us enough to let us do this exciting research without knowing what effects this may have. Archaeometry and archaeological science will only go forward with the support and engagement of people like yourself. Further thanks and acknowledgments to ANSTO for considering our proposal and deeming it a worthy study.

References
[1] David, A., 1997. Disease in Egyptian mummies : the contribution of new technologies. *The Lancet*, 349, pp.1760–63. https://doi.org/10.1016/S0140-6736(96)10221-X

[2] Ginsburg, L., 1999. Les chats momifiés de Bubasteion de Saqqarah. *Annales du Service des Antiquités de l'Égypte*, 74, pp.183–191.

[3] Ikram, S., 2011. Experimental Archaeology: From Meadow To Em-Baa-Lming Table. In C. Grave-Browns, ed. *Experiment and Experience*. Cardiff: University of Wales Press, pp. 53–74. https://doi.org/10.2307/j.ctvvnbgg.8

[4] Ikram, S. & Dodson, A., 1988. *The Mummy in Ancient Egypt: Equipping the Dead for Eternity.*, London: Thames and Hudson.

[5] Van Neer, W. et al., 2014. More evidence for cat taming at the Predynastic elite cemetery of Hierakonpolis (Upper Egypt). *Journal of Archaeological Science*, 45(1), pp.103–111. https://doi.org/10.1016/j.jas.2014.02.014

[6] Raymond, C.A., 2017. *Mummification unwrapped: investigating an Egyptian votive mummy using novel, non-invasive archaeometric techniques*. Macquarie University. Available: http://hdl.handle.net/1959.14/1268641.

[7] Raymond, C.A. et al., 2019. Recycled Blessings: an Investigative Case Study of a Re-wrapped Egyptian Votive Mummy using Novel and Established 3d Imaging Techniques. *Archaeometry*. https://doi.org/10.1111/arcm.12477

[8] Shaw, I., 2004. *Ancient Egypt: A Very Short Introduction*, Oxford: Oxford University Press. https://doi.org/10.1093/actrade/9780192854193.001.0001

[9] Smith, S.T., 2016. Unwrapping the Mummy: Hollywood Fantasies, Egyptian Realities. In J. M. Schablitsky, ed. *Box Office Archaeology: Refining Hollywood's Portrayals of the Past*. Routledge, p. 256.

Neutron Radiography - WCNR-11
Materials Research Proceedings 15 (2020) 256-261

Materials Research Forum LLC
https://doi.org/10.21741/9781644900574-40

Neutron Imaging, a Key Scientific Analytical Tool for the Cultural Heritage Project at ANSTO - Investigation of Egyptian Votive Mummies

Filomena Salvemini[1, a *], Constance Lord[2, b], Candace Richards[2, c]

[1] Australian Nuclear Science and Technology Organization (ANSTO), Lucas Heights, NSW (AU)

[2] Nicholson Museum, University of Sydney, NSW (AU)

[a]filomena.salvemini@ansto.gov.au, [b]constance.lord@sydney.edu.au, [c]candace.richards@sydney.edu.au

Keywords: Neutron Tomography, Non-Invasive Analysis, Cultural Heritage

Abstract. Neutron imaging has been instrumental in the development of the Cultural Heritage project at ANSTO with many successful applications reported to date. In this study we focus on the investigation of Egyptian votive mummies conducted by combining neutron and X-ray tomography. Two wrapped ibis bundles produced during the Roman Period (30 BC - 364 AD) were studied to inspect their content and ascertain the state of conservation. A new insight in the manufacturing method and composition of the mummified packages was gained in a non-invasive way. Advantages and limitations of the neutron tomographic method were also explored.

Introduction

A strategic scientific research project *Cultural Heritage* has been initiated at the Australian Nuclear Science and Technology Organization (ANSTO). The project aims to promote the access to the suite of nuclear methods available across the organization, and the use of a non-invasive analytical approach in the field of cultural-heritage, archaeology, and conservation science.

Neutron imaging, in particular, has become a valuable means for research in these fields. The fundamental properties of the neutron — no electric charge, deep penetration power into matter, and interaction with the nucleus of an atom rather than with the diffuse electron cloud —make this sub-atomic particle the ideal probe to non-invasively survey the bulk of a variety of heritage materials, such as metals, pottery, paintings, etc. [1]

In collaboration with Australian museum institutions and universities, and international experts, a series of archaeometric studies involving the neutron imaging beamline DINGO [2] at the Australian Centre for Neutron Scattering (ACNS) has been conducted so far [3] [4] [5]. In this paper, we present a recent tomographic examination of Egyptian votive mummies.

Egyptian mummies have always fascinated the modern world and the ancient Egyptians did not just mummify humans but a vast range of animal species as well [6]. In fact, animal mummies make up the largest category of objects we have from ancient Egypt and yet, until recently, are one of the least researched. In general terms, animal mummies can be divided into four main categories; food or victual mummies, pets, sacred animals and votive (offering) mummies. It is the latter of these that a visitor is most likely to encounter during a visit to an antiquities museum.

Votive mummies are a feature of the first millennium BCE, gaining great popularity in the Late Period and increasing even more so in the Ptolemaic and Roman Periods [7]. In much the same way as conventional offerings for the gods, votive mummies were produced for a range of

Neutron Radiography - WCNR-11
Materials Research Proceedings **15** (2020) 256-261

Materials Research Forum LLC
https://doi.org/10.21741/9781644900574-40

budgets, from simply wrapped figures with no decorative features for the poorer or the more frugal, up to lavishly wrapped mummies complete with cartonnage masks and mummified using an array of expensive embalming products for wealthier members of society or perhaps for a particularly urgent or important request [8].

As a non-invasive scientific method, tomographic investigation of mummies from ancient Egypt has been used since 1896 and provided information on mummification techniques, age and state of health of the individual at the time of death as well as any post-mortem damage to the mummy [9]. Medical Compute Tomography (CT) is commonly employed for these studies, but proved to have limited capabilities in differentiating between skeletal and desiccated soft tissues due to the limitation imposed by inbuilt safety mechanisms concerning radiation dose in human CT.

Neutron tomography, as a complementary method, can expand the investigation capabilities of standard analytical methods. In comparison to X-ray, neutrons feature a different contrast mechanism in the interaction with matter. The neutron attenuation coefficient is high for hydrogen and thus for organic materials. These characteristics make neutron tomography the ideal tool to study soft tissues, embalming substance and bandage of the mummies which mandatory require a non-invasive investigation approach. Moreover, thanks to the high penetration power of probing particles, neutron imaging can provide a unique insight, with high spatial resolution, into the bulk of dense materials, such as most metals and pottery. This enables to investigate the mummified bundle when encased into ceramic containers that is not always achievable with X-ray tomography [10] [11].

Experimental
Samples
The investigation was conducted on ibis mummies from the Egyptian antiquities collection of the Nicholson Museum in Sydney, since little scientific data are currently reported on this typology of votive offering.

The samples set consisted of two wrapped ibises and three jars said to contain ibis remains in the museum record. Examination of the content will enable further the scholarly discussion on the purpose of ancient Egyptian fake mummies as well as slightly narrowing the date-range of production as Ptolemy VI Philometer (180 – 168 BC), during an investigation into corrupt practices of the ancient manufacturers of ibis mummies, decreed that, from that time, only one bird per bundle was permitted.

The study is still in progress; here we present preliminary results about the two wrapped ibises (Figure 1-a) based on tomographic analyses, and refer the reader to future publications for a more detailed coverage of the entire research that will include the examination of the jars.

Analytical methods
Neutron tomography and medical CT were combined to characterize the contentment of the two bundles and experimentally explore advantage and limitation of each analytical method.

The neutron tomographic measurements were performed on DINGO. The measurements were conducted in the high-resolution acquisition mode (with the ratio of collimator-detector length L to inlet collimator diameter D equals to 1000) with a pixel size of 100 µm by setting a 200x200 mm^2 field of view with 50 mm lens coupled with a 100µm thick scintillation screen. Projections were acquired with an equiangular step of 0.1°over 180°with an exposure time of 15 seconds each. Since the length of the samples exceed the size of the maximum field of view available on DINGO (sample NM62.584 is 360x180x70 mm^3 and sample NM62.585 is 328x130x60 mm^3), 2 consecutive scans with vertical displacement will be acquired for each sample.

Neutron Radiography - WCNR-11   Materials Research Forum LLC
Materials Research Proceedings **15** (2020) 256-261   https://doi.org/10.21741/9781644900574-40

The medical CT analysis with X-rays was conducted at the University of Sydney by setting the X-ray source at 70kV and acquiring 397 projection equiangularly spaced over 180° with a pixel size of ˂400 μm.

The data were processed using the Octopus code for tomographic reconstruction [12]. The obtained slices were then recomposed and evaluated using the AVIZO software [13]. Neutron and X-ray data were registered and segmented on the base of the *correlation histogram* module.

Results and discussion

The combined application of neutron and X-ray tomographic analyses enable to investigate different aspect of the samples. The content and composition of the mummified packages were clarified; this includes determining the minimum number of individuals wrapped in a single package, and confirmation and identification of any ancient fakes within the set of investigated samples. Identification of species, age, pathologic abnormalities and cause of death can be also inferred on the base of the anatomical features of the animal remains (Figure 1).

Furthermore, more information on the mummification materials and methods can be also obtained as well as evaluation of specimen condition.

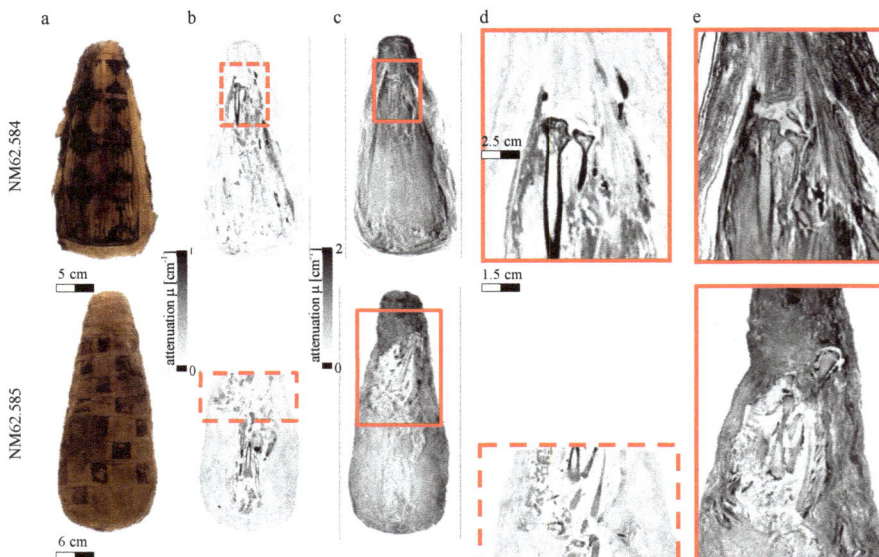

Figure 1: Each sample is identified using its museum registration numbers: NM62.584, NM62.585. a) Photographic images of the wrapped ibises; sections across the b) X-ray and c) neutron tomographic reconstructions and corresponding detailed view of the regions of interested d) and c) are shown. The scale bar at the left of each tomographic section indicates the colour code for the attenuation coefficient. Please, note that the X-ray CT for sample NM62.585 was acquired partially. The provided cross sections highlight the external wrapping enveloping the remains of the ibises, visible in the core area, and the difference in contrast that neutron and X-rays can provide for such components.

The combined tomographic datasets demonstrate the presence of skeletal remains with anatomical features typical of the ibis species in both items. Sample NM62.584 only contains few integral bones in association with several fragmented bones (Figure 2). The wrapping appears quite loose and some linen bandage might have been secured by a pin that was supposedly inserted by previous museum curators in more recent time. The bandage appears to be impregnated with resinous material mostly localised around bones, at the top and the bottom top, and in the outer portions of the wrapping.

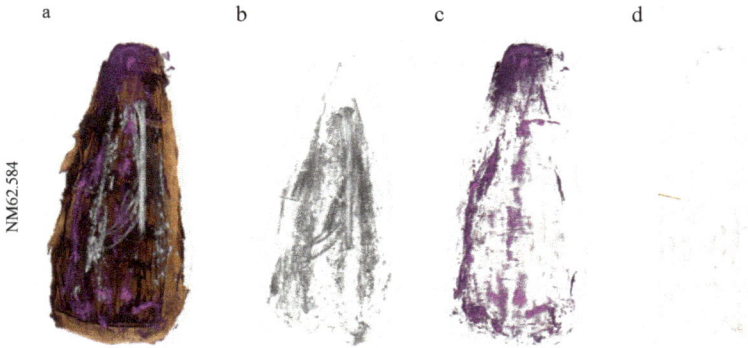

Figure 2: a) The photographic image of the museum item is overlapped to the segmentation based on neutron and X-ray data. The animal remains, distribution of the embalming material in the bandage, and a pin are rendered as separated volume in b), c) and d).

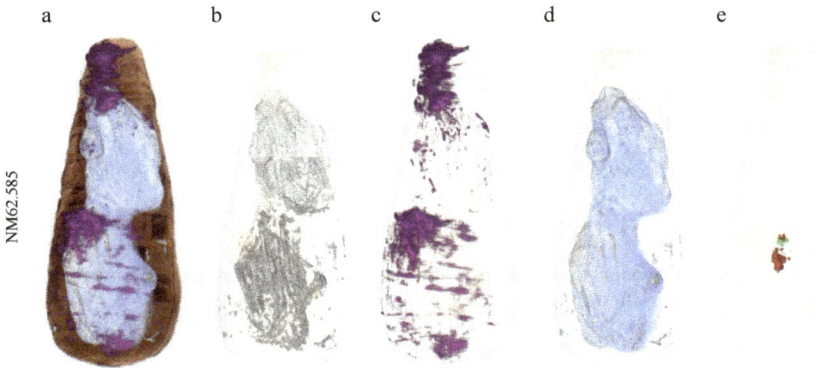

Figure 3: a) The photographic image of the museum item is overlapped to the segmentation based on neutron and X-ray data. The animal remains, distribution of the embalming material in the bandage, feathers, and the stomach content are rendered as separated volume in b), c), d) and e).

In the case of sample NM62.585 (Figure 3), the complete skeletons of two individual ibises were identified. Some details such as feathers and the stomach content were also reconstructed. The linen bandage was in better state of conservation and more tightly wrapped. Also in this case, the embalming component was mapped with a distribution similar to sample NM62.584.

The study is still ongoing including consultation with the Australian Museum in Sydney to identify the species and age of ibis.

Some considerations on the analytical method can be also made Figure 1. Neutron tomography demonstrated to be better suitable to detect the presence of feathers, otherwise not visible from the X-ray CT. The resinous balm could be more distinctly mapped due to the high attenuation cross-section of this hydrogen-rich material. Similarly, individual layers of the linen bandage can better be distinguished by neutrons; the tread of the textile and the direction of folding can be discerned to gain a better understanding of the wrapping method. However, limitations of the neutron tomographic method were also evidenced. Especially in the thickest region of the artefact, the neutron beam interacted with very high attenuating materials, such as the balm and several layers of textile; as predicable, in such circumstance, the low counting statistic due to the limited transmission through the sample affected the quality of the tomographic reconstruction. In these portions the tomographic slices are quite nosy and feature of the bulk can be discerned upon treatment of the data - 3D median and anisotropic diffusion filters were applied. X-ray CT has been always the technique of choice to investigate the content of votive offering; its value as analytical tool is hardly arguable as demonstrated by the high contrast offered, especially in the detection of the skeletal remains, and the superior penetration power through the thickest portions of the item. Most of the internal components were visible, despite a lower contrast is observed in the case of balm and textile compared to the neutron tomography. A limitation, intrinsic to the machine adopted for this study rather than the method itself, is the size of the sample that can scan. Figure 1-b shows that only a half of item NM62.585 was investigated since the length of the wrapping exceeded the size of the sample-chamber.

Conclusion

This study explores the application of a combined approach based on neutron and X-ray tomography to investigate the content of Egyptian votive mummy's part of the collection of the Nicholson Museum in Sydney.

While X-ray is indisputably the analytical method of choice to investigate the contest of such material in a non-invasive way, neutron tomography has demonstrated some advantages. Neutrons offered a better contrast to detect the presence of feathers, to map the distribution of the resinous balm, and to identify individual layers of the linen bandage. On the other hand, the penetration power was quite limited in the thickest portion of the item, thus affecting the image quality.

The investigation provided the museum with valuable information about the content and composition of the mummified packages. The condition of the artefacts was also ascertained; this will greatly assist in the museum's ongoing care and conservation of each of the collection items. The Museum's complete animal mummy collection will be analysed with the same instrumentation to achieve consistency across the collection and allow for comparative analysis between samples where appropriate. Furthermore, it is foretold that the tomographic data will be exploited for the development of 3D visualizations of each specimen for augmentation of museum displays while possible 3D printing of facsimiles will aid further (ongoing) study.

Acknowledgments

The authors would like to thank the team of Sydney Imaging at the University of Sydney for assistance with the X-ray CT analysis.

References

[1] N. Kardjilov and G. Festa, Neutron Methods for Archaeology and Cultural Heritage, Springer, 2017. https://doi.org/10.1007/978-3-319-33163-8

[2] U. Garbe, T. Randall, C. Hughes, G. Davidson, S. Pangelis and S. Kennedy, "A new Neutron Radiography / Tomography / Imaging Station DINGO at OPAL," in *10 World Conference on Neutron Radiography 5-10 October 2014*, 2015.

[3] F. Salvemini, V. Luzin, F. Grazzi, S. Gatenby and M. Kim, "Structural characterization of ancient Japanese swords from MAAS using neutron strain scanning measurements," in *Materials Research Forum*, 2016.

[4] F. Salvemini, K. Sheedy, S. R. Olsen, M. Avdeev, J. Davis and V. Luzin, "A multi-technique investigation of the incuse coinage of Magna Graecia," *Journal of Archaeological Science: Report,* pp. 748-755, 2018. https://doi.org/10.1016/j.jasrep.2018.06.025

[5] F. Salvmeini, V. Luzin, F. Grazzi, S. Olsen, K. Sheedy, S. Gatenby, M.-J. Kim and U. Garbe, "Archaeometric investigations on manufacturing processes in ancient cultures with the neutron imaging station DINGO at ANSTO," *Physics Procedia,* vol. 88, pp. 116-122, 2017. https://doi.org/10.1016/j.phpro.2017.06.015

[6] L. Bruno, The Scientific Examination of Animal Mummies in Soulful Creatures, E. Bleiberg, Y. Barbash and L. Bruno, Eds., Brooklyn: GILES, 2013.

[7] S. Ikram, Divine Creatures in Divine Creatures: Animal Mummies in Ancient Egypt, S. Ikram, Ed., Cairo: The American University in Cairo Press, 2005. https://doi.org/10.5743/cairo/9789774248580.001.0001

[8] S. Buckley, K. Clark and R. Evershed, "Complex organic chemical balms of pharaonic animal mummies," *Nature,* no. 431, p. 294–9, 2004. https://doi.org/10.1038/nature02849

[9] J. Adams and C. Alsop, Imaging Egyptian Mummies in Egyptian Mummies and Modern Science, R. David, Ed., 21-42, 2008. https://doi.org/10.1017/CBO9780511499654.004

[10] E. Abraham, M. Bessou, L. Szentmiklósi, B. Maryelle and A. Ziéglé, "Terahertz, X-ray and neutron computed tomography of an Eighteenth Dynasty Egyptian sealed pottery," *Applied physics. A,* no. Materials science processing, 2014. https://doi.org/10.1007/s00339-014-8779-3

[11] C. A. Raymond, J. J. Bevitt, Y. Tristant, R. K. Power, A. W. Lanati, C. J. Davey, J. S. Magnussen and S. M. Clark, "Recycled Blessings: an Investigative Case Study of a Re-wrapped Egyptian Votive Mummy using Novel and Established 3d Imaging Techniques," *Archaeometry,* 2019. https://doi.org/10.1111/arcm.12477

[12] M. Dierick, B. Masschaele and L. Van Hoorebeke, "Octopus, a fast and user-friendly tomographic reconstruction package developed in LabView," *Measurement Science and Technology,* vol. 15, no. 7, pp. 1366-1370, 2004. https://doi.org/10.1088/0957-0233/15/7/020

[13] "FEI," [Online]. Available: https://www.fei.com/software/amira-avizo/] .

[14] L. Evans L, M. Hartley and C. Lord , Animals in Ancient Egypt: Roles in Life and Death in The Museum of Ancient Cultures Catalogue, Y. Tristant, Ed., Macquarie University Press, in production.

Neutron Radiography - WCNR-11
Materials Research Proceedings 15 (2020) 262-267

Materials Research Forum LLC
https://doi.org/10.21741/9781644900574-41

Evaluation of Motion Blur in High-Speed Neutron Imaging at Kyoto University Research Reactor

Daisuke Ito[1,a*] and Yasushi Saito[1,b]

[1]Institute for Integrated Radiation and Nuclear Science, Kyoto University
2-1010 Asashiro-nishi, Kumatori-cho, Sennan-gun, Osaka, Japan

[a]itod@rri.kyoto-u.ac.jp, [b]ysaito@rri.kyoto-u.ac.jp

Keywords: High-Speed Neutron Imaging, Motion Blur, Rotating Disc

Abstract. The rapid multiphase flow phenomena such as the flow with vaporization and condensation must be clarified for the safety analysis and severe accident analysis for light water reactors. To understand the multiphase flows experimentally, measurement technique with high temporal resolution is required. In addition, the multiphase flow has spatial distributing characteristics, thus two- or three-dimensional visualization techniques are suitable for the understanding of the flow structure. In this study, temporal resolution of the neutron imaging technique was enhanced for the observation of rapid multiphase flow behavior. The existing imaging system was upgraded to improve the frame rate, and imaging with a frame rate of 10,000 fps could be achieved at B-4 port in Kyoto University Research Reactor (KUR) at 5 MW operation. Then, the imaging results were evaluated by using a rotational disc system. The relation between the rotational speed and motion blur was investigated in the high-speed neutron imaging.

Introduction

Multiphase flows appear in many industrial applications like power reactor and chemical reactor, and are phenomena which show large temporal fluctuation and characteristic spatial distribution. Therefore, the flow structure is very complicated and difficult to understand in detail. However, the multiphase flow should be clarified for the safety analysis and severe accident analysis for light water reactors. Especially, in the nuclear severe accident, several rapid multiphase flows might be seen, for example, steam explosion and re-flooding etc. Although it is difficult to investigate these phenomena experimentally and analytically, their understandings are essential for safety analysis of the nuclear reactors.

Usually, optical imaging technique using a high-speed camera has been applied to rapid flow observation. Very fast imaging (>1,000,000fps) can be performed using the high-speed camera [1] and the bubble shape and motion in bubbly flow can be detected easily. However, it is difficult to measure the flow behavior inside metal vessel which is generally used in the experiments at high temperature and high pressure conditions. The radiation imaging technique is very effective for such two-phase flow measurement in opaque vessel. In particular, neutron transmission imaging which has high sensitivity to the water and transparency to the metal has been applied various multiphase flow studies. The interaction of molten metal and water [2] and the bubble behavior in liquid metal pool [3] were visualized by thermal neutron imaging with the frame rate of 500 fps in JRR-3M. Kureta et al. also applied the neutron imaging to study the subcooled flow boiling in a narrow rectangular channel [4]. Two-phase flow behavior in porous media was observed at the frame rate of 200 fps in KUR B-4 port [5]. In addition, air-water two-phase flow was visualized at 800 fps using cold neutrons at ICON beam line in PSI [6] and boiling of pentane inside a steel tube was observed at 154 fps using NEUTROGRAPH in ILL [7]. High-speed neutron imaging with a 10µs exposure time was also performed at ANTARES facility in FRM II [8]. However, the

Neutron Radiography - WCNR-11   Materials Research Forum LLC
Materials Research Proceedings **15** (2020) 262-267   https://doi.org/10.21741/9781644900574-41

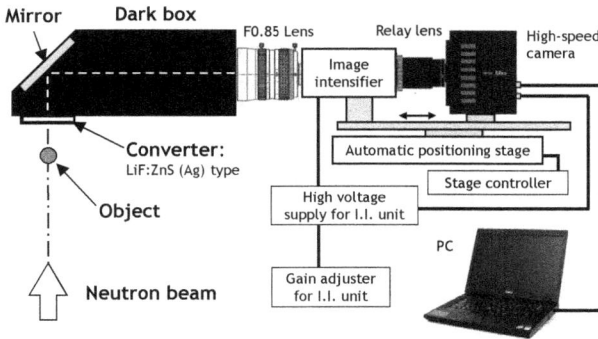

Fig. 1 High speed neutron imaging system

improvement of temporal resolution in neutron transmission imaging is still required for rapid multiphase flow observation. For high-speed flow observation using neutrons, high sensitivity imaging system and high neutron flux source are required. Since the construction of new high-power neutron source has large difficulty, the imaging system was upgraded in this study.

In addition, the image distortion and degradation are influenced by a number of system components in neutron imaging experiments [9]. To perform accurate imaging, these factors should be clarified. Generally, converter unsharpness, object scattering and geometric unsharpness cause the image degradation in the neutron imaging and, they depend on the system and facility. In high-speed neutron imaging, motion unsharpness and statistical system noise are important and their effect on acquired image must be understood. Thus, the purpose of the present study is to improve the high-speed neutron imaging system and to evaluate the image quality degradation in high-speed neutron imaging.

High speed neutron imaging system
To enhance the temporal resolution in the neutron imaging, the imaging system was upgraded. The high speed neutron imaging system used in this study consists of a high-speed camera, an optical image intensifier, optical lens and a converter. The present major improvements are the uses of a high sensitivity high-speed camera (Photron AX-50, ISO 40,000), an ultra-high sensitivity lens (50mm F0.85, VS Technology VS-50085/C) and a thick scintillator (RC Tritec 6LiF/ZnS:Ag 200μm). The improved imaging system is shown in Fig.1. The neutrons transmitted through the imaging object converted to the optical light by the scintillator. The light enters an optical image intensifier via the lens and intensified. Finally, the high-speed camera records the image. The focus of the camera was adjusted by a positioning stage. In the previous imaging system, the frame rate was 200 fps [10], however, the present imaging system could achieve the frame rate of 10,000 fps at B-4 Port in KUR at operation power of 5MW.

Experimental setup
To evaluate the image quality in the high-speed imaging, the experiments using a rotating indicator were performed. The rotating disc system is shown in Fig.2. The gadolinium (Gd) plate which has high neutron attenuation characteristics was used as the indicator. The shape of the used indicator is shown in Fig.3. This simulates small bubble in water, because water has large attenuation

Neutron Radiography - WCNR-11

Materials Research Forum LLC

Materials Research Proceedings **15** (2020) 262-267

https://doi.org/10.21741/9781644900574-41

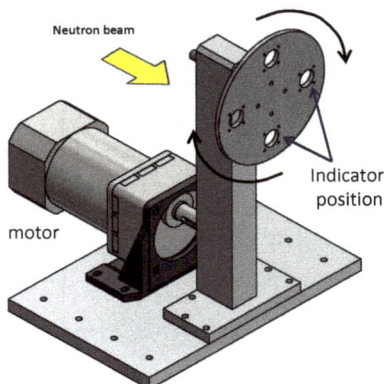

Fig. 2 Rotating disc system

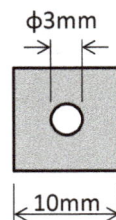

Fig. 3 Gd indicator plate

coefficient for thermal neutrons. The thickness of Gd plates is 0.2 mm. Four indicators can be fixed on the rotating disc equiangularly (circular holes in Fig. 2). The radial location of the indicators on the disc is 50 mm. The rotational speed of the disc was varied up to 960 rpm which is corresponding to the velocity of 5 m/s in the indicator position. The neutron imaging experiments were conducted at B-4 experimental room in Kyoto University Research Reactor (KUR). The thermal neutron flux at the beam exit is 8.5×10^7 n/cm^2 s. The beam size is 75 mm height and 10 mm width at beam exit. The distance between the indicator and the converter was 150 mm and the L/D in the present imaging was 150. The pixel resolution of the acquired image in the present experiments is 0.4 mm/pixel. The frame rate of the high-speed camera is 10,000 fps, and so the exposure time is 0.1 ms. The gate time of the image intensifier is also 0.1 ms.

Results and discussion
The neutron transmission image of a static indicator is shown in Fig. 4. 2,000 instantaneous images acquired with a 0.1ms exposure time were averaged. The neutrons are attenuated by the Gd plate, and the small hole could be observed. However, there is the blurring at the edge of the indicator and this may be caused by the converter thickness, object scattering and beam divergence. The neutron transmission profile along the vertical line in the center of the indicator is shown in Fig.5. The solid line denotes the transmission profile and the dashed line denotes the designed value of the indicator shown in Fig.3. The edge of the indicator is blurred even in the static observation.

The averaged images of the rotating indicator are represented in Fig. 6. The center of mass of the indicator in each image was estimated and those images were rotated around the center of rotation disc so as to be the same indicator location with the image in Fig.4. Then, 70~180 images at the same position are averaged. The rotational speed was changed from 0.6 m/s to 5 m/s. In these images, the indicator was moved from top to bottom. The difference between static and rotating indicators is obvious and the image degradation due to the indicator motion becomes strongly as the rotation speed increases. Especially, the rear side edge is more blurred. In addition, it is hard to recognize the circular hole in the indicator in 5 m/s motion. As a result, the effect of motion blur in high-speed neutron imaging corresponds to the rotating speed.

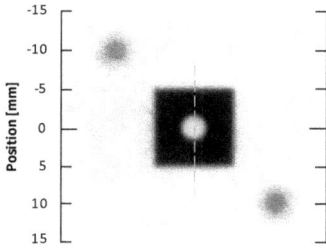

Fig. 4 Neutron transmission image of static indicator

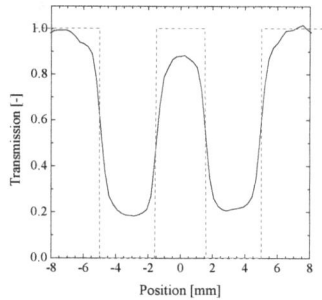

Fig. 5 Neutron transmission profile along the vertical line shown in Fig.4

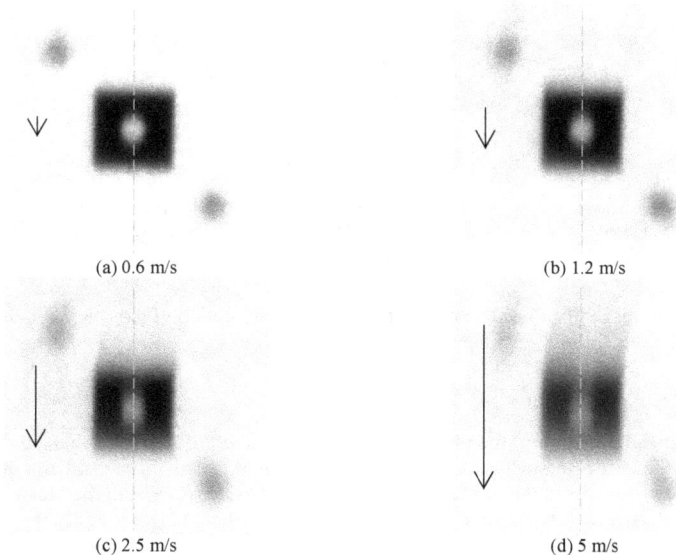

(a) 0.6 m/s

(b) 1.2 m/s

(c) 2.5 m/s

(d) 5 m/s

Fig. 6 Neutron transmission images of indicator rotated at different rotation speed

The neutron transmission profiles along the center line were made from the images in Fig. 6, as shown in Fig. 7. As the rotation speed increases, the edges become smoothly. In addition, the decay time characteristics of the scintillator might be one of the causes of the image degradation. Therefore, the smoothed edge characteristics due to the motion should be clarified and the image restoration method must be established by considering the motion blur and the afterglow. In this study, the image restoration using a point spread function (PSF) was applied to investigate the effect of the motion blur.

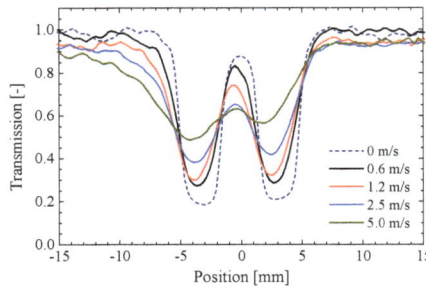

Fig. 7 Neutron transmission profile of rotating indicator

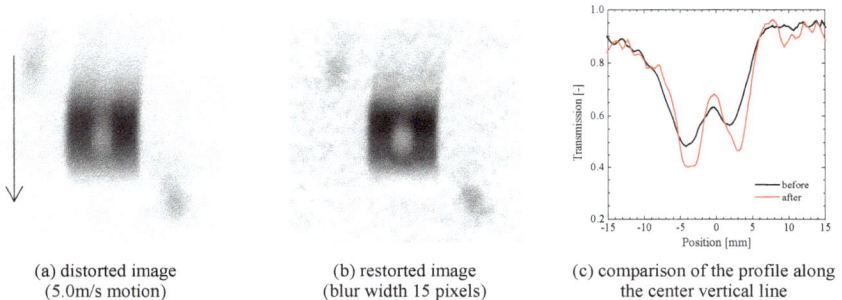

| (a) distorted image
(5.0m/s motion) | (b) restored image
(blur width 15 pixels) | (c) comparison of the profile along
the center vertical line |

Fig. 8 Blurred image restoration.

The distorted image $g(x,y)$ can be represented by the convolution of the original image $f(x,y)$ and the $PSF(x,y)$, as follows.

$$g(x,y) = PSF(x,y) * f(x,y) \tag{1}$$

If the width and direction of the motion blur are known, the PSF can be estimated. However, it is not easy to evaluate such parameter from only the distorted images. In this study, the parameters are assumed and the image was restored using Winner filter with the blur width of 15 pixels. The restored result is shown in Fig. 8(b). The motion blur can be reduced by this process. In addition, the comparison of the profile before and after the filter is show in Fig. 8(c). Although the filtering parameters were assumed manually, the edge of the indicator could be enhanced. However, the blur could not be removed completely, because it might be affected by the decay characteristics of the scintillator. Therefore, the thickness of the scintillator and gate time of the image intensifier should be considered to establish the image restoration processing.

Summary
In this study, the high-speed neutron imaging system was upgraded to improve the temporal resolution for rapid multiphase flow observation. As a result of the use of high sensitivity high speed camera and high sensitivity lens, the frame rate of 10,000 fps could be achieved at B-4 port, KUR. The applicability to rapid multiphase flow observation would be extended by the present improvement. In addition, the image quality in the high-speed neutron imaging was

Neutron Radiography - WCNR-11   Materials Research Forum LLC
Materials Research Proceedings **15** (2020) 262-267   https://doi.org/10.21741/9781644900574-41

evaluated by the rotating disc system and the image degradation due to the rapid motion was confirmed. The image distortion might be affected by not only the motion blur but also the decay characteristics of the scintillator and image intensifier. The image restoration processing method will be developed by considering the image distortion and the effect of the thickness of the scintillator and gate time of the image intensifier will be evaluated in the future.

References

[1] T. Alghamdi, S.T. Thoroddsen, J. F.Hernández-Sánchez, Ultra-high speed visualization of a flash-boiling jet in a low-pressure environment, Int. J. Multiphase Flow 110 (2019) 238-255. https://doi.org/10.1016/j.ijmultiphaseflow.2018.08.004

[2] K. Mishima, et al., Visualization study of molten metal–water interaction by using neutron radiography, Nucl. Eng. Design 189 (1999) 391-403. https://doi.org/10.1016/S0029-5493(98)00263-5

[3] Y. Saito, et al., Velocity field measurement in gas–liquid metal two-phase flow with use of PIV and neutron radiography techniques, Applied Radiation and Isotopes 61 (2004) 683-691. https://doi.org/10.1016/j.apradiso.2004.03.110

[4] M. Kureta, et al., Study on point of net vapor generation by neutron radiography in subcooled boiling flow along narrow rectangular channels with short heated length, Int. J. Heat Mass Trans. 46 (2003) 1171-1181. https://doi.org/10.1016/S0017-9310(02)00398-8

[5] D. Ito and Y. Saito, Visualization of bubble behavior in a packed bed of spheres using neutron radiography, Physics Procedia 69 (2015) 593-598. https://doi.org/10.1016/j.phpro.2015.07.084

[6] R. Zboray, P. Trtik, 800 fps neutron radiography of air-water two-phase flow, MethodsX 5, (2018) 96-102. https://doi.org/10.1016/j.mex.2018.01.008

[7] A. Hillenbach, M. Engelhardt, H. Abele, R. Gähler, High flux neutron imaging for high-speed radiography, dynamic tomography and strongly absorbing materials, Nuclear Instruments and Methods in Physics Research A 542 (2005) 116-122. https://doi.org/10.1016/j.nima.2005.01.290

[8] B. Schillinger, E. Calzada, K. Lorenz, M. Muhlbauer, Continuous neutron radioscopy with 1000fps and 10 microsecond time resolution, NEUTRON RADIOGRAPHY, DEStech Publications (2008) 234-239.

[9] A.A. Harms and D.R. Wyman, Mathematics and physics of neutron radiography, Springer Netherlands (1986). https://doi.org/10.1007/978-94-015-6937-8

[10] D. Ito, Y. Saito, Y. Kawabata, Hybrid two-phase flow measurements in a narrow channel using neutron radiography and liquid film sensor, Physics Procedia 69 (2015) 570-576. https://doi.org/10.1016/j.phpro.2015.07.081

Neutron Radiography - WCNR-11
Materials Research Proceedings **15** (2020) 268-273

Materials Research Forum LLC
https://doi.org/10.21741/9781644900574-42

Simultaneous Measurements of Water Distribution and Electrochemical Characteristics in Polymer Electrolyte Fuel Cell

Hideki Murakawa[1,a] *, Syun Sakihara[1,b], Katsumi Sugimoto[1,c], Hitoshi Asano[1,d], Daisuke Ito[2,e] and Yasushi Saito[2,f]

[1]Graduate School of Engineering, Kobe University, 1-1 Rokkodai, Nada, Kobe 657-8501 Japan

[2]Institute for Integrated Radiation and Nuclear Science, Kyoto University, 2 Asashiro-Nishi, Kumatori-cho, Sennan-gun, Osaka 590-0494 Japan

[a]murakawa@mech.kobe-u.ac.jp, [b]171t327t@stu.kobe-u.ac.jp, [c]sugimoto@mech.kobe-u.ac.jp, [d]asano@mech.kobe-u.ac.jp, [e]itod@rri.kyoto-u.ac.jp, [f]ysaito@rri.kyoto-u.ac.jp

Keywords: PEFC, EIS, Water Content, Neutron Radiography

Abstract. In this study, neutron radiography and electrochemical impedance spectroscopy (EIS) were simultaneously used to evaluate the relation between the water amount and the electro chemical characteristics in a polymer electrolyte fuel cell (PEFC). Two-dimensional water distributions in the through-plane direction of the proton exchange membrane (PEM) were measured every 60 s during the PEFC operation. The results were compared with ionic and the polarization resistances obtained from EIS. The ionic conductivity through the PEM increased with an increase in the liquid-water content in the membrane. The effects of water content on the ionic conductivity were much smaller in comparison to the Springer's model at a water content was less than 1. The polarization resistance increased with an increasing in liquid-water accumulation in the gas diffusion layer.

Introduction

Water management is a key topic of a polymer electrolyte fuel cell (PEFC). If condensed water exists in the gas diffusion layer (GDL) and the gas channel, it may depress the gas diffusion as flooding. However, the generated water must be appropriately supplied to the proton exchange membrane (PEM) for proton conduction. Hence, water management is significantly important for PEFC performance, and clarification of the water-transport mechanisms between the PEM, GDL, and gas channels is of great concern. Several investigations on water management have been carried out concerning water movement inside the PEM, water flooding in the GDL, and water plugging in the channel [1, 2].

Loss in electric power generation in the PEFC is due to resistances such as ionic, activation, concentration, etc. Many of these resistances are related to the mass transport in the PEFC. To evaluate the resistances in a PEFC, electrochemical impedance spectroscopy (EIS) has been widely used [3]. However, the effect of water in the PEFC on the resistances has not been fully understood. Aim of this study is to clarify the effect of water in the PEM and the GDL on the resistances. Simultaneous measurements of the water distribution and the electrical impedance were carried out by using neutron radiography and EIS. Changes in water accumulation in the PEM and the GDL were compared with the resistances.

Experimental setup and method

A schematic diagram of the PEFC is shown in Fig. 1. A proton exchange membrane was sandwiched between the gas diffusion layers and the gas channels. The separators were made of gold-plated aluminum having nine-parallel gas-channels with a cross-sectional area of 1×1

mm^2. The length of the channel was 10 mm. Nafion® NR-212 was used as the PEM with a thickness of approximately 90 μm, having catalyst layers on both the anode and the cathode sides. The electrode area was 10 × 19 mm^2. The GDL was carbon paper (Toray Ind.) with thickness of 190 μm at the cathode side and 280 μm at the anode side. The porosity of the GDL was approximately 78%.

Two-dimensional water distributions were measured using neutron radiography during PEFC operation. The neutron radiography facility at B-4 port in the Kyoto University Reactor (KUR) was used in this study. The total neutron flux at the beam exit was approximately 5 × 10^7 /cm^2 s with a nominal thermal output of 5 MW [4]. Neutrons entered from the side-view direction and were attenuated by the PEFC including the accumulated water. The transmitted neutrons were converted to visible rays using a scintillator screen, and the 16-bit gray-scale radiographs were taken using a cooled CCD camera (PIXIS 1024, Princeton Instruments) with an array of 1024 × 1024 pixels. Exposure time for obtaining an image was set at 1 min. Obtained water-distribution represents the average value over 1 min exposure time. The details of the measurement methods are described in reference [2].

The equivalent electric circuit for the EIS analysis used is shown in Fig. 2. As the polarization resistance of the cathode is much higher than that of the anode, associated elements of the anode can be neglected [5]. C_{dl} is the double-layer capacitance, R_{PEM} is the ionic resistance through the PEM, and R_{pol} is the polarization resistance consisting activation and concentration polarization resistances at the cathode. Impedance analysis took approximately 40 s for one measurement with a frequency range of 0.5 to 30 kHz. Therefore, an impedance measurement was complete within the time taken for a neutron radiograph measurement. Comparisons of the water distribution and the electro chemical characteristics were carried out over 1 min. Impedance analysis can be applied for steady state conditions, however, if the water distribution changes rapidly, results might be affected. Therefore, to change the water distribution gradually, the experiments were carried out at a relatively lower current density, i = 158 mA/cm^2 and 316 mA/cm^2. The air and hydrogen flow rates were 66 Ncc/min and 28 Ncc/min, respectively. The oxygen utilization was 7.5% at i =158 mA/cm^2 and 15% at i =316 mA/cm^2 and a hydrogen utilization was 7.5% at i =158 mA/cm^2 and 15% at i =316 mA/cm^2. The experiments were carried out at room temperature, while the temperature of the PEFC varied between 25 – 30 °C.

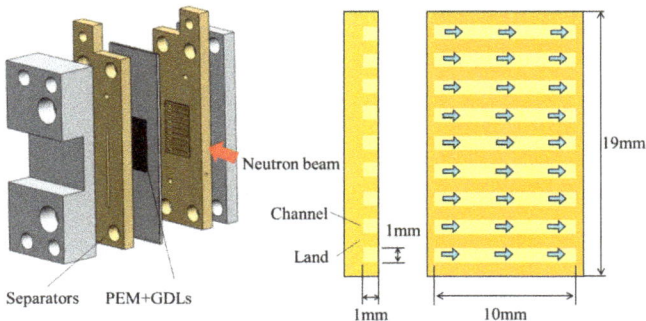

Fig. 1. Schematic diagram of the polymer electrolyte fuel cell and geometry of the gas channel.

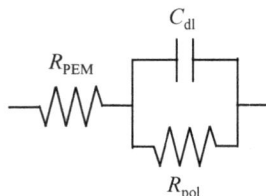

Fig. 2. *Equivalent electric circuit.*

Evaluation of the ionic conductivity

Springer et al. [6] proposed a correlation between ionic conductivity of the PEM, σ, and the water content in the PEM as

$$\sigma(T_{cell}) = (a\lambda + b)\exp\left[1268\left(\frac{1}{303} - \frac{1}{273 + T_{cell}}\right)\right] \qquad \text{for } \lambda > 1 \qquad (1)$$

Where λ [-] is the water content of the electrolyte and is defined as mol-H_2O / mol-SO_3^- ; T_{cell} [°C] is the cell temperature. The correction factors were proposed as $a = 0.5139$ and $b = -0.326$. To evaluate Eq. (1), λ was calculated using the following equation [7].

$$\lambda = \frac{t_w h t_{PEM} \rho / M_{H_2O}}{mS / EW}. \qquad (2)$$

Where t_w [m] is the average water thickness inside the PEM, h [m] is the electrode height, t_{PEM} [m] is thickness of the PEM, ρ [g/m^3] is density of the liquid water, M_{H_2O} [g/mol] is the molar mass of water, EW [kg/mol] is the mass of PEM per mol of the sulfonic acid group, m [kg/m^2] is the mass per area of PEM, and S [m^2] is the area of PEM. Ionic conductivity from the impedance measurement results can be expressed by the following equation, using R_{PEM} and the active area of A [m^2].

$$\sigma = \frac{A}{R_{PEM} t_{PEM}}. \qquad (3)$$

The PEM properties are listed in Table 1.

Table 1. *PEM properties*

PEM thickness, t_{PEM}	[m]	: 90×10^{-6}
Electrode height, h	[m]	: 19×10^{-3}
Liquid water density, ρ	[kg/m^3]	: 997.04
Molar mass, M_{H_2O}	[g/mol]	: 18
Equivalent weight, EW	[kg/mol]	: 1100
Basic weight of Nafion® NR-212, m	[g/m^2]	: 100
Area of PEM, S	[m^2]	: 0.00168

Anode   Cathode

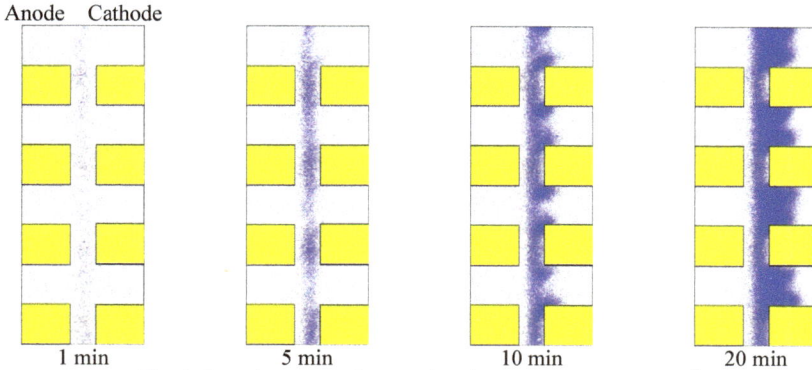

| 1 min | 5 min | 10 min | 20 min |

Fig. 3. *Two-dimensional water distributions at i =158 mA/cm².*

Two-dimensional water distributions

Two-dimensional water distributions at i =158 mA/cm² are shown in Fig. 3. Blue color represents the liquid water. Water accumulation in the PEM and the GDL increases with the PEFC operation time. The amount of water accumulation in the GDL under the lands is greater than that under the channels at 5 min. It indicates that water accumulation in the GDL at the cathode started under the lands. The liquid water, then reaches the channel and gets concentrated at the land corners. Liquid water droplets formed in the channels grow at the interface between the channels and the GDL along the channel walls.

Relation between water saturation and the resistances

Two-dimensional water distributions measured using neutron radiography expressed the relative water accumulation based on the initial condition. In this study, changes in λ ($\Delta\lambda$) and changes in σ ($\Delta\sigma$) from the initial condition are evaluated. Fig. 4 represents the relation between $\Delta\lambda$ and $\Delta\sigma$ at all times in each current density. It was confirmed that $\Delta\sigma$ increases with $\Delta\lambda$. However, effect of water contents on the ionic conductivity was much less than that proposed by Springer et al. [6]. It is difficult to determine the absolute value of λ in the PEM precisely using the neutron radiography. However, it was confirmed that a small change of the water amount in the PEM with supplying dry nitrogen was confirmed before each experiment. Therefore, it is expected that λ in the PEM before operation is less than 1. These results show that change in the ionic conductivity with respect to the water content becomes small in low water content range.

Fig. 5 shows the relationship between water saturation, s, in the GDL and the polarization resistance. The water saturation was calculated from the average two-dimensional water distribution in the GDL area. The results show that R_{pol} increases with the water saturation in the GDL. Tendencies at i = 158 mA/cm² and 316 mA/cm² are almost the same at s < 0.2. However, R_{pol} increases rapidly at s > 0.2. This indicates a blockage in the gas supply due to flooding in the GDL and the channel significantly degraded the power generation.

Neutron Radiography - WCNR-11   Materials Research Forum LLC
Materials Research Proceedings **15** (2020) 268-273   https://doi.org/10.21741/9781644900574-42

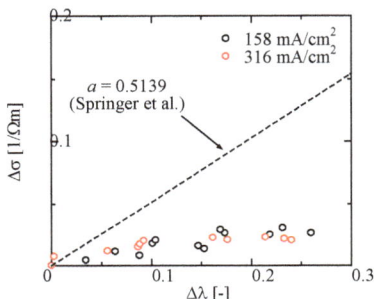

Fig. 4. *Relation between water content in MEA and the ionic conductivity.*

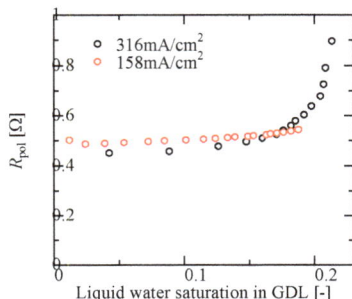

Fig. 5. *Relation between liquid saturation in GDL and the reaction resistance.*

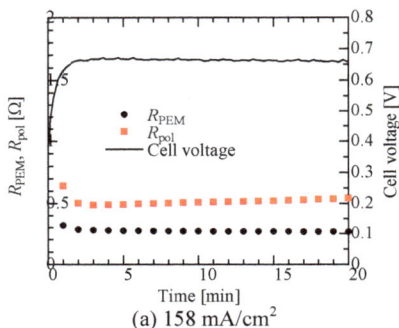

(a) 158 mA/cm^2

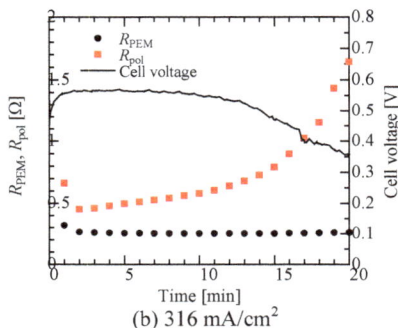

(b) 316 mA/cm^2

Fig. 6. *Relation between the cell voltage and the resistance.*

Fig. 6 shows time series of the cell voltage, the polarization and the ionic resistances. The large periodical fluctuations in the cell voltage were caused by EIS. For an easier understanding, a moving average was applied to the original cell voltage data during the impedance measurements, resulting in a smooth change in the cell voltages, as shown in Fig. 6. The cell voltage rapidly recovers after start of the power generation and is approximately constant at i = 158 mA/cm^2. The polarization resistance increases slightly, and the ionic resistances decreases slightly, with the operation time. Alternatively, the cell voltage rapidly decreases due to an increase in the polarization resistance after 10 min at i = 316 mA/cm^2. The tendency of the ionic resistance at i = 316 mA/cm^2 is approximately same as at i = 158 mA/cm^2. Because of the power generation, water accumulates in the PEM and the GDL. Increase in water accumulation in the PEM leads to a slight decrease in the ionic resistance. However, difference in the ionic resistance was not significant at 158 mA/cm^2 and 316 mA/cm^2. As discussed in Fig. 4, the effect of water content on the ionic conductivity was not significant. Thus, water accumulation influenced the polarization resistance and affected the cell performances.

Summary
In this study, simultaneous measurements of water distribution and the electrical impedance were carried out by using neutron radiography and electro-chemical impedance spectroscopy. The relation between the water amount in the PEFC and the electro chemical characteristics was evaluated.

Ionic conductivity increased with an increase in the membrane water content. However, effects of the water content on the ionic conductivity was much less than proposed by Springer et al. Springer's model cannot be applied for $\lambda < 1$. Polarization resistance increased with an increasing in liquid water in the GDL. This effect was dominant when water saturation reached 0.2. This indicates a blockage in the gas supply due to flooding in the GDL and the channel significantly degraded power generation. The polarization resistance, consisting of activation and concentration polarization resistances had a dominant effect on the degradation of the power generation if water accumulation in the membrane and the GDL was high.

Acknowledgement
The authors acknowledge financial support from The Iwatani Naoji Foundation. This work has been carried out in part under the Visiting Researchers Program of Kyoto University Institute for Integrated Radiation and Nuclear Science.

References

[1] H. Li, Y. Tang, Z. Wang, Z. Shi, S. Wu, D. Song, J. Zhang, K. Fatih, J. Zhang, H. Wang, Z. Liu, R. Abouatallah and A. Mazza, A review of water flooding issues in the proton exchange membrane fuel cell. J. Power Sources 178 (2008) 103-117. https://doi.org/10.1016/j.jpowsour.2007.12.068

[2] H. Murakawa, K. Sugimoto, N. Kitamura, H. Asano, N. Takenaka, Y. Saito, Visualization and Measurement of Water Distribution in Through-Plane Direction of Polymer Electrolyte Fuel Cell during Start-Up by Using Neutron Radiography, J. Flow Control, Measurement & Visualization 3 (2015) 122-133. https://doi.org/10.4236/jfcmv.2015.33012

[3] E. Antolini, L. Giorgi, A. Pozio and E. Passalacqua, Influence of Nafion loading in the catalyst layer of gas-diffusion, J. Power Sources 77 (1999) 136-142. https://doi.org/10.1016/S0378-7753(98)00186-4

[4] Y. Saito, S. Sekimoto, M. Hino and Y. Kawabata, Development of neutron radiography facility for boiling two-phase flow experiment in Kyoto University Research Reactor, Nucl. Inst. and Methods in Physics Research A 651, (2011) 36–41. https://doi.org/10.1016/j.nima.2011.01.103

[5] X. Changjun, Q. Shuhai, Drawing impedance spectroscopy for fuel cell by EIS, Procedia Environmental Sciences 11 (2011) 589-596. https://doi.org/10.1016/j.proenv.2011.12.092

[6] T.E. Springer, T.A. Zawodzinski, S. Gottesfeld, Polymer electrolyte fuel cell model, J. Electrochem. Soc. 138 (1991) 2334-2342. https://doi.org/10.1149/1.2085971

[7] T.J. Mason, J. Millichamp, T.P. Neville, P.R. Shearing, S. Simons and D.J.I. Brett, A study of the effect of water management and electrode flooding on the dimensional change of polymer electrolyte fuel cells, Journal of Power Sources 242 (2013) 70-77. https://doi.org/10.1016/j.jpowsour.2013.05.045

Neutron Radiography - WCNR-11
Materials Research Proceedings **15** (2020) 274-280

Materials Research Forum LLC
https://doi.org/10.21741/9781644900574-43

Visualization and Measurement of Boiling Flow Behaviors in Parallel Mini-channel Heat Exchanger by Neutron Radiography

Hitoshi Asano[1,a] *, Hideki Murakawa[1,b], Ryosuke Moriyasu[1,c],
Katsumi Sugimoto[1,d], Yohei Kubo[2], Kazuhisa Fukutani[2], Daisuke Ito[3,e]
and Yasushi Saito[3,f]

[1]Department of Mechanical Engineering, Graduate School of Engineering, Kobe University
1-1 Rokkodai, Nada-ku, Kobe 657-8501, Japan

[2]Mechanical Engineering Research Laboratory, Kobe Steel, Ltd.
1-5-5, Takatsukadai, Nishi-ku, Kobe 651-2271, Japan

[3]Institute for Integrated Radiation and Nuclear Science, Kyoto University
2 Asashiro-Nishi, Kumatori-cho, Sennan-gun, Osaka 590-0494, Japan

[a]asano@mech.kobe-u.ac.jp, [b]murakawa@mech.kobe-u.ac.jp, [c]189t373t@stu.kobe-u.ac.jp,
[d]sugimoto@mech.kobe-u.ac.jp, [e]itod@rri.kyoto-u.ac.jp, [f]ysaito@rri.kyoto-u.ac.jp

Keywords: Compact Heat Exchanger, Parallel Channel, Evaporator, Void Fraction, HFC134a, Neutron Radiography

Abstract. Boiling two-phase flows in cross-flow type mini-channel evaporator were visualized by neutron radiography. Refrigerant vertically upward flow of HFC134a $[CH_2FCF_3]$ in 21 parallel channels with the hydraulic diameter of 1.47 mm was heated by the heating medium of fluorocarbon FC3283. Void fraction distributions of evaporating two-phase flows were measured by neutron radiography. The effect of the inlet orifices for each refrigerant channel to prevent flow instability was evaluated. The refrigerant was supplied to the test section as subcooled liquid with the mass flux of 50 to 100 kg/(m^{2}s). The heating medium was supplied with the mass flux of 460 and 920 kg/(m^{2}s) and at the inlet temperature of 47.1 to 67.2 °C. As the result, it is shown that void fraction close to the inlet of heating medium was relatively higher. Flow instability of boiling flows and the effect of the inlet orifices were not observed.

Introduction

Compactness and decrease in temperature difference between two fluids are still important issues in developing heat exchangers. Especially, decrease in the temperature difference is required for the improvement in performance of energy systems, such as waste heat recovery, refrigerating and air-conditioning systems. One approach is the increase in heat transfer area density. On the other hand, for refrigeration and air-conditioning machines, reduction of the refrigerant charging amount is strongly required to satisfy the Kigali agreement of Montreal Protocol, because popular refrigerants have high global warming potential (GWP). The reduction in the channel diameter is effective way for the both requirements.

Microchannel compact heat exchanger manufactured by diffusion bonding process is developed. Smaller refrigerant channel diameter leads to larger pressure loss of refrigerant flows. Larger pressure loss requires larger pumping power. For an evaporator and condenser, the larger pressure loss causes temperature decrease, because the refrigerant is under the saturation condition. Therefore, microchannel heat exchanger has many parallel channels, and refrigerant flow is distributed into the parallel channels to decrease the mass flow rate through each channel. For a liquid or gas single-phase flow application, uniform flow distribution among channels is easily obtained by appropriate design of the header for flow distribution. However, in the case where the refrigerant flows with evaporation or condensation, maldistribution often occurs, regardless of the inlet flow condition. The maldistribution causes deterioration in the heat transfer performance.

Such parallel microchannel systems are investigated for the application of the cooling device of electro equipment. Huang and Thome [1] conducted boiling flow experiments in 67 parallel microchannel cooling equipment using three types of refrigerant. The channel hydraulic diameter was 0.1 mm. As a result; it was shown that boiling flows could be stabilized by an inlet orifice for each channel, and pressure loss through the straight section except of the inlet orifice decreased. Dário, et al. [2] experimentally evaluated the pressure drop of HFC-134a boiling flows through 9 parallel microchannels with the hydraulic diameter of 0.77 mm. Frictional and acceleration contributions to the pressure drop were analyzed. In these studies, parallel channels were symmetrically heated with constant heat flux. Kuroki, et al. [3] conducted HFC-32 boiling flow experiments in 2 parallel channels with the diameter of 1.0 mm. The effect of non-uniformly heating on heat transfer coefficient was evaluated. It was reported that under the non-uniformly heating condition, the heat transfer coefficient of the flow with higher heating load became lower, and overall heat transfer coefficient of the parallel channels decreased by non-uniform heating. The deterioration in heat transfer coefficient might be caused by refrigerant maldistribution.

In this study, boiling two-phase flows in cross-flow type mini-channel evaporator manufactured by diffusion bonding were visualized by neutron radiography. Liquid refrigerant was heated by a heating medium. The effects of mass flow rate of the refrigerant, mass flow rate and inlet temperature of the heating medium on heat transfer rate and void fraction distribution were evaluated.

Experimental setup and method
A schematic diagram of experimental apparatus is shown in Fig. 1. HFC-134a, whose chemical formula is CH_2FCF_3, was used as the refrigerant, and fluorocarbon FC-3283 was used as the heating medium. The attenuation of neutron is quite low for FC3283 because FC3283 does not include hydrogen. Subcooled liquid refrigerant was supplied to the test section through a pre-heater by a gear pump. The refrigerant inlet temperature was measured by a K-type sheathed thermocouple inserted in the channel at the downstream of a mixing section. A tank was connected to the upstream of the pump to absorb volume change due to boiling. Refrigerant pressures were measured at the inlet and outlet of the test section. On the other hand, the heating medium was supplied from a temperature controlled bath. Inlet and outlet temperatures were measured by inserted thermocouples. Mass flow rates of both fluids were measured by mass flow meter.

A schematic of tested heat exchanger is shown in Fig. 2. Refrigerant and heating medium channels were formed on a thin stainless steel sheet by chemical etching process, individually. Then, the sheets was stacked by diffusion bonding process. The cross-sectional shape of channels was semi-circular with the hydraulic diameter of 1.47 mm as shown in Fig. 2 (b). The same channel shape was applied to both fluids. The number of channels was 21 for the refrigerant and 20 for the heating medium. The channels were set in crossflow arrangement as shown in Fig. 2(a). The temperature of the heating medium would be higher at the right channel near the heating medium inlet. The test section was placed vertically to form upward boiling flows. Two types of refrigerant channel were used, namely, Type I with an inlet orifice for each channel and Type II without inlet orifice.

Neutron radiography experiments were carried out at B4 port of Kyoto University Research Reactor. The neutron beam was irradiated from the front face. The exposure time was set to 30

Neutron Radiography - WCNR-11
Materials Research Proceedings 15 (2020) 274-280

Materials Research Forum LLC
https://doi.org/10.21741/9781644900574-43

seconds. Since the width of the radiation field was narrower than that of the heat transfer area, radiograph was obtained for 5 sections by moving horizontally. Figure 3 shows an original visualized image of the center section. The pixel size was 87.9 μm/pixel. Void fractions, α, were measured from three radiographs of two-phase flows, the liquid single-phase flow and the vapor single-phase flow.

Assuming that the brightness of a visualized image is proportional to the beam intensity on a scintillation converter and neglecting the attenuation term due to a gas phase, the brightness of two-phase flow image $S_{TP}(x,y)$, the image of liquid single-phase flow $S_0(x,y)$, i.e., $\alpha(x,y)=0$, and the image of vapor single-phase flow $S_1(x,y)$, i.e., $\alpha(x,y)=1$, are expressed as the following equations [4].

$$S_{TP}(x, y) = G(x, y) \exp[- \rho_w \mu_{mw} t_w(x, y) - \{1 - \alpha(x, y)\} \rho_L \mu_{mL} t_c(x, y)] + O_{TP}(x, y) \quad (1)$$

$$S_0(x, y) = G(x, y) \exp[- \rho_w \mu_{mw} t_w(x, y) - \rho_L \mu_{mL} t_c(x, y)] + O_0(x, y) \quad (2)$$

$$S_1(x, y) = G(x, y) \exp[- \rho_w \mu_{mw} t_w(x, y)] + O_1(x, y) \quad (3)$$

where ρ and μ_m are the density and mass attenuation coefficient, respectively. t_w and t_c is thickness of wall and channel along a neutron beam, respectively. $G(x, y)$ is the gain and depends on the position due to a non-flatness of the initial beam intensity and of the sensitivity in a imaging system. $O_{TP}(x, y)$, $O_0(x, y)$, $O_1(x, y)$ are the offset value in brightness. In this study, the dark current $O(x, y)$ was used as the offset values. Using equations (1) to (3), a two-dimensional void fraction distribution can be expressed as

$$\alpha(x, y) = \frac{\ln[(S_{TP}(x, y) - O(x, y))/(S_0(x, y) - O(x, y))]}{\ln[(S_1(x, y) - O(x, y))/(S_0(x, y) - O(x, y))]} \quad (4)$$

Experimental conditions are shown in Table 1. Experiments were carried out for varied refrigerant mass flux of 50 and 100 kg/(m²s), heating medium mass flux of 460 and 920 kg/(m²s), and heating medium inlet temperature of 29.0 to 67.2 °C.

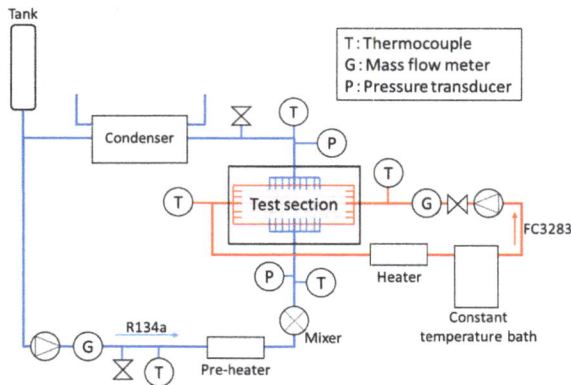

Fig. 1. *Schematic diagram of experimental apparatus.*

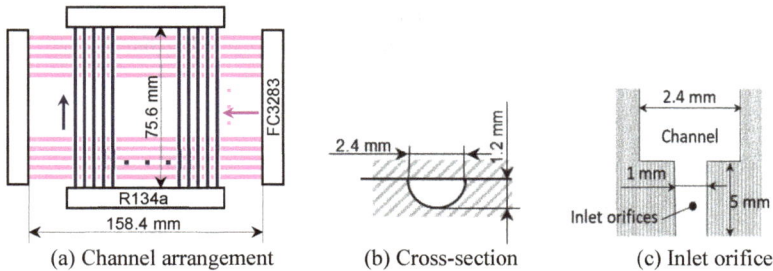

| (a) Channel arrangement | (b) Cross-section | (c) Inlet orifice |

Fig. 2. Test section.

Table 1. Experimental condition.

		Type I (with inlet orifices)	Type II
Refrigerant R134a	Mass flow rate (Mass flux)	142.5, 285 g/min (50, 100 kg/(m²s))	
	Inlet subcooling degree	3.14~4.66 K	0.60~2.93 K
	Inlet pressure (Saturation temperature)	0.632~0.659 MPa (23.3~24.7 °C)	0.570~0.619 MPa (19.9~22.6 °C)
Heating medium FC3283	Mass flow rate (Mass flux) Re for liquid flow	1.25, 2.5 kg/min (460, 920 kg/(m²s)) 450, 900	
	Inlet temperature	47.3~48.5 °C 65.1~67.2 °C	47.1~48.5 °C 64.7~67.2 °C

Experimental results and discussion
Void fraction distribution

Void fraction can be measured for each pixel. Spatial average void fractions were calculated in the area of 11 by 11 pixels. Eleven pixels were equivalent to 0.97 mm. Then, one-dimensional void fraction distributions were obtained by moving average to the flow direction for each channel. Void fraction distributions are shown in Figs. 4 by color scale. Figures 4 (a) and (b) show the results for Type I with inlet orifices and Type II without orifices, respectively. The heating medium was flowing from the right to left side.

Fig. 3. Original visualized image of the center section of the heat transfer

Void fraction became higher to the downstream due to boiling. The distributions were seemed to be continuous among channels. For two-phase flows in parallel channels, a maldistribution of refrigerant flow sometimes occurs and produces a channel whose mass flow rate and exit vapor quality is quite different form the other channels. Since the distribution was continuous, such flow instability might not occur in this system. Void fraction was higher the closer to the right and left edge. Mass flow rate might be higher the closer to the center channel, because the inlet and outlet ports were at the horizontal center of the header. Void fraction on the right side close to the heating medium inlet was higher than the left side. The difference might be caused by the difference in temperature difference between the refrigerant and heating medium. On the other hand, horizontal distributions of void fraction tend to become uniform for the higher mass flow rate of the heating medium, because the temperature change of the heating medium became smaller due to the higher heat capacity flow rate.

In comparison of the test sections, the void fraction of Type I was lower than Type II. The difference was caused by the difference in the operating pressure of refrigerant.

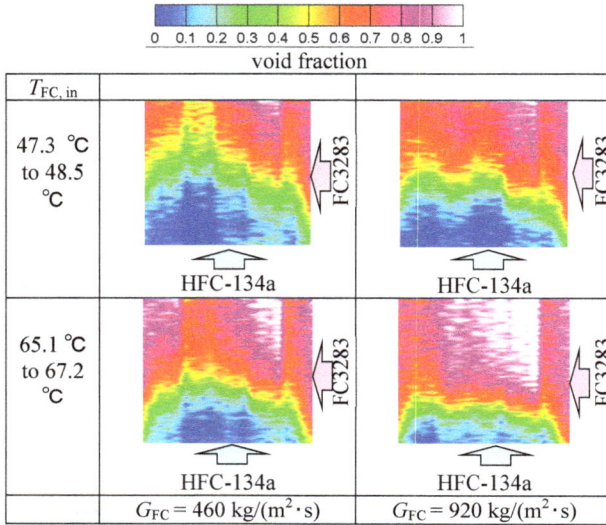

(a) Type I (with inlet orifices)

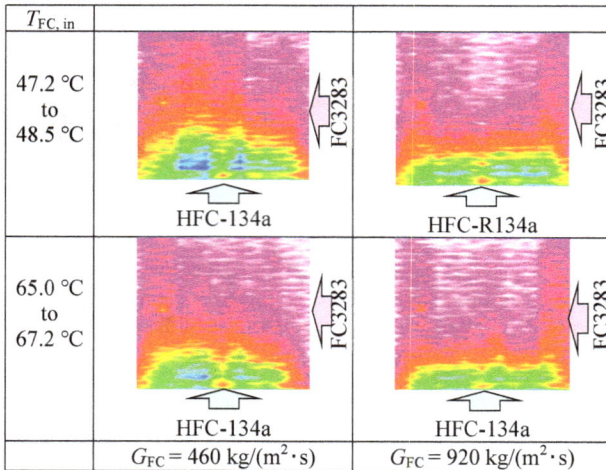

(b) Type II (without inlet orifice)

Fig. 4. *Void fraction distribution ($G_R = 100$ kg/($m^2 \cdot s$))*

Heat transfer rate

Heat transfer rate was calculated from temperature change and heat capacity flow rate of the heating medium. Measured heat transfer rates are shown in Fig. 5. The measured heat transfer rate increased with increasing inlet temperature of the heating medium due to the increase in temperature difference of heat exchange. While the effect of mass flow rate of the refrigerant was little, the effect of mass flow rate of the heating medium was noticeable. Higher mass flow rate produced larger heat transfer rate. There are two possible reasons. One is the decrease in temperature change of the heating medium due to the increase in heat capacity flow rate. The decrease leads to the increase in temperature difference between fluids. The other is the improvement of heat transfer in heating medium. Total thermal resistance of two fluids can be expressed as the summation of thermal resistance in the refrigerant, the stainless steel wall, and the heating medium. Since the heat transfer coefficient of boiling flow is quite higher than that of liquid single-phase flow, total thermal resistance might be dominated by the thermal resistance in the heating medium flow.

As shown in void fraction distribution, heat transfer rate of Type II was larger than that of Type I with the inlet orifices. The difference might be caused by the difference in refrigerant pressure. In the experiments for Type II, flow instability was not observed. Since the inlet orifices are used to stabilize the boiling flows, the orifices might not have an effect on the heat transfer performance.

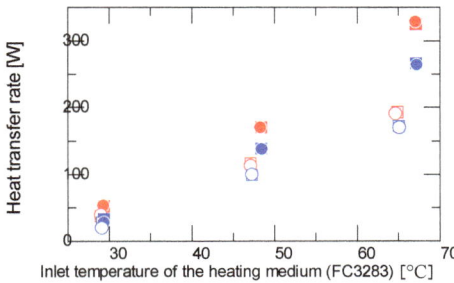

	mass flux R134a $[kg/(m^2 \cdot s)]$	mass flux FC3283 $[kg/(m^2 \cdot s)]$	
		460	920
Type I	50	○	●
	100	□	■
Type II	50	○	●
	100	□	■

Fig. 5. Heat transfer rate.

Summary

In this study, boiling two-phase flows of HFC-134a in 21 parallel mini-channels were visualized by neutron radiography, and void fraction distributions were measured from radiographs. The refrigerant was heated by fluolocarbon FC3283 in 20 parallel mini-channels. As a result, it was shown that void fraction was higher the closer to the inlet of the heating medium. Horizontal distributions of void fraction tend to become uniform by increasing mass flow rate of the heating medium. Flow instability was not confirmed from the visualized results.

Acknowledgement

This work has been carried out in part under the Visiting Researchers Program of Kyoto University Institute for Integrated Radiation and Nuclear Science.

Neutron Radiography - WCNR-11
Materials Research Proceedings **15** (2020) 274-280

Materials Research Forum LLC
https://doi.org/10.21741/9781644900574-43

References

[1] H. Huang, J. R. Thome, An Experimental Study on Flow Boiling Pressure Drop in Multimicrochannel, Experimental Thermal and Fluid Science, 80 (2017) 391–407. https://doi.org/10.1016/j.expthermflusci.2016.08.030

[2] E. R. Dário, J. C. Passos, M. L. Sánchez Simón, L. Tadrist, Pressure Drop during Flow Boiling inside Parallel Microchannels, International Journal of Refrigeration, 72 (2016) 111–123. https://doi.org/10.1016/j.ijrefrig.2016.08.002

[3] K. Kurose, K. Miyata, Y. Hamamoto, H. Mori, Characteristics of Flow Boiling Heat Transfer in Non-Uniformly Heated Parallel Mini-Channels, Trans. of the JSRAE, 35(2) (2018), 101–108. https://doi.org/10.1615/TFEC2018.che.021663

[4] N. Takenaka, H. Asano, T. Fujii, M. Matsubayashi, "A Method for Quantitative Measurement by Thermal Neutron Radiography", Nondestructive Testing and Evaluation, 16 (2001), 345–354. https://doi.org/10.1080/10589750108953089

Neutron Radiography - WCNR-11
Materials Research Proceedings **15** (2020) 281-286

Materials Research Forum LLC
https://doi.org/10.21741/9781644900574-44

3D Velocity Vector Measurements in a Liquid-metal using Unsharpness in Neutron Transmission Images

Yasushi Saito[1,a][*], and Daisuke Ito[1,b]

[1]Institute for Integrated Radiation and Nuclear Science, Kyoto University
2 Asashiro-Nishi, Kumatori-cho, Sennan-gun, Osaka 590-0494 Japan

[a]ysaito@rri.kyoto-u.ac.jp, [b]itod@rri.kyoto-u.ac.jp

Keywords: Liquid Metal Flow, Image Unsharpness, Velocimetry, Neutron Radiography

Abstract. To measure the two/three-dimensional velocity vectors in a liquid metal flow, a new neutron imaging method was applied. By using a conventional transmission neutron imaging (neutron radiography) only two-dimensional information of tracer particles can be obtained. In this study, three-dimensional position was measured by analyzing the image unsharpness (blur) of tracer particles depending on the distance between the particle and the imaging screen in the neutron beam direction. Neutron imaging has been performed at the Kyoto University Research Reactor and the variation of image unsharpness of particles was investigated.

Introduction

Detailed modelling of liquid-metal two-phase flow is needed not only for safety analysis of liquid-metal fast breeder reactors (LMFBRs) [1] but also for development of accelerator-driven system (ADS) that makes use of lead bismuth as a spallation target and coolant. Liquid-metal two-phase flow has a larger liquid-to-gas density ratio and a higher surface tension in comparison with those of ordinary two-phase flows such as air-water flow. In order to predict the flow behavior of a gas-molten metal mixture in a pool precisely, it is essential to examine the applicability of the existing model to the gas-molten metal mixture pool, and if necessary, to propose suitable constitutive relations in the momentum exchange between phases. From this point of view, present authors [2,3] performed study on the flow characteristics of nitrogen gas-molten lead bismuth eutectic (LBE) mixture in a rectangular pool and measured the void fraction distribution and time averaged liquid velocity field using neutron imaging and particle image velocimetry techniques [4].

However, only two-dimensional positions of tracer particles can be obtained from the conventional neutron imaging method. Three-dimensional position of particles can be measured by using more than two neutron beams. The purpose of this study is to measure the distance between tracer particles and the imaging screen (scintillator) by analyzing the image unsharpness of the neutron transmission imaging.

Experimental setup and method

Figure 1 shows the schematic of experimental setup. The test section consists of a Newton alloy (Pb:50%, Bi: 25%, Sn:25%) rod and a rotating table. Four cadmium particles of 2mm in diameter were cast in the Newton alloy rod with 18 mm in diameter. The imaging system consists of a scintillator screen (LiF/ZnS:Ag), mirrors, and a CCD camera (Princeton Instruments, PIXIS 1024B). Experiments were performed at the B-4 port of Kyoto University Research Reactor. Figure 2 shows the neutron image of the test section and the cadmium particles, where the coordinate system x, y, z and rotating angle θ of the particle are also shown. Neutron imaging was performed with rotating step of 6 deg. And the exposure time was 10 s.

Neutron Radiography - WCNR-11
Materials Research Proceedings **15** (2020) 281-286

Materials Research Forum LLC
https://doi.org/10.21741/9781644900574-44

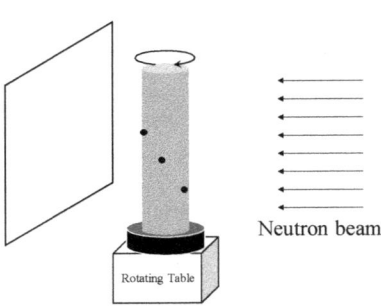

***Fig. 1** Experimental setup.*

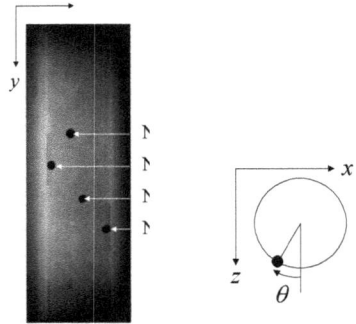

***Fig. 2.** Coordinate system and particle number.*

Evaluation of the image unsharpness

Two-dimensional position of tracer particle can be easily obtained from the neutron transmission images. Applying Particle Tracking Velocimetry (PTV) to the successive images of moving tracer particles, two-dimensional velocity vector can also be estimated. However, the remaining position, the distance between the particle and the scintillator screen, cannot be directly obtained from them. In this study, the image unsharpness information of tracer particles are analyzed to obtain the distance (z in Fig.2).

The divergent incident neutron beam generates geometric unsharpness in the neutron transmission images as shown in Fig.3. The image unsharpness U_g due to the divergence of the incident beam is simply expressed by

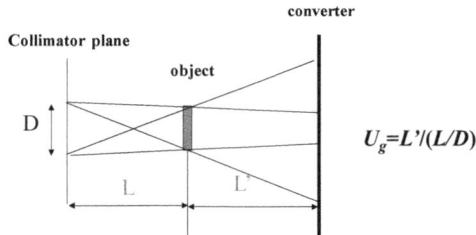

***Fig. 3** Geometric unsharpness arising from divergent neutron beam.*

$$U_g = L'/(L/D)$$ (1)

where L' is the distance between the object and the scintillator screen. L and D denote the length and the inlet aperture diameter of the collimator, respectively. Thus, the image unsharpness includes the information of the distance between the object and the screen. As it is known, the measured profile can be expressed by the following convolution [4]:

$$G'(x, y) = G(x, y) * h(x, y)$$ (2)

Neutron Radiography - WCNR-11   Materials Research Forum LLC
Materials Research Proceedings **15** (2020) 281-286   https://doi.org/10.21741/9781644900574-44

where $G'(x,y)$ is the measured profile of the blurred image, $G(x,y)$ the original profile, and $h(x,y)$ a point spread function. The point spread function can be modeled by Gaussian form:

$$h(x, y) = 1/\sqrt{2\pi\sigma^2} \exp\left(-\left(x^2 + y\right)/2\sigma^2\right) \tag{3}$$

where σ is the standard deviation of the Gaussians as shown in Fig.4.

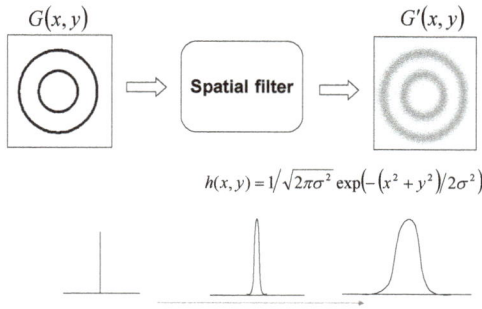

Fig. 4 *Mathematical expression of image blur due to beam divergence.*

Thus, if the original image and the blurred image is known, the standard deviation in the point spread function can be calculated. However, the shape of the tracer particles employed in this study was not perfectly spherical. Therefore, the binarized image was calculated and the convolution in Eq.(2) was conducted with point spread function Eq.(3) to obtain the optimum value of standard deviation of the point spread function as shown in Fig.5. Then the standard deviation can be converted to the image unsharpness U_g or the distance between the object and the screen, L 'or z. Once x, y, and z of each tracer particle are known, three-dimensional velocity vector of tracer particles can be measured using PTV method.

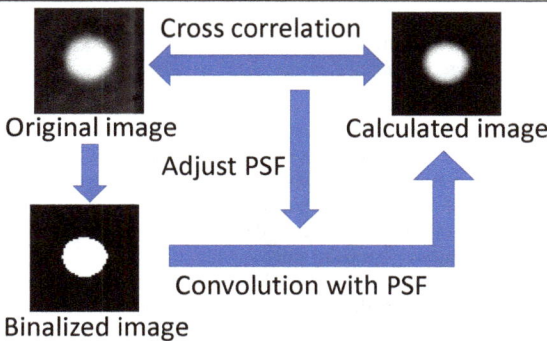

Fig. 5 *Mathematical expression of image blurr due to beam divergence.*

Neutron Radiography - WCNR-11   Materials Research Forum LLC
Materials Research Proceedings 15 (2020) 281-286   https://doi.org/10.21741/9781644900574-44

Results and Discussions

Figure 6 shows the two-dimensional position of Cd particles obtained from the transmission images for each Cd particle (No.1~4). Position of Cd particle was calculated by using a cross correlation method of tracer image. Since the Newton alloy rod was rotated around the vertical axis, the y-position of each tracer particle was kept constant during the rotation. Then, the image unsharpness of the tracer images was calculated for each particle, as shown in Fig. 7. As shown in these figures, the image unsharpness, the standard deviation, corresponds to the rotation angle for each Cd particle.

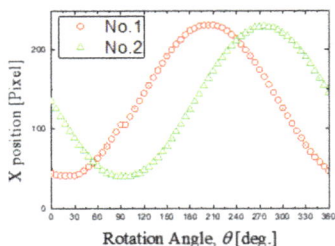

(a) Results for No.1 and No.2 particle   (a) Results for No.3 and No.4 particle

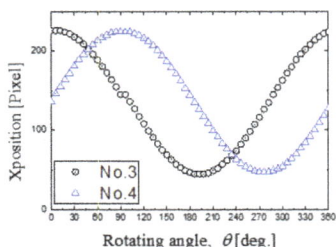

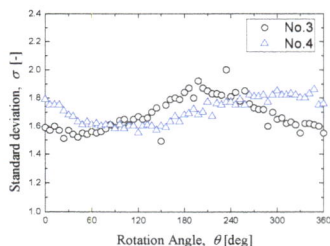

Fig. 6. *Variation of x-position with rotating angle.*

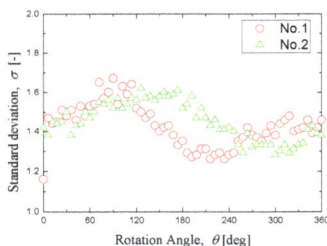

(a) Results for No.1 and No.2 particle   (a) Results for No.3 and No.4 particle

Fig. 7. *Variation of image unsharpness with rotating angle.*

Figure 8 shows the variation of image unsharpness with x-position of particles. As can be seen, variation of the standard deviation shows two tendency. When the particle stays near the scintillator, the standard deviation, image unsharpness, shows smaller value. In contrast, when the particle stays at the opposite side, the standard deviation takes larger value even at the same x-position. From such tendency, we can roughly detect the position of particle, which side the particle stays, near the scintillator or opposite side.

If the image unsharpness is caused only by the beam divergence, the experimental results should show a circle shape, however, the shape of trajectory in Fig. 8 is clearly distorted. Such distortion may be caused by the additional image unsharpness due to the imaging system itself such as lens aberration, image distortion through the image intensifier and so on. Therefore, z distance can be estimated by modifying the image unsharpness information by taking the distortion due to the imaging system. The solid lines in Fig.7 show the average tendency of the

Neutron Radiography - WCNR-11   Materials Research Forum LLC
Materials Research Proceedings **15** (2020) 281-286   https://doi.org/10.21741/9781644900574-44

image unsharpness along the x direction, which may denote the image unsharpness due to the image unsharpness. If the image unsharpness due to the imaging system can be assumed to be constant depending on the location in the image plane, z position of each Cd particle can be correlated to the imaging system.

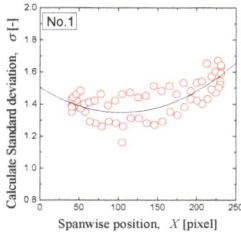

(a) Results for No.1   (b) Results for No.2

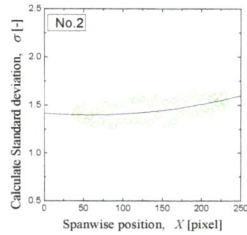

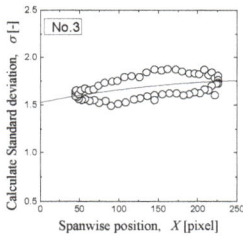

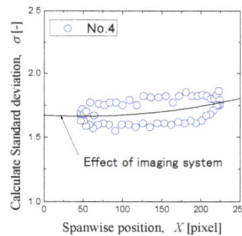

(c) Results for No.3   (d) Results for No. 4

Fig. 8. *Variation of image unsharpness with x-position.*

Figure 8 shows the comparison of calculated z position from the image unsharpness and that measured from the x position of the particles. As shown in this figure, good agreement can be achieved between calculated and measured z position.

Summary

To develop three-dimensional velocity vector in a liquid-metal flows, neutron imaging was applied and the image unsharpness was estimated by changing the distance between tracer particles and the imaging screen. From the preliminary experiments, the variation of image unsharpness was clearly observed depending on the distance. However, to estimate the z position of particles accurately, the other origin of image unsharpness due the imaging system should be taken into account. In addition, image blur due to the moving object should be considered to establish the 3D measurements of tracer particles in the actual applications.

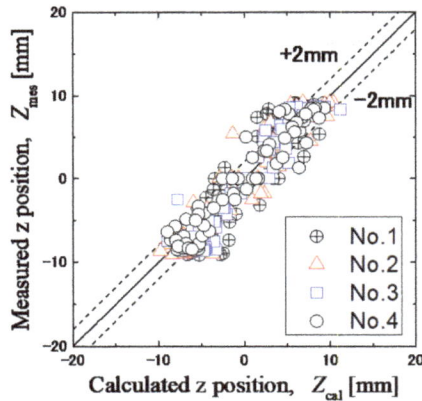

Fig. 9. *Variation of image unsharpness with x-position.*

References

[1] Kondo S., Brear D.J., Tobita Y., Morita K., Maschek W., Coste P., and Wilhelm D., Status and Achievement of Assessment Program for SIMMER-III, A Multicomponent Code for LMFR Safety Analysis, Proc. 8th International Topical Mtg. on Nuclear Reactor Thermal-Hydraulics (NURETH-8), 3, Kyoto, Japan, 1997.

[2] Mishima K., Hibiki T., Saito Y., Tobita Y., Konishi K., and Matsubayashi M., Visualization and measurement of gas-liquid metal two-phase flow with large density difference using thermal neutrons as microscopic probes, Nucl. Instrum. Methods Phys. Res. A, 424, pp.229-234, 1999. https://doi.org/10.1016/S0168-9002(98)01300-X

[3] Saito Y., Morimoto T., and Mishima K., Development of bubble measurements by using 4-sensor probe, Japanese J. Multiphase Flow, 24(5), pp.673-680, 2011.(in Japanese). https://doi.org/10.3811/jjmf.24.673

[4] Saito Y., Mishima K., Tobita Y., Suzuki T., Matsubayashi M., Lim I.C., and Cha J.E., Application of high frame-rate neutron radiography to liquid-metal two-phase flow research, Nucl. Instrum. Methods Phys. Res. A, 542, pp.168-174, 2005. https://doi.org/10.1016/j.nima.2005.01.095

Neutron Radiography - WCNR-11
Materials Research Proceedings **15** (2020) 287-291

Materials Research Forum LLC
https://doi.org/10.21741/9781644900574-45

Investigation of SINQ (Lead/Zircaloy) Spallation Target Structures by Means of Neutron Imaging Techniques

M. Wohlmuter[2], S. Dementjevs[2], P. Vontobel[1], J. Hovind[1], P. Trtik[1], E.H. Lehmann[1,a,*]

[1]Laboratory for Neutron Scattering & Imaging, Paul Scherrer Institut, CH-5232 Villigen PSI

[2]Large Research Facilities Division, Paul Scherrer Institut, CH-5232 Villigen PSI

[a]eberhard.lehmann@psi.ch

*corresponding author

Keywords: Spallation Neutron Source, Target Technology, Melting Point Lead, Neutron Imaging, Gamma Dose Rate, Dy Converter Technique, Imaging Plates, NEURAP

Abstract. We report here on neutron imaging material studies of lead/zircaloy spallation target structures that are used to operate the Swiss Spallation Neutron Source (SINQ) facility at Paul Scherrer Institut (PSI). Lead provides a relatively high yield when exposed by the high intense proton beam delivered by the 590 MeV cyclotron [1]. With the melting point of 327.5°C, this target material gets liquid inside the Zircaloy cladding during the proton beam exposure of about 1 MW thermal power. Short beam interruptions and other shut-down phases result in the solidification of the lead structures. In order to investigate the long-term behavior of this relevant material during proton exposure methods of neutron imaging have been used: (i) on-line studies in thermal cycling time sequences during external heating of dummy rods and (ii) inspection of "spent" rods after dismounting from the rod bundle. While the first type of studies can be performed in a normal neutron imaging setup, the second kind of investigations needs a well-shielded configuration (NEURAP) together with an imaging method, insensitive to the accompanying gamma background from the highly activated rod samples.

Introduction

The Paul Scherrer Institut (PSI, Switzerland) successfully operates the spallation neutron source SINQ [2] since 1997. It is the most powerful national neutron source in Switzerland and the basis for neutron research, also accessible by the international user community via semiannual open calls for scientific proposals.

Lead was found the most useful material for the spallation process with its high neutron yield and the low neutron absorption cross section. The SINQ targets consists today of Pb filled Zircaloy-2 tubes of about 10 mm diameter and 20 cm length arranged in a hexagonal rod bundle, placed perpendicular to the proton beam direction. Colloquially, the Pb-filled Zircaloy-2 tubes are called "Cannellonis".

The "Cannellonis" bundle is cooled via a forced cross flow of heavy water (D_2O). At the highest power level, the target operation has to allow for local melting of the Pb inside the Zicaloy-2 cladding rod. In cases of proton beam trips or other shutdown of the heating power the Pb freezes and starts shrinking immediately. For these cases of this thermal cycling a reduced filling of the Cannelloni's by 10% is foreseen during the configuration process.

In the last years, SINQ has been operated in competitive mode with another spallation target station – ultra cold neutron (UCN) source [3]; the full proton beam power is diverted to the UCN source for a few seconds only with a duty cycle of up to 3%. This short period of missing beam

Neutron Radiography - WCNR-11   Materials Research Forum LLC
Materials Research Proceedings **15** (2020) 287-291   https://doi.org/10.21741/9781644900574-45

power is enough to result in solidification of the molten Pb for the time of the beam kick to the UCN source.

The consequences for the target reliability have been studied in different manners. Next to thermo-dynamic simulations we performed two kinds of neutron imaging studies: in-situ studies for the melting/freezing process with heating/cooling elements and a mock-up target rod setup [2]; inspection of samples from used targets with the NEURAP technique [4]. Neutron imaging is a very useful approach since the target materials (Pb, Zr) are very transparent whereas common X-ray studies would fail due to the limited transmission. This report describes the applied techniques and shows preliminary results of selected examples.

The spallation neutron source SINQ and its target technology

Based on the already existing high intense proton accelerator at PSI, the spallation neutron source SINQ (Fig. 1) went into operation in 1997. Protons with 590 MeV and a current of up to 2 mA are sent to the target where the conversion to neutrons happens by the spallation process in the lead target. Because the heating goes beyond the melting point of lead, the Zircaloy cladding provides the real barrier during high power operation.

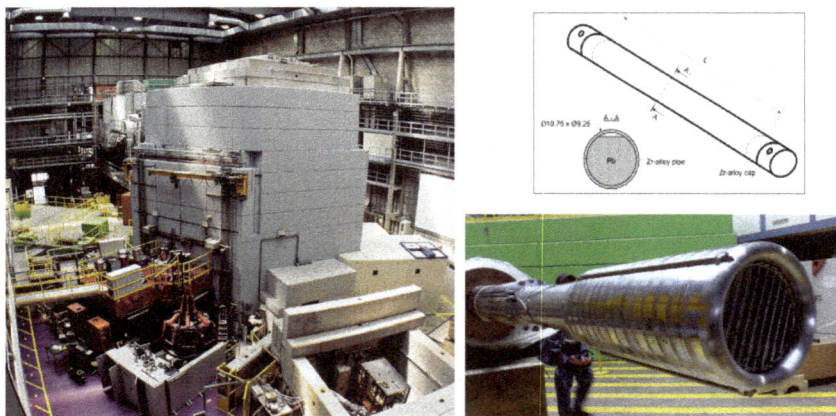

Fig. 1:   *The spallation neutron source SINQ at PSI (left) and its lead rod target (right); the diameter is approximately 25 cm only, active length 60 cm. It is positioned vertically in the source center while the proton beam is injected from the bottom*

In-situ study of the melting/solidification process (inactive) of target rods

Standard target rods were used for a study of the lead distribution during cycling above and below the melting point by means of an external heating device using a hot air according to a thermal cycling regime given in Fig. 2. At the same time, neutron images were taken with a common setup with a digital neutron imaging detector (scintillator-camera combination) at the NEUTRA beamline [5]. In order to tolerate the volume expansion during melting, the initial lead filling of Zircaloy cladding tubes was about 50%.

Some characteristic results are shown in Fig. 3: the sample before cycling and after 357 and 658 cycles, respectively.

Neutron Radiography - WCNR-11   Materials Research Forum LLC
Materials Research Proceedings **15** (2020) 287-291   https://doi.org/10.21741/9781644900574-45

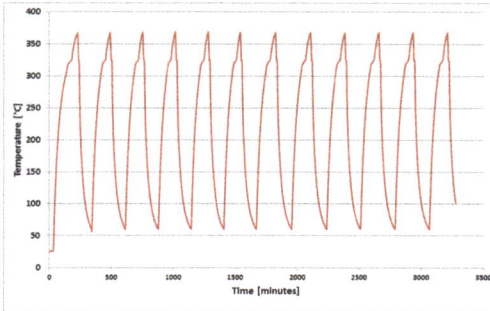

Fig. 2: *Temperature evolution at the central point of the target rod shown in Fig. 1 on the right side.*

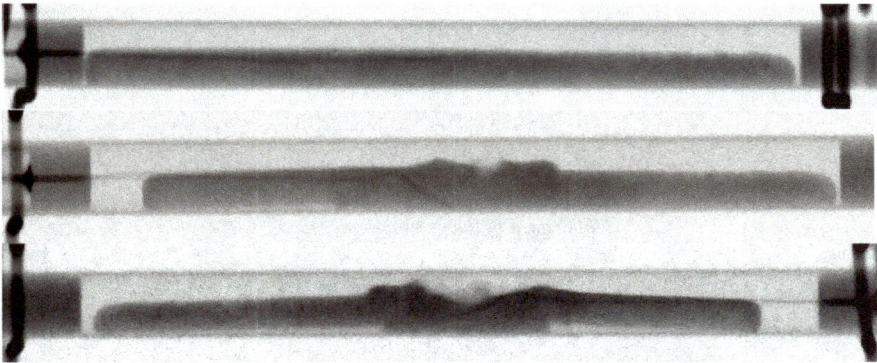

Fig. 3: *Dummy test rods with 50% lead filling in the virgin state (top); the same sample after 357 and 658 thermal cycles, respectively. A thermocouple is used to determine the temperature very precisely (see the data in Fig. 1)*

As images shown in Figure 3, prove the lead distribution within Zircaloy cladding tube changes without any mechanical loading. This is particularly apparent in the central part where melting and freezing takes place all time during the cycling. The external parts of lead stay solid, but a certain amount is redistributed towards the center and increases the lead amount there. Because the filling of the investigated rods was only 50% in volume no risk for cladding damage occurred. However, the real filling in the target rods is 90% - and some cladding interaction might happen.

Therefore, further, more realistic tests are currently under preparation.

Inspection of hot target rods after «normal» long-term proton exposure (active samples)
The SINQ target rods are designed for a long-term operation of about 2 years with proton exposure of about 1.5 mA, corresponding to an accumulated proton current of more than 6 Ah. However, there is no material study possible during the operation cycle before the target is exchanged, removed therefore and investigated. One of the non-destructive inspections has been routinely done with a special neutron imaging technique, based on avoidance of the gamma background from the highly activated target material. It is based on the activation of Dy inside

imaging plates and has to be used in a dedicated regime [6]. Furthermore, the target rod samples have to be extracted under hot-lab conditions and can be only transported to SINQ within a well shielded container. The same container is later be used to manipulate the sample in the beam and to position it towards the Dy-doped imaging plate.

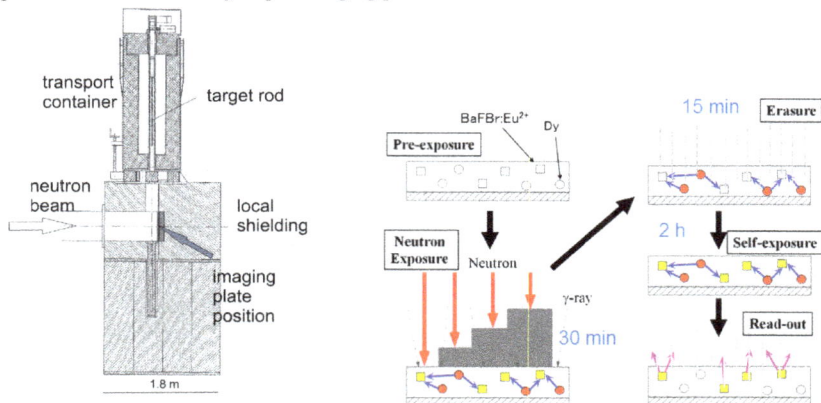

Fig. 4: *Setup for the shielded environment for the study of highly-activated samples (left); procedure for using Dy-doped imaging plates to study the neutron induced effects only – withdrawal of the gamma-background*

Fig. 5: *The "virgin" Pb rod (top) is quite transparent and uniform – a gap at the upper surface of the rod is visible due to the filling with only 90% of the volume; (bottom) a long-term exposed rod, containing a central thermal couple, is shown below. The neutron contrast is increased due to the accumulation of highly absorbingly spallation products, such as mercury.*

Investigation of target rods after damage (active)

A target rod failure occurred in SINQ-target No. 8 (operated during 12 month in 2009 and 2010, exposed to 12.15 Ah proton beam). One of the Zircaloy rods failed and lead was thus accidentally released into the coolant flow. The operation of SINQ had to be stopped and the reasons of this failure be assessed.

One of the applied methods was neutron imaging with the option of the suppression of gamma radiation by the Dy-doped imaging plates (as described above). The Figure 6 shows the results together with the photo of the damaged target rod with its long crack in the middle. The neutron imaging data clearly show remaining parts of the lead filling at the edges, fitting well at the right side, but moved to the center on the left side. Some residual material is found diluted in the central part. With the help of the digital information of the image data and the known geometry of the rod, it was possible to determine the missing part of lead in the damaged rod. Quantitatively, 38.2 volume per cents of the lead, corresponding to 30.2 g was missed in the damaged rod and was hence distributed into the coolant circuit.

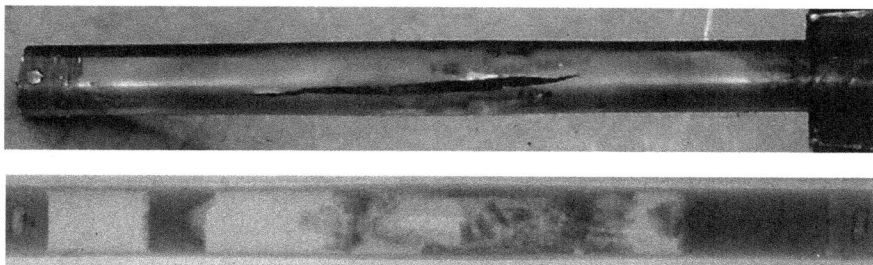

Fig. 6: *Photo and neutron image of a broken target rod with lead filling; some of the material is missing and lost by the gap in the cladding*

Fortunately, such target rod failures happened very seldom during the 20 years of SINQ operation. However, in order to avoid such target rod failures, it is important to study the material behavior BEFORE some damage happens. Next to the volume expansion of Pb, the hydrogen accumulation in the Zr cladding during proton exposure has to be seen as a risk because it is accompanied by the risk to embrittlement. Because hydrogen provides a high contrast in neutron imaging data, already small amounts can be defected efficiently.

Summary
The herewith described neutron imaging techniques were found very useful for non-destructive investigations of target structures containing Pb and Zr. Both materials are rather transparent for neutrons and very little material changes can be observed sensitively. In the future, the high sensitivity of neutron imaging w.r.t. hydrogen determination will be exploited by means of a high resolution setup, the neutron microscope [8], for the distribution analysis of hydrides.

References

[1] https://www.psi.ch/science/large-research-facilities

[2] https://www.psi.ch/sinq/sinq

[3] https://www.psi.ch/ucn/ucn-source

[4] https://www.psi.ch/sinq/neutra/sample-environment

[5] https://www.psi.ch/sinq/neutra/

[6] M. Tamaki et al., Dy-IP characterization and its application for experimental neutron radiography tests under realistic conditions, Nuclear Instruments and Methods in Physics Research A 542 (2005) 320–323. https://doi.org/10.1016/j.nima.2005.01.156

[7] W. Gong et al, J. of Nucl. Materials, 2018, submitted.

[8] P. Trtik et al., Improving the Spatial Resolution of Neutron Imaging at Paul Scherrer Institut The Neutron Microscope Project, Physics Procedia 69 (2015) 169 – 176, https://doi.org/10.1016/j.phpro.2015.07.024

Neutron Radiography - WCNR-11
Materials Research Proceedings **15** (2020) 292-298

Materials Research Forum LLC
https://doi.org/10.21741/9781644900574-46

Reactivation of the Transient Reactor Test (TREAT) Facility Neutron Radiography Program

Shawn R. Jensen[a,*], Aaron E. Craft[b], Glen C. Papaioannou[c],
Wyatt W. Empie[d], and Blaine R. Ward[e]

Idaho National Laboratory, Idaho Falls, ID 83415, United States of America

[a]Shawn.Jensen@inl.gov, [b]Aaron.Craft@inl.gov, [c]Glenn.Papaioannou@inl.gov,
[d]Wyatt.Empie@inl.gov, [e]Blaine.Ward@inl.gov

Keywords: Transient Reactor Test (TREAT) Facility, TREAT, Neutron Radiography

Abstract. The TREAT radiography system is used to perform neutron radiography of fuels, experiments, and other specimens before and after irradiation within the TREAT reactor. The TREAT neutron radiography facility performed approximately 5,000 radiographs by the spring of 1977. Originally built in 1958, the TREAT Facility was in operation until it was placed in a shutdown status in 1994. Following the Fukushima incident and seeing a need for enhanced accident tolerant fuels, the United Stated Department of Energy decided to restart the TREAT facility and resume transient operations. In November 2017, the TREAT reactor was successfully restarted and is currently performing operational testing in preparation for initial experiment irradiations and transient testing. This paper discusses efforts to reactivate the TREAT neutron radiography facility. To characterize the neutron beam, gold foil activation measurements were made to determine an average neutron flux and flux profile. An open beam image provides the information about variations in the beam profile. A series of system qualification radiographs have been acquired to determine the effective image acquisition parameters, resulting image quality, and the relationships between the two.

TREAT Neutron Radiography Program and Facility Description

Neutron radiography of irradiated nuclear fuel provides more comprehensive information about the internal condition of irradiated nuclear fuel than any other non-destructive technique to date. Neutron radiography has seen significant use for nuclear applications since the capability was first developed, and continues to prove its value still today [1]. The TREAT radiography system is used to perform neutron radiography of fuels, experiments, and other specimens before and after irradiation at the TREAT reactor [2]. Neutron radiographs are acquired using transfer method with dysprosium conversion foils and image plates [3], similar to the process used at INL's Neutron Radiography (NRAD) Reactor [4]. The system includes the radiography facility where radiography cassettes are exposed to the neutron beam and a Radiography Room where radiographs are subsequently processed. Figure 1 shows a schematic of the TREAT Neutron Radiography Facility. The neutron beam originates in the graphite reflector of the reactor. A collimator routes the neutron beam from the reflector to the radiography stand, and consists of inner and outer collimators and a beam shutter. A carriage system remotely positions a radiography cassette behind the sample position. Highly-radioactive objects can be moved to the radiography stand using a cask. The radiography stand contains radiation shielding materials to protect personnel from harmful amounts of radiation. A Radiography Room provides a location for processing the radiographs. Activated foils are coupled to an image plate in a special cassette, which are then placed in a shielded cabinet during the decay process. The radiography room includes environmental and dust control systems for a clean and consistent image processing environment.

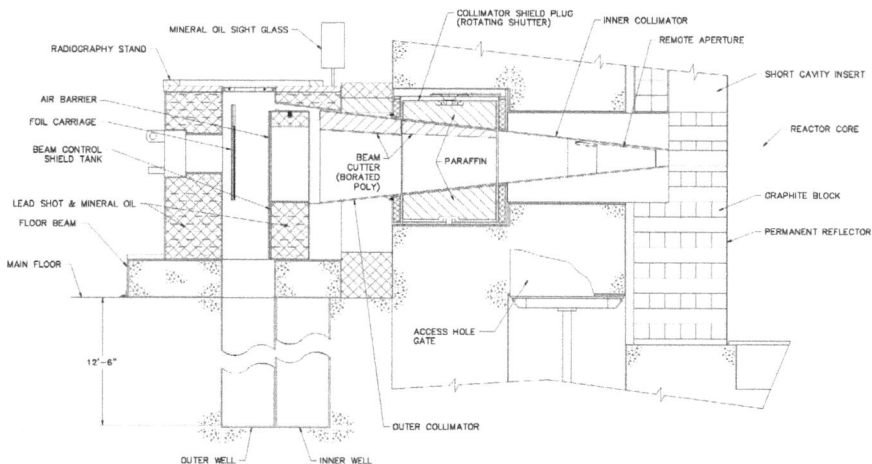

Figure 1. A side-view depiction of the TREAT neutron radiography facility.

Purpose of Neutron Radiography at TREAT

Neutron radiography at TREAT has three main purposes. The first purpose is for operational safety. Often, specimens are complex assemblies and/or are highly radioactive which makes looking inside of an experiment not a valid option. Numerous safety barriers and containments are designed into the experiments and must be intact prior to irradiation. Radiographs are taken to provide visual indication the specimen is constructed as designed and that the components remained in position during shipment, which ensure that the reactor and experiment remain safe in all conditions and testing situations.

The second purpose is for post-irradiation test. Transients are frequently designed to cause intentional failure of the specimen. What happened during a transient? Did the specimen remain intact? What is the condition of the containment vessel? The ability to quickly obtain a preliminary answer to those questions is important. This information can then be transferred to the experimenters and to the facility that will be performing subsequent detailed analyses and examinations.

The third purpose is related to non-irradiated items. The industry is seeing uses for neutron radiography in fields such as archeology, historical artifacts, biology, etc. Since neutron radiography is complimentary to x-ray radiography, additional information can be obtained that could not otherwise be seen in a traditional x-ray radiograph. Taking neutron radiographs of these types of items will most likely be a minor activity, but the capability is available.

History of the TREAT Neutron Radiography Program

Initial experiments using the TREAT reactor for neutron radiography were conducted in 1964 [5]. The original radiography facility was installed in the mid-1960s with the first production radiograph being generated in 1967. By 1977, over 5,000 radiographs had been acquired [6]. This system worked well but had some deficiencies, namely: insufficient structural support for larger handling vessels, poor alignment mechanism, manual operation of the shield door and foil carriage, and lack of precise specimen positioning. The system was upgraded in 1975-76 with a new stand designed to remedy the aforementioned deficiencies. The manual functions of opening/closing the shield door and inserting/removing the foil carriage was replaced with remotely-operated motorized actuators. The remote system substantially reduced the radiation dose to personnel working around the stand. The new stand also provided additional internal shielding to further reduce the exposure. In 1984 and in 1986 additional upgrades enhanced the load capacity of the stand and installed an aperture upgrade.

Restart and Status of the TREAT Neutron Radiography Facilities
System Readiness. Since the radiography facility had not been used for 20 plus years and normal maintenance was not performed during that time, it was essential to perform a detailed inspection of the system prior to operating. The system readiness plan included activities to establish configuration management, baseline the current condition of the equipment, ensure system operability, repair identified defects, identify and procure critical spare parts, and establish a preventative maintenance program. For components that could not be readily accessed, a borescope was used to inspect and observe the operating components. In spite of sitting idle for over 20 years and the overall age of the components, the radiography facility functioned remarkably well. Additional system enhancements have been installed over the past few years. The shielding inside the stand is composed of lead shot and mineral oil. Due to the original design of the stand, oil leakage has occurred. A new oil reclamation system was recently installed to capture any leaking oil and return it to the stand along with visual confirmation of the oil level. A picture of the radiography facility is shown in Figure 2. External shielding consisting of borated-poly and concrete has also been installed around the exterior of the radiography facility along with additional structural bracing.

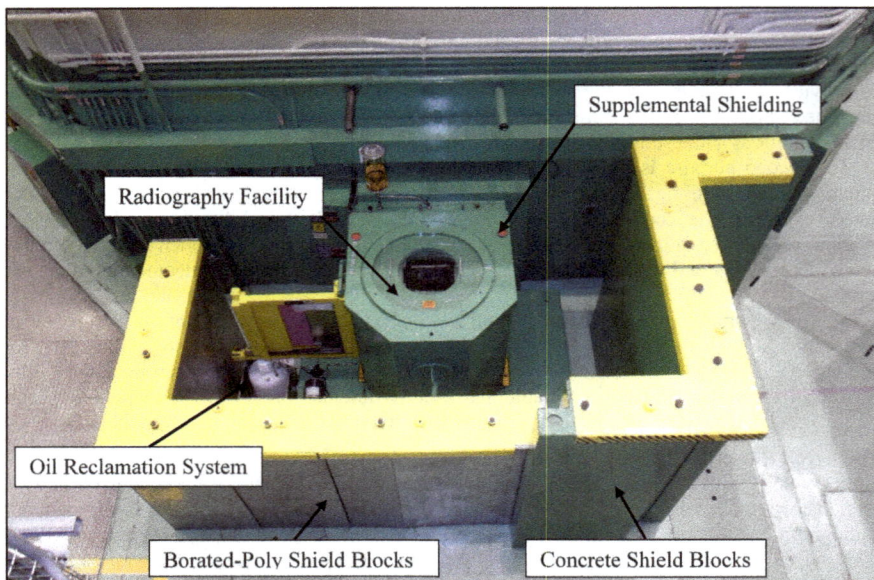

Figure 2. Picture of the TREAT neutron radiography facility layout.

New exposure cassettes were fabricated to provide consistent foil placement while still allowing for easy transfer of foils into decay cassettes. The original electro-mechanical exposure timer has also been replaced with a digital timer which has increased repeatability and accuracy for shot exposures. A neutron radiography room was constructed to provide a dust, climate, and lighting controlled workspace for equipment storage, radiography operations, image processing, and a shielded decay safe. A vacuum system for handling activated foils significantly reduces radiation dose to radiographers while transferring irradiated foils from exposure cassettes to decay cassettes.

Upgrade from Film to Computed Radiography
Traditionally, all radiographs taken at TREAT were based on film. This required substantial time for decay and film processing. Radiographs were not available for viewing until the following day. Film is becoming more difficult to procure and the chemicals have environmental concerns. TREAT has elected to produce radiographs via the CR process, which utilizes PSP image plates and a digital scanner [1]. Image plates are more sensitive than film, and thus require less exposure time than previous operations with film. CR uses no chemicals, can be much quicker than film, and can achieve similar resolution to film. CR images also have a linear dose response, compared to the S-curve for film, along with greater latitude, allowing for simpler image processing and interpretation. Another benefit of CR is that it directly produces a digital image.

Beam Characterization
The neutron beam flux was measured using foil activation methods [7]. An array of 21 gold foils was activated in the TREAT neutron beam for 3 hours with the reactor power of 80 kW. The resulting activity was measured using a calibrated high-purity germanium detector, then the activity at the end of exposure in the neutron beam was calculated based on this measurement, the decay constant, and the time between the measurement and end of exposure [7][8]. The measured average thermal equivalent neutron flux at the image plane is $8.25 \times 10^6 \pm 1.89 \times 10^5$ n/cm^{2}s with the reactor power at 80 kW. The neutron beam uniformity was measured using a neutron radiograph of the open beam with no sample. The resulting peak-to-average ratio is 1.006 in the horizontal direction and 1.011 in the vertical direction.

System Qualification
The purposes of TREAT radiography include pre-transient operational safety shots and post-transient detail shots. The shot types can vary based on the information requested and the image quality desired. As such, it was necessary to determine exposure times and decay times for the various shot types. The first shot type is an information-only quick-shot which has lower resolution but a very short turnaround time. Higher-quality radiographs for programmatic use need to be of higher image quality (resolution and signal-to-noise ratio, SNR) to provide greater detail and clarity. A system qualification plan was implemented to determine the required settings for each shot type. A newly designed Resolution Test Piece (RTP) was utilized as the specimen, containing an American Society for Testing and Materials (ASTM) Sensitivity Indicator and Beam Purity Indicator [9], a 2.5 cm square foil of gadolinium, and a 2.5 cm square foil of hafnium. These ASTM indicators and foils provided the necessary data for image resolution calculation to be performed.

Three radiography campaigns were performed using a range of exposure times and decay times to determine the optimum settings for each shot type. The first campaign varied exposure times to establish this value for further tests. The second campaign varied decay time to establish this value for programmatic-quality radiography. The third campaign also varied decay time, but for much shorter times, to establish the decay time needed for quick shots. All shots were taken

Neutron Radiography - WCNR-11
Materials Research Proceedings 15 (2020) 292-298

Materials Research Forum LLC
https://doi.org/10.21741/9781644900574-46

with the reactor operating at 80 kW steady state and using dysprosium converter foils. All image plates were scanned using a Carestream HPX-1 scanner.

The first campaign contained eleven radiographs. Seven radiographs were acquired with nominal exposure times of 20, 12, 10, 8, 6, 4, and 2 minutes. The foils were then placed on Carestream general purpose (GP) image plates in individual decay cassettes and left to decay overnight. Four additional radiographs were acquired with nominal exposure times of 20, 14, 12, and 10 minutes. The foils were then placed on Carestream high resolution (HR) image plates in individual decay cassettes and also left to decay overnight. The intent was to determine exposure times and scanner settings that produced the highest quality images based on signal (i.e. pixel value), SNR, and spatial resolution. The effective spatial resolution seems unaffected by exposure time. The data from the SNR and peak pixel values show diminishing returns beyond 10 minute exposure time, which is the basis for choosing 10 minutes as the exposure time for further tests.

A second set contained five radiographs with an exposure time of 10 minutes and varying decay times, including: 240, 180, 120, 60, and 30 minutes. Following exposure, the foils were placed on GP image plates in decay cassettes and left to decay. This campaign with longer decay times sought to identify the decay time required for high quality radiographs for programmatic use.

A third radiography campaign contained six radiographs with varying decay times, including: 60, 50, 40, 30, 20, and 10 minutes. This campaign was designed to identify parameters for information-only quick-shot radiographs. The exposure time was set at 10 minutes each based on the data obtained from the first set. A decay time of only 20 minutes provided 60% the SNR of a radiograph with 240 minute decay time, and thus was chosen as a reasonable decay time for information-only quick-shots. It is notable that the SNR for radiographs with 240 minute and overnight decay times are both 194, indicating that decay times beyond 240 minutes may not provide additional benefit for SNR. A decay time of 120 minutes provided 94% of the peak SNR, and thus was chosen as the decay time for high-quality radiography for programmatic use. The effective spatial resolution was also measured as a function of decay time and shows that spatial resolution is unaffected by decay time.

Results of Initial System Demonstration Shots
Based on the data, current settings for high-quality radiography for programmatic use are 10 minute exposure time and 120 minute decay time using a GP image plate. The HR plates provided higher SNR and effective spatial resolution, which may lead to their eventual use for radiography operations, but additional measurements are needed first to verify their performance. Figure 3 shows a neutron radiograph of an ASTM sensitivity indicator acquired using a GP plate (left) with 10 minute exposure time and overnight decay time, and a corresponding line profile (right). The smallest shim in the sensitivity indicator is visible in both the GP and HR radiographs.

Future of TREAT Neutron Radiography
As part of a 5-year improvement plan, upgrades are in the early design stages that will support enhanced capability for the TREAT radiography program. The primary focus is to design and install a new radiography stand and control panel to support installation of camera-based digital neutron imaging systems and subsequent development of neutron computed tomography (CT). CT involves taking multiple shots from various angles along the entire length of the specimen and then mathematically combining them into a reconstructed 3-dimensional image. Neutron CT is an active area of development at NRAD, where novel equipment and techniques are being

Neutron Radiography - WCNR-11   Materials Research Forum LLC
Materials Research Proceedings **15** (2020) 292-298   https://doi.org/10.21741/9781644900574-46

evaluated for neutron CT of highly-radioactive objects. This is an ongoing effort in collaboration with partners from universities, industry, and national and international research centers.

Another area for potential new research is flash neutron radiography, which would take advantage of the exceptionally bright neutron beam produced during transients. The peak transient power at TREAT is 19,000 MW, which produces a calculated neutron beam flux of 1.95×10^{12} n/cm^2s for ~100 ms, which is ~1,000 times brighter than the current brightest neutron beams in the world. The total energy available in a transient is limited to ~2,500 MJ, but the intensity and duration of transient are highly customizable with TREAT's advanced reactivity control system.

TREAT neutron radiography is prepared to accept and radiograph the newest generation of experiments and specimens. The first of such experiments is scheduled to begin irradiations in September 2018. As new experiments are designed, new radiography support equipment may be required to accommodate these experiments. The TREAT radiography program is poised to remain a valuable asset to the TREAT irradiation program for many years to come.

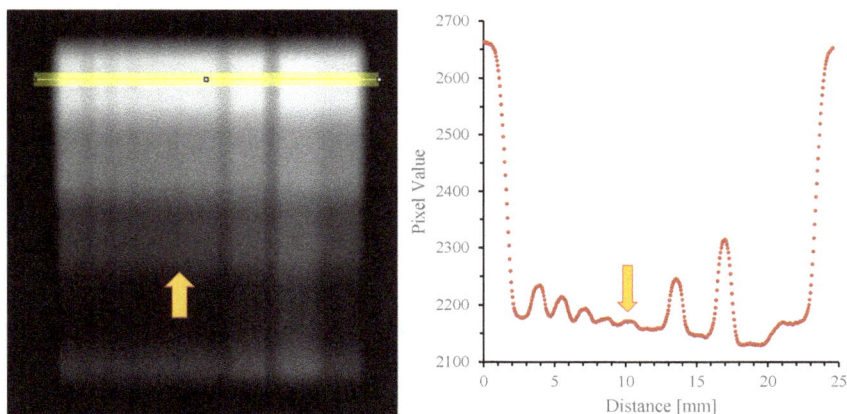

Figure 3. A neutron radiograph of an ASTM sensitivity indicator acquired using a GP plate (left) with 10 minute exposure time and overnight decay time, and a corresponding line profile (right).

References

[1] A.E. Craft and J.D. Barton, "Applications of neutron radiography for the nuclear power industry," Physics Procedia 88 (2017) 73-80. https://doi.org/10.1016/j.phpro.2017.06.009

[2] L.J. Harrison, *"TREAT neutron radiography facility,"* In: Barton J.P., von der Hardt P. (eds) Neutron Radiography, 251-256, 1983. https://doi.org/10.1007/978-94-009-7043-4_30

[3] A.E. Craft, G.C. Papaioannou, D.L. Chichester, and W.J. Williams, *"Conversion from film to image plates for transfer method neutron radiography of nuclear fuel,"* Physics Procedia 88 (2017b) 81-88. https://doi.org/10.1016/j.phpro.2017.06.010

[4] A.E. Craft, D.M. Wachs, M.A. Okuniewski, D.L. Chichester, W.J. Williams, G.C. Papaioannou, and A.T. Smolinski, *"Neutron radiography of irradiated nuclear fuel at Idaho National Laboratory,"* Physics Procedia 69 (2015) 483-490. https://doi.org/10.1016/j.phpro.2015.07.068

[5] D.C. Cutforth and J.F. Boland, *"Request for use of TREAT,"* INL 5163270, Apr.16,1964.

[6] L.J. Harrison, R.M. Conant and R.W. Mouring, *"Improvements in neutron radiography equipment at TREAT,"* 25th Conf. on Remote Systems Tech., Am. Nucl. Soc., 1977.

[7] A.E. Craft, B.A. Hilton, and G.C. Papaioannou, "Characterization of a neutron beam following reconfiguration of the Neutron Radiography Reactor (NRAD) core and addition of new fuel elements," Nuc. Eng. & Tech. 48 (2016) 200-210. https://doi.org/10.1016/j.net.2015.10.006

[8] R.L. Murri and D.G. Vasilik, "A Method to Determine Fast and Thermal Neutron Fluxes by Foil Activation Analysis," DOW Chem. Co., Rocky Flats Division, RFP-1466, 1971. https://doi.org/10.2172/4022263

[9] ASTM International, 2014. "Standard Test Method for Determining Image Quality in Direct Thermal Neutron Radiographic Examination," ASTM E545-14.

Neutron Radiography - WCNR-11   Materials Research Forum LLC
Materials Research Proceedings **15** (2020) 299-304   https://doi.org/10.21741/9781644900574-47

Fission Neutron Tomography of a 280-L Waste Package

T. Bücherl *, Ch. Lierse von Gostomski, T. Baldauf

ZTWB Radiochemie München, Technische Universität München, Walther-Meißner-Str. 3, 85748 Garching, Germany

* thomas.buecherl@tum.de

Keywords: Fission Neutron, Transmission Tomography, Emission Tomography, NECTAR, Radioactive Waste, Non-Destructive Characterization

Abstract. Based on a recent feasibility study, where it is demonstrated that fission neutron radiography of 200-l (radioactive) waste drums is possible at NECTAR, these investigations are extended to fission neutron tomography in transmission and emission mode. As sample, a 280-l drum is used containing inactive waste being typical for radioactive waste drums. For emission tomography, an AmBe-neutron source is measured separately and at different positions in the drum.

Introduction

For the non-destructive characterization of radioactive waste packages for the declaration or verification of their radioactive inventory, well-established passive and active methods are applied. These are mainly based on gamma-spectroscopic emission measurements (segmented gamma scanning), gamma-transmission measurements (e.g. radiography and tomography) using intense external radionuclide sources (e.g. ^{60}Co) or accelerators, neutron emission counting with time correlation analysis to distinguish between neutrons originating from spontaneous fission and (α,n) events, and neutron interrogation techniques inducing fission events. Tomography using fission neutrons, in transmission and emission mode, is to the best of our knowledge not applied on waste packages, yet.

In a feasibility study[1] it was demonstrated that fission neutron radiography of 200-l (radioactive) waste drums is possible at NECTAR, a neutron radiography and tomography facility using fission neutrons [2]. In a subsequent step, the study was extended on tomographic investigations of 200-l and 280-l mock-up waste drums both in transmission and emission mode.

Experimental

NECTAR Facility. The measurements are performed at the NECTAR facility at the research reactor FRM II of the Technical University Munich (TUM) using an ANDOR iKon L 4 Megapixel back illuminated CCD camera operating at -100 °C with a Nikkor 100mm f/2.8 lens, resulting in a field of view of 291mm x 288mm (width x height). The conversion of fission neutrons to blue light (450 nm) is performed by a PP/ZnS:Ag (30%) scintillator with dimensions 2.4 mm x 400 mm x 400 mm (depth x width x height) distributed by RC TRITEC, Switzerland, attached to the light tight housing of the detector box by a sliding mechanism. The L/D-value used for all transmission measurements is 132 ± 12 [3].

Sample. The sample is a 280-l drum with 100 cm in height, 71 cm in diameter and a mass of about 500 kg. It houses a 200-l drum containing two supercompacted pellets on the bottom and then filled up with metallic and plastic scrap (Fig. 1). A cylindrical container made of polyethylene (PE) is used for housing additional neutron sources.

Neutron Radiography - WCNR-11
Materials Research Proceedings **15** (2020) 299-304

Materials Research Forum LLC
https://doi.org/10.21741/9781644900574-47

Fig. 1: Left: Open 200-l drum during assembling with two pellets covered by scrap. The white PE-cylinder can ingest neutron sources. Right: 280-l drum on the sample manipulator of NECTAR.

External Neutron Sources. For the feasibility tests of neutron emission tomography, two certified AmBe-sources with neutron emission rates of $2.2 \cdot 10^5$ s^{-1} and $2.2 \cdot 10^6$ s^{-1} are available. Their mean neutron energy is about 2.7 MeV with a significant amount (about 23%) of neutrons below 1 MeV having a mean energy of 400 keV. The compacted mixture of AmO_2- and Beryllium-metal-powder is housed in a sealed container of 6cm height and 3cm diameter.

Measurements. The first series is focused on transmission measurements. The 280-l drum is positioned as close as possible to the detector still enabling undisturbed rotation, i.e. at a distance of 38 cm between the axis of rotation and the scintillator screen. As the FOV of the detector system is smaller than the diameter of the drum, the sample is translationally scanned, i.e. one projection is setup by three transmission images differing in position by 23 cm each. This results in an overlap of successive images of 6.1 cm. The integration time for each image is 60 s. In total 491 angular projections for 360° are measured applying the golden ratio method [4]. To minimize the deadtime of the system due to mechanical movements, the translational scanning is performed by reciprocating motion. The total measuring time is about 1.5 days, Dark images and open beam images are measured additionally.

In the second series, four runs in total are investigating the feasibility of the NECTAR facility for emission tomography. The first run is performed without drum, positioning the strong AmBe-source in the axis of rotation of the turntable at a distance of 3 cm to the center of the fission neutron scintillator screen. The measurement is using the golden ratio method with 207 projections and an integration time of 3500 s for each projection, a total measuring time of almost 9 days. The next runs are performed with drum, positioned identically like in the transmission measurement. First, the weak AmBe-neutron source is ingested in the PE-container, then it is replaced by the strong AmBe-neutron source in the second run. In the last run this source is positioned close to the outer side of the PE-container avoiding moderation by the PE. The parameters for the last three runs are the same like for the transmission measurements, except the fission neutron beam of the reactor is closed. Additional test measurements with different scintillators being sensitive to thermal neutrons are performed, too.

Data Evaluation
Preprocessing. All images are filtered [5] to remove artefacts caused by gamma-radiation and/or scattered neutrons hitting the CCD sensor directly, followed by dark-image correction and normalization [6].

Neutron Radiography - WCNR-11
Materials Research Proceedings **15** (2020) 299-304

Materials Research Forum LLC
https://doi.org/10.21741/9781644900574-47

The stitching of images to get a projection is performed by using a minimization routine, i.e. by superimposing the left and right parts of two images, estimating the minimum of the sums of the absolute differences of all pixels as a function of the width of superimposition. As the intensity of the neutron beam is differing between the right and left side of each image and thus the uncertainties of each pixel value of the normalized image, an appropriate weighing matrix is taken into account. Fig. 2 shows as an example three normalized images for one projection and the resulting stitched projection.

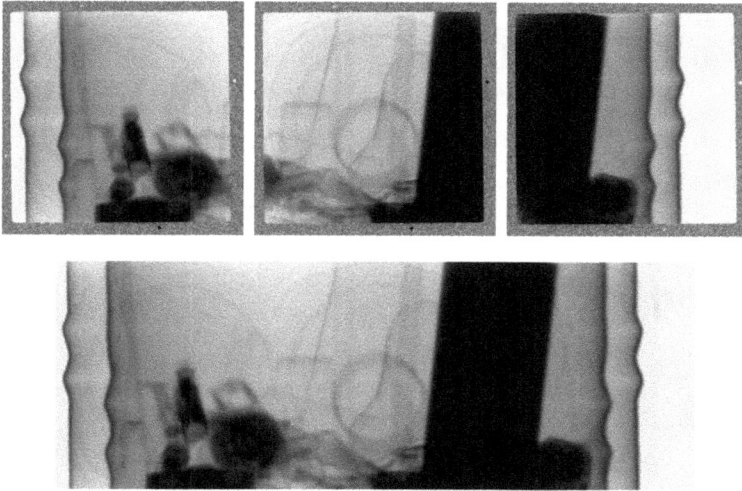

Fig. 2: Upper row: Three normalized radiographs making up one projection. Note, that the images are not cropped, i.e. the area not covered by the scintillator but seen by the CCD is visible. Bottom: Stitched projection estimated from the cropped images in the upper row.

Reconstruction Algorithms. For reconstruction of the transmission data, the filtered backprojection (FBP) method provided by the program Octopus [7] is applied, assuming parallel beam geometry. The emission data are reconstructed applying the FBP algorithm with ramp-filter and the maximum likelihood expectation maximization (MLEM) [8], using python scripts. Here, the projection data is binned by five pixels to minimize computation time for this feasibility study.

Results and Discussion

Transmission Measurements. The reconstruction of the transmission data does not result in utilizable data, as the tomograms are blurred and no details can be identified (Fig. 3). This is based on two effects, the low dynamic range of the data of 660 grey values, and more problematic, the effect of beam divergence. The latter is demonstrated by comparing the stitching parameters of projections at $52.52°$ and $327.54°$ with an estimated overlap of two images of 189 and 143 pixels, respectively (Fig. 4). For the assumed parallel beam geometry, the value of overlap must be constant, while indicating a change of 46 pixels, i.e. a shift of about 7

Neutron Radiography - WCNR-11   Materials Research Forum LLC
Materials Research Proceedings **15** (2020) 299-304   https://doi.org/10.21741/9781644900574-47

mm, due to a maximum beam divergence of 2.1° [9]. Thus, an improved iterative method for a combined stitching and reconstruction process has to be developed, taking into account beam divergence and changing uncertainties in the pixel values due to non-homogeneous areal beam intensity distribution.

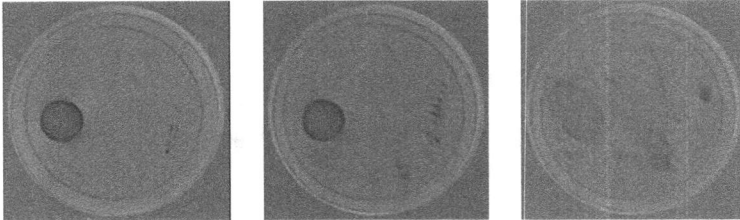

Fig. 3: Three tomograms at different height positions of the drum. The borders of the 280-l- and 200-l-drums are visible as well as the PE-container, but no further details.

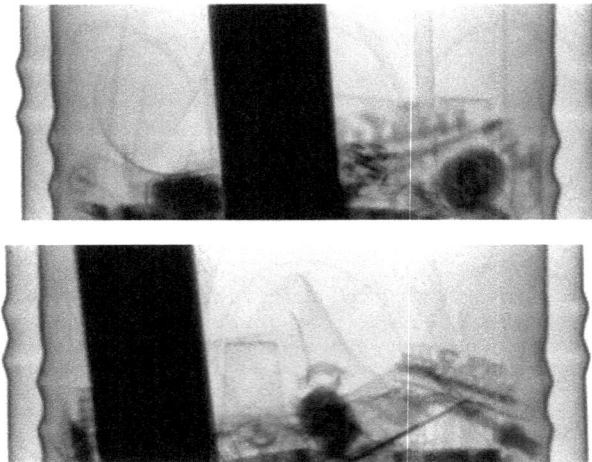

Fig. 4: Projection at 52.52°(top) and 327.54° (bottom) with an overlap of two adjacent images of 189 and 143 pixels, respectively. A shift of the axis of rotation is present, too.

A first idea of such an algorithm is sketched in Fig. 5. For the presented measurement, three identical neutron sources at distances d being the translational shift of the manipulator of successive image positions contribute to one projection, taking into account the beam divergences and the overlap. Using an ART algorithm modified for the use of multiple external sources, an improved image quality may be achieved. This is the work of a future R&D project.

Neutron Radiography - WCNR-11   Materials Research Forum LLC
Materials Research Proceedings **15** (2020) 299-304   https://doi.org/10.21741/9781644900574-47

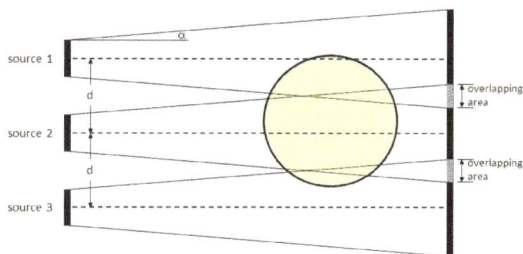

Fig. 5: Sketch of the basic idea for an improved combined stitching and reconstruction algorithm taking into account a beam divergence α.

Emission Measurements. Fig. 6 shows the neutron activity distributions of the first run using the strong AmBe-source evaluated by MLEM and FBP. For data evaluation a binning of 8x8 pixels is applied. The horizontal line profiles through the center of the distributions indicate that the influences of scattering and absorption in the sample are not taken into account completely and an improved data evaluation has to be performed.

The images of the three other emission measurement runs with the samples in the drum show no significant signals in the detector for all scintillators used, indicating that the detection efficiency at NECTAR is too low. A rough calculation for the AmBe-source emitting $2.2 \cdot 10^6$ s^{-1} neutrons positioned in the center of a 200-l barrel, i.e. at 30 cm distance from the detector plane, results in about 900 cm^{-2}s^{-1} or 0.02 pixel^{-1}s^{-1} neutrons in this plane. These values do not consider attenuation and moderation effects. In combination with the actual detection efficiency for fission neutrons of less than 1%, their detection is unlikely within reasonable integration times. Thus an improvement in detection efficiency is one of the most important topics for transmission and emission measurements.

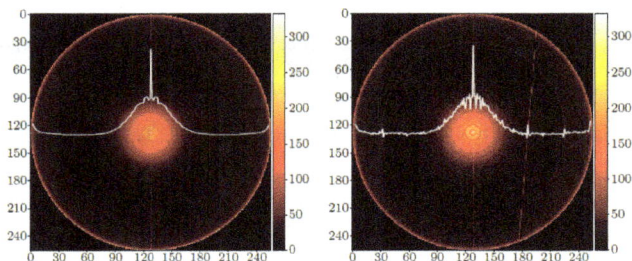

Fig. 6: MLEM-reconstruction after 30 iterations and smoothing filter applied (left) and FBP (right). The axes are given in pixels (i.e. 256 x 256 pixels), with one pixel corresponding to about 0.78mm in size. The line profiles show the horizontal distributions of the neutron activity.

Summary
The results of the study show, that transmission tomography using fission neutrons on large sized samples like 280-l drums is feasible, but requires further developments in a combined stitching

and reconstruction algorithm as well as for increased detection efficiency. A possible toehold for an algorithm is sketched. The applicability will strongly depend on density, thickness and distribution of the different materials in the drums. It is demonstrated that neutron emission tomography is possible at the NECTAR facility, although only applicable in praxis on large samples for extremely high neutron emission rates. Here passive neutron assay systems like SANDRA [10] are still in favor due to their much higher detection efficiency, but not giving the activity distribution in the volume of the sample. Developments for improved detection efficiencies for fission neutrons can overcome these limitations.

References

[1] T. Bücherl, O. Kalthoff, Ch. Lierse von Gostomski, A feasibility study on reactor based fission neutron radiography of 200-l waste packages, Physics Procedia 88 (2017) 64 – 72. https://doi.org/10.1016/j.phpro.2017.06.008

[2] NECTAR: Heinz Maier-Leibnitz Zentrum. (2015). NECTAR: Radiography and tomography station using fission neutrons. Journal of large-scale research facilities, 1, A19. https://doi.org/10.17815/jlsrf-1-45

[3] T. Bücherl, Ch. Lierse von Gostomski, H. Breitkreuz, M. Jungwirth, F.M. Wagner, NECTAR – A fission neutron radiography and tomography facility, Nucl. Instr. Meth. Phys. Res. A 651(2011) 86-89. https://doi.org/10.1016/j.nima.2011.01.058

[4] L. G. Butler, E. H. Lehmann, B. Schillinger, Neutron Radiography, Tomography, and Diffraction of Commercial Lithium-ion Polymer Batteries, Physics Procedia, 43, 2013, 331-336, https://doi.org/10.1016/j.phpro.2013.03.039.

[5] K. Osterloh, T. Bücherl, Ch. Lierse von Gostomski, U. Zscherpel, U. Ewert, S. Bock, Filtering algorithm for dotted interferences, Nucl. Instr. Meth. Phys. Res. A 651(2011) 171-174. https://doi.org/10.1016/j.nima.2011.01.107

[6] A. C. Kak, M. Slaney, Principles of Computerized Tomographic Imagimg, IEEE Press, 1988, ISBN 0-87942-198-3.

[7] M. Dierick, B. Masschaele, L, Van Hoorebeke, Octopus, a fast and user-friendly tomographic reconstruction package developed in LabView, Measurement Science and Technology, Volume 15, Number 7, 2004. https://doi.org/10.1088/0957-0233/15/7/020

[8] http://campar.in.tum.de/twiki/pub/Main/MoritzBlume/EMPET.pdf (last call 07.11.2018).

[9] J. Guo, T. Bücherl, Y. Zou, Z. Guo, Study on beam geometry and image reconstruction algorithm in fast neutron computerized tomography at NECTAR facility, Nucl. Instr. Meth. Phys. Res. A, 651 (2011) 180-186. https://doi.org/10.1016/j.nima.2011.01.097

[10] http://www.en-trap.eu/doc/neutronsynopsis.pdf (last call 07.11.2018).

Keyword Index

www.ingramcontent.com/pod-product-compliance
Lightning Source LLC
Chambersburg PA
CBHW071326210326
41597CB00015B/1360